Yersinia enterocolitica

Editor

Edward J. Bottone

Director
Department of Microbiology
The Mount Sinai Hospital
New York, New York

CRC Press, Inc.
Boca Raton, Florida

Library of Congress Cataloging in Publication Data

Yersinia enterocolitica.

　Bibliography: p.
　Includes index.
　1. Yersinia enterocolitica. 2. Yersinia infections. I. Bottone, Edward J. [DNLM: 1. Yersinia. 2. Yersinia infections. QW138.5. Y3 Y47]
QR201.Y45Y47　　　616′.0145　　　80-19597
ISBN 0-8493-5545-1

This book represents information obtained from authentic and highly regarded sources. Reprinted material is quoted with permission, and sources are indicated. A wide variety of references are listed. Every reasonable effort has been made to give reliable data and information, but the author and the publisher cannot assume responsibility for the validity of all materials or for the consequences of their use.

All rights reserved. This book, or any parts thereof, may not be reproduced in any form without written consent from the publisher.

Direct all inquiries to CRC Press Inc., 2000 N.W. 24th Street, Boca Raton, Florida 33431.

© 1981 by CRC Press, Inc.

International Standard Book Number 0-8493-5545-1

Library of Congress Card Number 80-19597
Printed in the United States

PREFACE

The evolution of a microorganism to the current level of awareness by the science oriented fraternity is steeped in fascinating vignettes. At this juncture, as we simultaneously gaze retrospectively and prospectively, the unfolding of awareness of *Yersinia enterocolitica* is actually the chronicling of information initially sequestered in the laboratories of a few geographically dispersed investigators and presently, the focal point of intensive global activity.

In tracing the legacy of such an enigmatic microbial entity, one embarks on a microbiologic journey through the early literature in which every page is embellished with the author's perception of this fleeting microorganism. It is thus that we envision our predecessors penning according to their "state of the art" their observations of the evolving microorganism under descriptive terms ranging from "an unidentified microorganism pathogenic for man" which appeared in the earliest known description in 1939, through "*Bacterium enterocoliticum*" to the highly imaginative and provocative "*Le Germe X*" in 1964.

Early studies sought to establish the identifying microbiologic criteria and clinical syndromes associated with the microorganism now recognized as being distinct from the closely related *Y. pseudotuberculosis* and designated *Y. enterocolitica* (1964). In contrast, contemporary interest has kept pace with innovative trends in microbiology and medicine. Hence, in the ensuing pages each author has traversed time to give perspective to his presentation. In so doing, we are treated to the usage of modern day technology in the form of DNA-DNA hybridization studies to establish the taxonomic status of *Y. enterocolitica* and related species. The utility of such speciation is then translated into the provision of useful microbiologic, clinical, and epidemiologic data.

Germane to the understanding of the varied clinical and epidemiologic phenomena associated with *Yersinia*, is their recovery from specimens derived from clinical and environmental sources. As these species have long defied the microbiologists isolation attempts, procedures for their recovery and subsequent identification have been detailed. In the absence of isolation, diagnosis through a consideration of immunologic response as measured serologically is also brought into sharp focus through discussion of surface antigens of *Y. enterocolitica* and the host response engendered during the various clinical syndromes.

Linking the chapters dealing with the clinical manifestations of "yersiniosis enterocolitica" are those of investigators working to delineate the mode of pathogenesis of *Y. enterocolitica* and the laboratory approach to determining such parameters as invasiveness and enterotoxin production.

Epidemiologically, numerous questions remain still unanswered regarding the exact mode of transmission of yersiniae, the exact reservoir in nature of several of the more commonly occurring serotypes, e.g., O:8, associated with human disease, and the public health problem relative to their recovery from foods. In attempting to resolve such dilemmas, knowledge of the zoonotic host range, clinical manifestations in infected animals, and the potential for human transmission from zoonotic reservoirs becomes mandatory.

The significance of *Y. enterocolitica* in food products is still debated as only one major outbreak, occurring in the U.S., has been traced to contaminated chocolate milk. Despite this direct incrimination, species of *Yersinia*, especially the rhamnose fermenting *Y. intermedia*, are frequently recovered from food products without known production of disease. While not considered among the *Yersinia* species capable of causing "classic syndromes" such as mesenteric lymphadenitis, terminal ileitis, or septicemia, great interest in the pathogenic potential of *Y. intermedia* has been the subject of intensive investigation by food microbiologists. The nature of the *Yersinia* strains isolated in foods, and their control is a subject of great public health significance. In this regard the correlation of the biochemical, serologic, and invasive potential of food isolates has issued a new dimension in *Yersinia* research.

To be certain, the present work will not have answered several still outstanding questions regarding yersiniosis. Yet to be elucidated in terms of human disease are the roles of the various temperature dependent biologic attributes necessary for the expression of virulence. For instance, as surface topography (antigens), invasiveness, and enterotoxin production are all expressed at 22°C, little is known of the initial microbe-host interaction at the time of acquisition of the microorganism from its natural (cold?) reservoir. Evidence supporting the importance of the bacterium's surface antigens is suggested in the serologic response to infection in which sera from patients with yersinosis will agglutinate their own isolate to a greater titer when the strain is tested after growth at 22°C as compared to 37°C.

Although much remains to be expounded, it is hoped, that as our predecessors left us a legacy upon which the present work firmly rests, the efforts herein will traverse scientific time and serve as a foundation for all those who will turn these pages in their search of the truth underlying *Y. enterocolitica,* indeed, a multifaceted human pathogen.

Edward J. Bottone

THE EDITOR

Edward J. Bottone, Ph.D., is Director, Department of Microbiology, The Mount Sinai Hospital, and Associate Professor of Microbiology at Mount Sinai School of Medicine, New York, New York.

Dr. Bottone received his B. S. degree in Biology in 1965 from City College of City University of New York, his Master of Science in Bacteriology and Public Health in 1968 from Wagner College, New York, and his Ph.D. in Biology in 1973 from St. John's University, Jamaica, New York.

Dr. Bottone is a member of the National and New York City Branch of the American Society for Microbiology and serves on the National Taxonomy Subcommittee on Gram-negative, Facultatively Anaerobic Rods, and for the local branch served as Membership Chairman, Program Chairman (Clinical Section), and as Branch Councilor. Dr. Bottone is also a member of the New York Medical Mycology Society, is Director of an approved (American Academy of Medical Microbiology), Postdoctoral Program in Public Health and Medical Laboratory Microbiology, and is Co-Director of Workshops in Clinical Laboratory Sciences. He is a Diplomate of the American Board of Medical Microbiology.

He is a biographee in Who's Who in the East and American Men and Women of Science and is recipient of the Excellence in Teaching Award at the Mount Sinai School of Medicine.

Dr. Bottone has presented numerous invited lectures at National Meetings as well as guest lectures at various universities and institutes. He has published 70 papers to date, and is on the Editorial Board of the Journal of Clinical Microbiology. His major research interest include bacteriocin interactions, the study of unusual microbial species, and the epidemiology of infections caused by *Yersinia enterocolitica* and related species.

CONTRIBUTORS

Kimmo Aho, M.D.
Head, Department of Immunobiology
Central Public Health Laboratory
Helsinki, Finland

Paavo Ahvonen, M.D.
Microbiologist
Municipal Bacteriological Laboratory
Aurora Hospital
Helsinki, Finland

Marjorie L. Bissett, Ph.D.
Research Microbiologist
Microbial Diseases Laboratory
California Department of Health
Berkeley, California

Edward J. Bottone, Ph.D.
Director
Department of Microbiology
The Mount Sinai Hospital
New York, New York

Don J. Brenner, Ph.D.
Chief, Enteric Section
Center for Disease Control
Atlanta, Georgia

Philip B. Carter, Ph.D.
Associate Member
Trudeau Institute, Inc.
Saranac Lake, New York

Guy Cornelis, Dr. Sc. Ph.
Chef de travaux
Unite de Microbiologie
Universite Catholique de Louvain
Brussels, Belgium

James C. Feeley, Ph.D.
Chief, Special Pathogens Laboratory
    Section
Bureau of Epidemiology
Center for Disease Control
Atlanta, Georgia

Bengt Hurvell, D.V.M., Ph.D.
Professor
Head, Department of Bacteriology
National Veterinary Institute
Uppsala, Sweden

Gerald T. Keusch, M.D.
Professor of Medicine
Chief, Division of Geographic Medicine
Tufts University School of Medicine
Boston, Massachusetts

**H. J. Koornhof, M.B.,Ch.B., F.R.C.
    Path.**
Head, Department of Microbiology
University of the Witwatersand
South African Institute for Medical
    Research
Johannesburg, South Africa

Lucette Lafleur, M.D.
Associate Professor of Microbiology
Universite de Montreal
Hospital Sainte Justine
Montreal, Canada

Ossi Laitinen, M.D.
Professor of Rheumatology
University of Tampere
Tampere, Finland

Wei Hwa Lee, Ph.D.
Microbiologist
USDA-FSQS
Beltsville, Maryland

Marjatta Leirisalo, M.D.
Lecturer of Medicine
Second Department of Medicine
Helsinki University Central Hospital
Helsinki, Finland

Melvin I. Marks, M.D.
Associate Professor of Pediatrics
McGill University
Montreal, Quebec
Canada

Chik H. Pai, M.D., Ph.D.
Director, Department of Microbiology
Montreal Children's Hospital
Associate Professor of Microbiology
McGill University
Montreal, Quebec
Canada

A. R. Rabson, M.B., B.Ch., D.C.P.
Head, Department of Immunology
University of the Witwatersand
South African Institute for Medical
  Research
Johannesburg, South Africa

Roy M. Robins-Browne, M.B., B.Ch., F.F. Path.
Head, Department of Microbiology
University of Natal
Durban, South Africa

David B. Sachar, M.D.
Associate Professor of Medicine
The Mount Sinai School of Medicine
New York, New York

Mehdi Shayegani, Ph.D.
Director, Laboratories for Bacteriology
Division of Laboratories and Research
New York State Department of Health
Albany, New York

Louis Shenkman, M.D.
Associate Professor of Medicine
New York University Medical School
New York, New York

Norman J. Stern, Ph.D.
Microbiologist
Meat Science Research Laboratory
Beltsville, Maryland

Sandu Toma, M.D.,
Chief Bacteriologist
Ontario Ministry of Health
Public Health Laboratories
Toronto, Canada

Carl Vanderzant, Ph.D.
Professor of Food Microbiology
Department of Animal Science
Texas A & M University
College Station, Texas

Jacob S. Walfish, M.D.
Clinical Instructor of Medicine
Mount Sinai School of Medicine
New York, New York

Georges Wauters, M.D.
Professor of Microbiology
University of Louvain
Brussels, Belgium

Sten G. Winblad, M.D.
Professor of Clinical Bacteriology
University of Lund
Malmo, Sweden

Gary P. Wormser, M.D.
Chief of Infectious Diseases
Assistant Chief of Medicine
Bronx Veterans Administration
  Hospital
Assistant Professor of Medicine
The Mount Sinai School of Medicine
New York, New York

Hiroshi Zen-Yoji, Ph.D.
Professor of Microbiology
Kyorin University
Tokyo, Japan

DEDICATION

This endeavor is dedicated to S. Stanley Schneierson, M.D. (1906-1976) former Director, Department of Microbiology of the Mount Sinai Hospital whose teachings embraced a philosophy of professional ethics, awareness of life's amenities, and pursuit of academic achievement. For those of us whose paths were fortunate enough to intersect with Dr. Schneierson, his example is indelible, practiced in daily encounter and lives in all his professional children.

E. J. B.

TABLE OF CONTENTS

Chapter 1
Classification of *Yersinia enterocolitica* 1
Don J. Brenner

Chapter 2
Isolation Techniques for *Yersinia enterocolitica* 9
James C. Feeley

Chapter 3
Yersinia enterocolitica: An Approach to Laboratory Identification with Reference to Deoxyribonucleic Acid Hybridization Groups and Biochemical Characteristics 17
Edward J. Bottone

Chapter 4
Microbiological Aspects of *Yersinia pseudotuberculosis* 31
Marjorie L. Bissett

Chapter 5
Antigens of *Yersinia enterocolitica* 41
Georges Wauters

Chapter 6
Antibiotic Resistance in *Yersinia enterocolitica* 55
Guy Cornelis

Chapter 7
Human *Yersinia enterocolitica* Infection: Laboratory Models 73
Philip B. Carter

Chapter 8
Yersinia enterocolitica: Clinical Observations 83
Gary P. Wormser and Gerald T. Keusch

Chapter 9
Yersinia enterocolitica Gastroenteritis in Children and Their Families 95
Melvin I. Marks, Chik H. Pai, and Lucette Lafleur

Chapter 10
Yersinia enteritis and Crohn's Disease 105
Jacob S. Walfish and David B. Sachar

Chapter 11
Arthritis Associated with *Yersinia enterocolitica* Infection 113
Kimmo Aho, Paavo Ahvonen, Ossi Laitinen, and Marjatta Leirisalo

Chapter 12
Erythema Nodosum Associated with Infection with *Yersinia enterocolitica* ... 125
Sten Winblad

Chapter 13
The Occurrence of Antibodies to *Yersinia enterocolitica* in Thyroid Diseases 135
Louis B. Shenkman and Edward J. Bottone

Chapter 14
Zoonotic *Yersinia enterocolitica* Infection: Host Range, Clinical Manifestations, and Transmission Between Animals and Man 145
Bengt Hurvell

Chapter 15
The Occurrence of *Yersinia enterocolitica* in Foods 161
Wei Hwa Lee, Carl Vanderzant, and Norman Stern

Chapter 16
Epidemiological Aspects of *Yersinia enterocolitica* in New York State with Emphasis on a Recent Food-Borne Outbreak 173
Mehdi Shayegani

Chapter 17
Yersinia enterocolitica Infections in Canada 183
Sandu Toma and Lucette Lafleur

Chapter 18
Yersinia enterocolitica in South Africa 193
Roy M. Robins-Browne, A. R. Rabson, and H. J. Koornhof

Chapter 19
Epidemiologic Aspects of Yersiniosis in Japan 205
Hiroshi Zen-Yoji

Index 217

Chapter 1

CLASSIFICATION OF *YERSINIA ENTEROCOLITICA*

Don J. Brenner

TABLE OF CONTENTS

I.	Introduction	2
II.	Biochemical Characterization of *Y. enterocolitica* and *Y. enterocolitica*-Like Organisms	2
III.	Guanine Plus Cytosine Content	5
IV.	DNA Relatedness in *Y. enterocolitica* and *Y. enterocolitica*-Like Organisms	5
V.	Classification	6
	References	7

I. INTRODUCTION

The genus *Yersinia* was proposed by Van Loghem[26] in honor of A. J. E. Yersin to accommodate *Yersinia pestis* and *Y. pseudotuberculosis* which were previously classified in the genus *Pasteurella*. Thal[25] first proposed that *Yersinia* be placed in the family *Enterobacteriaceae*. The Gram-negative, fermentative organism *Y. enterocolitica* was first reported by Schleifstein and Coleman[22] who called their isolates *Bacterium enterocoliticum*. Hassig et al.[14] were the first to isolate *Y. enterocolitica* from humans in Europe. Their isolates were named *Pasteurella pseudotuberculosis rodentium*. Daniels and Goudzwaard[9] referred to similiar isolates as *Pasteurella* X.

Frederiksen[13] examined 55 of these strains, including isolates from Schleifstein and Coleman, Hassig et al., and Daniels and Goudzwaard. Biochemical reactions were similar in all of the strains. Their antibiotic susceptibility patterns were also similar. Three different O antigens were present, none of which reacted with *Y. pseudotuberculosis*. Frederikson concluded: "The characteristics of this group of bacteria are sufficiently distinct to separate them from *Y. pseudotuberculosis*. Yet they resemble *Y. pseudotuberculosis* sufficiently to justify their classification in the genus *Yersinia* (within the family *Enterobacteriaceae* as a separate species: *Y. enterocolitica*."

II. BIOCHEMICAL CHARACTERIZATION OF *Y. ENTEROCOLITICA* AND *Y. ENTEROCOLITICA*-LIKE ORGANISMS

Y. enterocolitica strains, unlike *Y. pseudotuberculosis*, show considerable biochemical variability. The biochemical profiles of *Y. enterocolitica* and *Y. enterocolitica*-like strains are quite dependent upon the temperature of incubation. Results obtained at 22°C to 28°C are often considerably different from those obtained at 36°C ± 1°C.[2,4,8,16,21,23] Furthermore, strains from different sources and/or geographical origins often have different biochemical reactions. It was, therefore, difficult to develop a biochemical definition and biochemical boundaries for *Y. enterocolitica*. Biochemical reactions obtained for *Y. enterocolitica* in several laboratories are given in Table 1.

Most strains identified as *Y. enterocolitica* are characterized and separated from *Y. pseudotuberculosis* by the reactions given in Table 2. *Y. enterocolitica* strains differ in reactions for lecithinase (lipase), indole, xylose, esculin, and salicin. These reactions became the basis for several biotyping schemes (Table 3). Several other biogroups of *Y. enterocolitica*-like strains were encountered. These included metabolically inactive strains isolated from hares.[19] The hare strains are trehalose-negative and are often sucrose-negative. Niléhn[21] and Wauters[28] included them as biogroup 5 of *Y. enterocolitica* (Table 3). Stevens and Mair[24] did a numerical taxonomic study of *Y. enterocolitica* and considered the trehalose-negative strains to be members of a separate species. A second sucrose-negative group was described by Fredericksen.[13] These organisms were trehalose-positive.

Two groups of rhamnose-positive *Y. enterocolitica*-like strains have been described.[1,4-6,17,28] One rhamnose-positive group is also positive in reactions for melibiose, raffinose, Simmons' citrate, and α-methyl-D-glucoside. These strains are substantially more active at 22°C. At 37°C these reactions, except for melibiose, may be delayed or negative. The second rhamnose-positive group apparently differs from typical *Y. enterocolitica* only in its ability to ferment rhamnose. Botzler et al.[6] described a sucrose-negative, rhamnose-positive group of *Y. enterocolitica*-like strains.

In addition to these *Y. enterocolitica*-like groups, there are a number of strains with various reactions atypical for *Y. enterocolitica*. These include lactose-positive, lactose

Table 1
BIOCHEMICAL REACTIONS OF *YERSINIA ENTEROCOLITICA*

Reaction	Smith and Thal[23] (1965) 35—37°C	Niléhn[21] (1969) 22—25°C	Niléhn[21] (1969) 35—37°C	Bejot et al.[2] (1975) 22—25°C	Bejot et al.[2] (1975) 35—37°C	Darland et al.[10] (1975) 22—25°C	Darland et al.[10] (1975) 35—37°C	Brenner et al.[8] (1976) 35—37°C
Urease	+		99.7		+		91(7)	+
Indole	+		5		v	27	27	v
Voges-Proskauer	−	83	0	+	−	87	0	−
KCN	v					96	0	−
Motility	−	52(48)	0	+	−		0	−
Ornithine decarboxylase	+		93		+		91(7)	+
Lactose	−		0(2)			0(3)	0(5)	+
Sucrose	+		95(2)		+		96	+
Salicin	−		2				13(15)	v
i-Inositol	−		2(29)		v	9(32)	3(17)	+
D-Sorbitol	v		98(1)		+		99	+
L-Arabinose	+		96(4)		+		99(1)	+
Raffinose	−		0		−		7	−
L-Rhamnose							1	−
Maltose	+	96(4)	16(79)		d	99(1)	43(51)	+
D-Xylose	+		27		d		3(6)	v
Trehalose	+		93		+		89(9)	+
Cellobiose	+		100		+		88(12)	+
Glycerol	+		47(53)		+		65(28)	+
Esculin	−		6		v		15(21)	v
Amygdalin	d		0(87)				96(4)	+
β-Galactosidase		91	83(6)		+	93	24	+
Galactose	+		82(18)					+
Sorbose			95(2)		+			+
β-Xylosidase					v			
Arbutin								v
Christensen's citrate							11(21)	v
Jordan's tartrate							100	v
Acetate						7(44)	0	−
Lipase					d	72(20)	73(24)	+
NO$_3^-$ → NO$_2^-$			92		+		99	+
Deoxyribonuclease							3	v

Note: + = 90% or more positive; − = less than 10% positive; v = 10 to 89.9% positive; ( ) = delayed reaction, positive in 3 days or longer. The following tests done in one or more of the laboratories cited above were 100% negative: H$_2$S, Simmons' citrate, Koser's citrate, gelatin, lysine decarboxylase, arginine dihydrolase, phenylalanine deaminase, glucose (gas), dulcitol, adonitol, malonate, mucate, alginate, D-arabinose, α-methyl-D-glucoside, erythritol, melibiose, inulin, melizitose, α-methyl-xyloside, α-methyl-D-mannoside, glycogen, pigment, oxidase. The following tests done in one or more of the laboratories cited above were 100% positive: methyl red, glucose (acid), mannitol, mannose, dextrin, polypectate.[27]

Table 2
DEFINITIVE BIOCHEMICAL REACTIONS FOR *YERSINIA ENTEROCOLITICA* AND *YERSINIA PSEUDOTUBERCULOSIS*

Reaction	*Y. enterocolitica*	*Y. pseudotuberculosis*
Ornithine decarboxylase	+	−
Sucrose	+	−
Cellobiose	+	−
D-Sorbitol	+	−
L-Rhamnose	−	+
Melibiose	−	+
Raffinose	−	v
Indole	v	−
Motility 22 to 25°C	+	+
Motility 35 to 37°C	−	−
Urease	+	+
Methyl red	+	+
$NO_3^- \rightarrow NO_2^-$	+	+
D-Glucose	+	+
D-Mannitol	+	+
D-Mannose	+	+
Trehalose	+	+
Voges-Proskauer	−	−
H_2S	−	−
Arginine dihydrolase	−	−
Lysine decarboxylase	−	−
Phenylalanine deaminase	−	−
Simmons' citrate	−	−
Gelatin 22°C	−	−
Gas from glucose	−	−
KCN	−	−
Pigment	−	−
Oxidase	−	−
Dulcitol	−	−
Adonitol	−	−
Malonate	−	−
Mucate	−	−

Note: Tests were done at 35 to 37°C unless otherwise specified. + = 90% or more positive within 48 hr; − = less than 10% positive within 48 hr; v = 10.1 to 89.9% positive within 48 hr.

and raffinose-positive, urease-negative, methyl red-negative, and D-mannose-negative strains.

Knapp and Thal[15,16] expressed concern about the biochemical boundaries of *Y. enterocolitica*. There was no species definition for *Pasteurella* X.[9] Therefore, strains that differed from *Pasteurella* X in important reactions were included in *Y. enterocolitica*. Knapp and Thal considered their group 2 (biotypes 3 and 4 of Niléhn and of Wauters, Table 3), which is indole-positive, to be true *Y. enterocolitica*. They proposed the name *Y. enteritidis* for the indole-negative strains of their group 2. The strains in their groups 3 (positive for indole, xylose, esculin, and salicin) and 4 (including rhamnose-positive and sucrose-negative types) were not considered to be *Y. enterocolitica* or *Y. enteritidis*.[15,16] The subcommittee on the Taxonomy of Pasteurella, Yersinia, and Francisella was also concerned with the properties of *Y. enterocolitica*, and requested additional information on serology, genetics, and deoxyribonucleic acid (DNA) homology.[18]

Table 3
BIOTYPING SCHEMES FOR YERSINIA ENTEROCOLITICA

Reaction	Niléhn[21] (1969) 1 2 3 4 5	Wauters[28] (1970) 1 2 3 4 5	Knapp and Thal[16] (1973[a]) 1 2 3 4[b]
Lecithinase		+ − − − −	
Salicin	+ − − − −		− − + v
Esculin	+ − − − −		− − + v
Indole	+ + − − −	+(+) − − −	− + + v
D-Xylose	+ + + − −	+ + + − −	v + + v
Nitrate reduction	+ + + + −	+ + + + −	
Trehalose	+ + + + −	+ + + + −	
D-Sorbitol	+ + + + −		
Ornithine decarboxylase	+ + + + −	+ + + + −	
Voges-Proskauer (25°C)	+ + + + −		
β-Galactosidase (25°C)	+ + + + −	+ + + + −	
Sucrose	+ + + + −		
Sorbose	+ + + + −		

[a] Tests were done at 35 and 37°C unless otherwise specified.
[b] Group O:4 strains gave atypical results in reactions for lactose, rhamnose, sucrose, and ornithine decarboxylase. + = positive within 48 hr, (+−) = positive after 48 hr; − = negative; v = variable.

III. GUANINE PLUS CYTOSINE CONTENT

Domaradskij et al.[11] reported that *Y. enterocolitica* DNA contained somewhat more than 45% guanine plus cytosine (G+C). Moore and Brubaker[20] used thermal denaturation of DNA from 12 *Y. enterocolitica* strains to determine a GC range of between 47.0% and 48.5% with an average of 47.7%. Similar results were reported by Ewing et al.[12] Using buoyant density ultracentrifugation, Mandel found 49% GC in DNA of typical strains and rhamnose-melibiose-raffinose-positive strains.[29]

IV. DNA RELATEDNESS IN *Y. ENTEROCOLITICA* AND *Y. ENTEROCOLITICA*-LIKE ORGANISMS

Moore and Brubaker[20] used DNA hybridization to determine that 12 strains of *Y. enterocolitica* from biotypes 1 to 4 belonged to the same species. They questioned the inclusion of *Y. enterocolitica* in the genus *Yersinia* because it was only distantly related to *Y. pseudotuberculosis* (18 to 22%). Brenner et al.[8] found 40 to 60% relatedness between *Y. entercolitica* and *Y. pseudotuberculosis* on the basis of criteria closer to optimal for determining DNA reassociation. They concluded that *Y. enterocolitica* was a perfectly good species in the genus *Yersinia*.

Brenner et al.[8] reported three separate DNA-relatedness groups and a possible fourth group formed from *Y. enterocolitica* and *Y. enterocolitica*-like strains. The first group was composed of strains from Knapp and Thal groups O:1, O:2, and O:3; strains from Niléhn's groups O:1, O:2, and O:3 (Table 3); and strains that had the following patterns for indole, xylose, esculin, and salicin, respectively: −++−, −−++, −+−+, −−−+, −+++, ++−+, and +++−. There was no evidence of species-level differences between indole-positive and indole-negative strains. Other biochemically atypical strains also fell into this DNA relatedness group, including urease-negative, nitrate-negative, methyl red-

Table 4
PERCENTAGE OF DNA RELATEDNESS AMONG Y. ENTERCOLITICA

Source of labeled DNA	Source of unlabeled DNA[a]							
	Typ	Hare	Suc	RRM	R_a	R_b	Ypst	Ent
Typ	85	80	70	60	55	65	50	10—30
Hare	85	95	65	55	—	60	50	10—30
Suc	70	65	85	60	60	60	35	10—30
RRM	60	55	60	90	55	60	55	10—30
R_a	55	45	55	50	95	55	—	—
R_b	60	55	55	60	55	80	50	10—30

[a] Data taken from Brenner, Ursing, Bercovier, Steigerwalt, Fanning, Alonso, and Mollaret, in preparation.

[b] Typ = typical; Hare = hare strains; Suc = sucrose-negative strains; RRM = rhamnose-melibiose-raffinose-positive strains; R_a and R_b = rhamnose-positive only strains that fall into two DNA hybridization groups; Ypst = *Y. pseudotuberculosis*; Ent = species in other genera within *Enterobacteriaceae*.

negative, ornithine decarboxylase-negative, lactose-positive and/or raffinose-positive, and D-mannose-negative. The second DNA relatedness group was composed of rhamnose-positive strains and the third of rhamnose-melibiose-rafinose-α-methyl-D-glucoside-positive strains. Their data suggested the existence of a fourth DNA relatedness group composed of sucrose-negative strains. They did not test hare strains from biogroup 5 of Niléhn and of Wauters (Table 3).

A comprehensive, collaborative study of *Y. enterocolitica* and *Y. enterocolitica*-like strains was recently completed by Bercovier, Alonso, and Mollaret (France); Ursing (Sweden); and Brenner, Steigerwalt, Fanning, and Carter (U.S.). Their investigation confirmed and extended the observations of Brenner et al.[8] They confirmed the existence of a DNA relatedness group containing sucrose-negative strains. The hare strains were shown to be in the same relatedness group as typical *Y. enterocolitica*. Rhamnose-positive only strains were in two separate DNA relatedness groups.

DNA relatedness among *Y. enterocolitica*, *Y. enterocolitica*-like groups, *Y. pseudotuberculosis*, and other *Enterobacteriaceae* is shown in Table 4. At conditions optimal for DNA reassociation, relatedness between strains of the same species is 70% or higher.[7] In addition, the percentage of relatedness between strains of the same species remains high (60% or more) when experiments are done at temperatures high enough to prevent reassociation of all but very closely related DNA strands. Each DNA relatedness group shown in Table 4 is composed of strains that are 80% or more related.

With one exception, relatedness between hybridization groups is 45 to 65%. The exception is that sucrose-negative strains are at the borderline of species relatedness (70%) with typical *Y. enterocolitica*. At higher temperatures the relatedness between sucrose-negative and typical *Y. enterocolitica* strains drops to between 35 and 55%. *Y. enterocolitica* and *Y. enterocolitica*-like strains are 35 to 55% related to *Y. pseudotuberculosis* and 10 to 30% related to other species of *Enterobacteriaceae*. These data are consistent with classifying *Y. enterocolitica* and *Y. enterocolitica*-like strains in the family *Enterobacteriaceae* and in the genus *Yersinia*.

V. CLASSIFICATION

On the bases of these genetic data and the biochemical data of Bercovier et al.,[3] *Y. enterocolitica* and *Y. enterocolitica*-like organisms have been classified as follows:

Table 5
BIOCHEMICAL CHARACTERISTICS OF *YERSINIA ENTEROCOLITICA* AND NEW SPECIES OF *YERSINIA*[3]

Reaction	*Y. intermedia*	*Y. frederiksenii*	*Y. kristensenii*	*Y. enterocolitica* (Wauters biotypes)				
				1	2	3	4	5
Lipase (Tween 80)	+	+	v	+	−	−	−	−
Deoxyribonuclease	−	−	v	−	+	+	+	+
Simmons' citrate	+	v	−	−	−	−	−	−
Voges-Proskauer	+	+	−	+	+	v	+	+
Ornithine decarboxylase	+	+	+	+	+	+	+	v
Indole	+	+	v	+	+	−	−	−
Acid from								
D-Xylose	+	+	+	+	+	+	−	−
Trehalose	+	+	+	+	+	+	+	−
Sucrose	+	+	−	+	+	+	+	v
Sorbose	+	+	+	+	+	v	+	v
i-Inositol	+	+	+	+	+	v	+	+
L-Rhamnose	+	+	−	−	−	−	−	−
Melibiose	+	−	−	−	−	−	−	−
Raffinose	+	−	−	−	−	−	−	−
α-CH$_3$-D-glucoside	+	−	−	−	−	−	−	−

Note: Reactions done at 28°C. + = positive; − = negative; v = variable.

1. *Y. enterocolitica:* Includes biotypes 1 to 5 of Niléhn and of Wauters, with biogroup 5 containing both sucrose-negative and sucrose-positive, trehalose-negative hare strains.[3a]
2. *Y. intermedia:* A new species composed of rhamnose-positive, melibiose-positive, raffinose-positive strains.[3a]
3. *Y. frederiksenii:* A new species composed of rhamnose-positive DNA relatedness groups R$_a$ and R$_b$ (Table 4) (Ursing, Brenner, Bercovier, Fanning, Steigerwalt, Alonso, and Mollaret, in preparation). At this time there are no simple biochemical tests with which to separate R$_a$ and R$_b$. If such tests become available, R$_a$ and R$_b$ may become separate species.
4. *Y. kristensenii:* A new species composed of sucrose-negative, Voges-Proskauer-negative strains.[3a]

Biochemical reactions useful in defining these species and differentiating between them are shown in Table 5. These species, especially *Y. intermedia*, are more active metabolically at 22 to 28°C than at 35 to 37°C. The data in Table 5 were generated at 28°C.

These four species, generated from what were considered to be all *Y. enterocolitica* or *Y. enterocolitica* and *Y. enterocolitica*-like organisms, also differ in serology, habitat, and pathogenicity.[1,4,5,16,17,19,21,23,28] The establishment of these species after detailed biochemical and genetic study provides an excellent example of how proper classification eliminates confusion for medical and nonmedical microbiologists.

REFERENCES

1. **Alonso, J. M., Bejot, J., Bercovier, H., and Mollaret, H. H.**, Sur un groupe de souches de *Yersinia enterocolitica* fermentant le rhamnose, *Med. Mal. Infect.,* 10, 490, 1975.
2. **Bejot, J., Alonso, J. M., and Mollaret, H. H.**, Le diagnostic bacteriologique des yersinioses humanines (infections à *Yersinia pseudotubercolosis* et *Yersinia enterocolitica), Med. Mal. Infect.,* 5, 233, 1975.
3. **Bercovier, H., Alonso, J. M., Bentaiba, Z. N., Brault, J., and Mollaret, H. H.**, Contribution à la definition et à la taxonomie de *Yersinia enterocolitica,* in press.
3a. **Bercovier, H., Brenner, D. J., Fanning, G. R., Alonso, J. M., Carter, P., Mollaret, H. H., Ursing, J., and Steigerwalt, A. G.,** *Curr. Microbiol.,* in press.
4. **Bottone, E. J.,** *Yersinia enterocolitica:* a panoramic view of a charismatic microorganism, *CRC Crit. Rev. Microbiol.* 5, 211, 1977.
5. **Bottone, E. J., Chester, B., Malowany, M., and Allerhand, J.,** Unusual *Yersinia enterocolitica* isolates not associated with mesenteric lymphadenitis, *Appl. Microbiol.,* 27, 858, 1974.
6. **Botzler, R. G., Wetzler, T. F., and Cowan, A. B.,** *Yersinia enterocolitica* and *Yersinia*-like organisms isolated from frogs and snails, *Bull. Wildl. Dis. Assoc.,* 4, 110, 1968.
7. **Brenner, D. J., Fanning, G. R., Skerman, F. J., and Falkow, S.,** Polynucleotide sequence divergence among strains of *Escherichia coli* and closely related organisms, *J. Bacteriol.,* 109, 953, 1972.
8. **Brenner, D. J., Steigerwalt, A. C., Falcao, D. P., Weaver, R. E., and Fanning, G. A.,** Characterization of *Yersinia enterocolitica* and *Yersinia psuedotuberculosis* by deoxyribonucleic acid hybridization and by biochemical reactions, *Int. J. Syst. Bacteriol.,* 26, 180, 1976.
9. **Daniels, J. J. H. M. and Goudzwaard, C.,** Enkle stammen van een op *Pasteurella pseudotuberculosis* Gelijkend niet Geidentifeceerd species, Geisoleerd bij knaagdieren, *Diergeneesk,* 88, 96, 1963.
10. **Darland, G., Ewing, W. H., and Davis, B. R.,** The Biochemical Characteristics of *Yersinia enterocolitica* and *Yersinia pseudotuberculosis,* Center for Disease Control Publication, Atlanta, 1975.
11. **Domaradskij, I., Marchenkov, V., and Shimanjuk, N.,** New data obtained in comparative studies of plaque agent and related organisms, *Contrib. Microbiol. Immunol.,* 2, 2, 1973.
12. **Ewing, W. H., Ross, A. J., Brenner, D. J., and Fanning, G. R.,** *Yersinia ruckeri* sp. nov., the Redmouth (RM) bacterium, *Int. J. Syst. Bacteriol.,* 28, 37, 1978.
13. **Frederiksen, W.,** A study of some *Yersinia pseudotuberculosis*-like bacteria *("Bacterium enterocoliticum"* and *Pasteurella* X"). *Scand. Congr. Pathol. Microbiol.,* 14, 103, 1964.
14. **Hässig, A., Karrer, J., and Pusterla, F.,** Uber pseudotuberkulose beim menschem, *Schweiz. Med. Wohenschr.,* 79, 971, 1949.
15. **Knapp, W. and Thal, E.,** Differentiation of *Yersinia enterocolitica* by biochemical reactions, *Contrib. Microbiol. Immunol.,* 2, 10, 1973.
16. **Knapp, W. and Thal, E.,** A simplified schema for *Yersinia enterocolitica* (synonym: *"Pasteurella* X") based on biochemical characteristics, *Zentralbl. Bakteriol. Parasitenkd. Infektionskr. Hyg. Abt. I Orig.,* 223, 88, 1973.
17. **Lassen, J.** *Yersinia enterocolitica* in drinking water, *Scand. J. Infect. Dis.,* 4, 125, 1972.
18. **Mollaret, H. H. and Knapp, W.,** International Committee on Systematic Bacteriology Subcommittee on the Taxonomy of *Pasteurella Yersinia* and *Francisella, Int. J. Syst. Bacteriol.,* 22, 401, 1972.
19. **Mollaret, H. H. and Lucas, A.,** Sur les particularites biochemiques des souches de *Yersinia enterocolitica* isolees chez les lievres, *Ann. Inst. Pasteur (Paris),* 108, 121, 1965.
20. **Moore, R. L. and Brubaker, R. R.,** Hybridization of deoxyribonucleatide sequences of *Yersinia enterocolitica* and other selected members of *Enterobacteriaceae, Int. J. Syst. Bacteriol.,* 25, 336, 1975.
21. **Nilehn, B.,** Studies on *Yersinia enterocolitica* with special reference to bacterial diagnosis and occurrence in human acute enteric disease, *Acta Pathol. Microbiol. Scand. Suppl.,* 206, 1, 1969.
22. **Schleifstein, J. and Coleman, M. B.,** An unidentified microorganism resembling *B. lignieresi* and *Pasteurella pseudotuberculosis,* pathogenic for man, *N.Y. State J. Med.,* 39, 1749, 1939.
23. **Smith, J. E. and Thal, E.,** A taxonomic study of the genus *Pasteurella* using a numerical technique, *Acta Pathol. Microbiol. Scand.,* 64, 213, 1965.
24. **Stevens, M. and Mair, N. S.,** A numerical taxonomic study of *Yersinia enterocolitica* strains, *Contrib. Microbiol. Immunol.,* 2, 17, 1973.
25. **Thal, E.,** Untersuchungen uber *Pasteurella pseudotuberculosis,* Berlingska Boktryckeriet, Lund, 1954.
26. **Van Loghem, J. J.,** The classification of plague-bacillus, *Antonie van Leeuwenhoek,* 10, 15, 1944.
27. **Van Riesen, V. L.,** Polypectate digestion by *Yersinia, J. Clin. Microbiol.,* 2, 552, 1975.
28. **Wauters, G.,** Contribution à l'étude de *Yersinia enterocolitica,* Ph.D. Thesis, Vander, Louvain, Belgium, 1970.
29. **Mandel, M.,** personal communication, 1979.

Chapter 2

ISOLATION TECHNIQUES FOR *YERSINIA ENTEROCOLITICA*

James C. Feeley

TABLE OF CONTENTS

I.	Introduction	10
II.	Collection and Transport of Specimens	10
	A. Clinical Specimens	10
	B. Environmental Specimens	11
	C. Food Specimens	11
III.	Laboratory Methods	11
	A. Sample Preparation	11
	1. Clinical Specimens	11
	2. Environmental Specimens	11
	3. Foodstuffs	11
	B. Inoculation of Selective Media	12
	1. Agar Media	12
	2. Broth Media	12
	C. Incubation of Selective Plating Media	12
	D. Examination of Selective Plating Media	12
	E. Selection of Suspect Colonies	13
	1. Biochemical Screening	13
	2. Presumptive Identification	13
	3. Definitive Identification	13
IV.	Conclusion	14
References		14

I. INTRODUCTION

Yersinia enterocolitica has been isolated worldwide from a variety of human clinical specimens such as wounds, mesenteric lymph nodes, abscesses, feces, and sputa.[1] It has also been recovered from numerous animal hosts such as the beaver, birds, cats, cows, deer, horses, oysters, and fish.[2] These findings and the detection of this bacterium in foodstuffs such as meat, ice cream, oysters, and water has resulted in an increased concern by public health officials and medical scientists (human and veterinary) for the role of this organism in human and animal disease.[3]

Although numerous procedures and media have been recommended for the isolation of *Y. enterocolitica*, those described here are used personally by the author[3] and they are listed according to the type of specimen being examined, namely, clinical, environmental, or foodstuffs.

II. COLLECTION AND TRANSPORT OF SPECIMENS

Successful isolation of *Y. enterocolitica* bacteria from any of the three specimen types requires that the sample: (1) be adequate in size, especially those to be cold enriched; (2) be protected from dehydration; and (3) be generally packed in wet ice during transport from the field to the laboratory. Additionally, specimens should be refrigerated in the laboratory while being held for examination.

A. Clinical Specimens

Both human and animal specimens should be handled according to their clinical nature.

Tissues, blood, CSF, pleural fluid, and wound exudates — These may contain *Y. enterocolitica*, if present, in relatively pure culture. Consequently, *Y. enterocolitica* may be easily isolated from these sources and such specimens can be collected and transported to the laboratory by procedures generally recommended and routinely practiced for these specimen types.

Feces and throat cultures — These cultures usually contain large numbers of other bacteria. If *Y. enterocolitica* is present, it may not be the predominant organism in the mixed flora. Consequently, isolation is more difficult. Therefore, these specimens must be collected, transported, and held in the laboratory in a manner to prevent the overgrowth of *Y. enterocolitica* by other microbial species. It is recommended that the specimen be held in $1/15\ M$ disodium phosphate buffer (PBS) pH 7.6 for 21 days at 4°C and subcultured at various intervals. The frequency of isolating *Y. enterocolitica* has been increased by this cold enrichment.[7,8] The rationale for this procedure results in the fact that *Y. enterocolitica* grows at 4°C while most other bacteria present in the mixed flora either do not grow or die.[9] In order for the procedure to be successful, a minimal but adequate amount of nutrient must be available. Generally the specimen itself provides this. This is true for rectal swabs that contain 0.5 g of feces and are placed in 5.0 mℓ of PBS (phosphate buffered saline). PBS-containing throat and rectal swabs that are not suspended to a 10% (wt/vol) suspension must be given supplemental nutrition; addition of 10% peptone broth (0.5 to 5.0 mℓ PBS) is recommended. Urine specimens should also be cold enriched after dilution to a 10% (vol/vol) concentration in 1% peptone PBS broth. Recently an all-purpose transport medium developed by Cary and Blair has been used successfully to transport and cold-enrich specimens from an outbreak of gastrointestinal illness caused by *Y. enterocolitica*.[6,10] While other thioglycolate-base media such as Stuart's and Amie's were not evaluated, they should also be suitable.

B. Environmental Specimens

Soil — Soil samples should be collected aseptically, placed in sealable plastic bags, packed in wet ice, and transported to the laboratory where on arrival they should be refrigerated.

Water — Water may be sampled in the field by filtering duplicate 100-mℓ samples through each of two 0.45-μm membrane filters. If this equipment is unavailable, specimens should be collected in sterile plastic bottles that contain sufficient sodium thiosulfate to neutralize residual chlorine. Both types of samples should be transported to the laboratory packed in wet ice. Each filter may be placed in a small volume of PBS to protect it from dehydration.

C. Food Specimens

Foodstuffs should be collected aseptically and be placed in sealed sterile containers to prevent dehydration. They should also be refrigerated or packed in wet ice during transport and until analyzed.

III. LABORATORY METHODS

A. Sample Preparation

1. Clinical Specimens

Clinical samples collected in transport media require no additional preparation and should be inoculated to selective media immediately and again after being held at 4°C for 21 days. Recommended media are described in a following section. Plating of cold-enriched specimens may not be necessary if other enteric pathogens or *Y. enterocolitica* are isolated by direct plating. The latter is generally true in countries like Canada where serotype O:3 *Y. enterocolitica* is prevalent and is easily isolated on direct plating of diarrheic stools.[11] Isolation of *Y. enterocolitica* serotypes other than O:3, such as O:8, that are prevalent in the U.S. generally requires cold enrichment in order for successful recovery. Specimens that are not inoculated directly to transport media should be reinoculated to such a medium and handled as previously described.

2. Environmental Specimens

Soil samples — Place 20 g of soil in 100 mℓ sterile distilled water containing 0.5% Tween 80® as a wetting agent and mix on a mechanical shaker for 1 hr at moderate speed. Leave undisturbed for 5 to 10 min to allow the large soil particles to settle and transfer 10 mℓ of the supernatant into 90 mℓ of 1% peptone PBS broth. Remove several 0.1-mℓ aliquots immediately and plate onto selective media. Repeat after 21 days of cold enrichment at 4°C.

Water samples — Specimens that were not passed through a 0.45-μm membrane filter on collection in the field should be filtered on arrival at the laboratory. Two membrane filters (MF) are recommended per sample (100 mℓ water/MF). One membrane is directly cultured on M. Endo broth and incubated at 25°C for 72 hr.[5,12] The second membrane is placed into 10 mℓ of Cooked Meat broth and held at 4°C for 21 days. If membrane filtering equipment is not available, two 100-mℓ aliquots of water should be centrifuged at 2900 × g for 30 min. The sediment is resuspended in 1.0 mℓ of sterile distilled water. Three 0.1-mℓ aliquots are plated immediately to selective media and the remainder is added to 10 mℓ Cooked Meat broth and held at 4°C for 21 days.

3. Foodstuffs

Liquid samples — Specimens that can be either membrane filtered or centrifuged should be treated as a water specimen. Specimens such as milk that cannot be filtered but have adequate nutrient to grow *Y. enterocolitica* should be plated to selective media immediately and after enrichment at 4°C for 21 days. Peptone broth (1%) should be added to other specimens requiring supplemental nutrition in a 9:1 proportion before they are plated and cold enriched. It should be noted that contaminated chocolate milk is the only food that has been proven both bacteriologically and epidemiologically linked to a major outbreak of yersiniosis.[13] Readers are referred to the article by Schiemann and Toma[14] for a more detailed description of available methods for culturing milk.

Nonliquid samples — Specimens that are solid or semisolid should be suspended in both PBS and 1% peptone broth by blending duplicate 25-g amounts of the food in 225 mℓ of each liquid. Both preparations should be plated and cold enriched as recommended and previously described for liquid food specimens.

B. Inoculation of Selective Media

Samples in transport media and those subsequently prepared in the laboratory should be inoculated immediately to selective plating agars and optionally to selective liquid broths followed by plating again to selective agar.

1. Agar Media

Y. enterocolitica has a tolerance to high concentrations of bile salts. Consequently most of the common enteric plating media have been used successfully to isolate *Y. enterocolitica*. Examples are MacConkey's agar, SS agar, Levine's eosin methylene blue (EMB) agar, Tergitol 7 agar, lactose sucrose urea agar, and desoxycholate citrate agar. Recently several new media have been developed that offer enhanced differential detection of *Y. enterocolitica*.[15-17] Additionally, *Y. enterocolitica* has been found to form characteristic black colonies on bismuth sulfite agar.[18]

The author prefers to use the two plating media, MacConkey's agar and SS (*Salmonella-Shigella*) agar adjusted to pH 7.4. Strains may vary in their growth on various media, but most will grow on MacConkey's agar. However, knowledge that a specific serotype can grow on a more selective medium should be used to advantage.[19]

2. Broth Media

A variety of selective broths such as magnesium chloride-malachite green-carbenicillin (MCMGC) broth,[19] selenite F broth,[2] and tetrathionate broth[20] have been used for isolating *Y. enterocolitica*. Selection of a specific medium should be according to the *Y. enterocolitica* serotype sought. For O:3 organisms, the MCMGC broth is very effective. However, this broth is not very good for isolating serotype O:8 bacteria and one of the other less restrictive broths should be used. When tetrathionate broth is used, it must not contain brilliant green because this dye is toxic to *Y. enterocolitica*. If a selective broth is used, transfer 0.1 mℓ of the cold-enriched sample into 10 mℓ of the selective broth and incubate at 25°C for 48 hr before inoculating selective plating media.

C. Incubation of Selective Plating Media

Although *Y. enterocolitica* is a member of the family *Enterobacteriaceae*,[21] it grows better than other enteric bacteria at 25°C than at 36°C. This offers a selective advantage for isolation and primary cultures ideally should be incubated at 25°C for 48 hr. However, some clinical laboratories may wish to initially incubate their cultures at 36°C for the first 24 hr to enhance the isolation of *Salmonella* and *Shigella* followed by incubation at 25°C for a second 24 hr in order to isolate *Yersinia*.

D. Examination of Selective Plating Media

Y. enterocolitica colonies on SS agar are round, opaque, and colorless, and are about 1 mm in diameter, whereas colonies on MacConkey's agar can be as large as 3 mm (after

Table 1
REACTIONS OF *YERSINIA ENTEROCOLITICA* AND OTHER CLOSELY RELATED BACTERIA

Tests	*Y. enterocolitica*	*Y. enterocolitica-like*[a]	*Y. pseudotuberculosis*	*Y. pestis*	*Vibrio cholerae*	*Aeromonas hydrophila*	*Serratia*	*Enterobacter*	*Citrobacter diversus*	*Klebsiella*	*Proteus morganii*	*Proteus rettgeri*	*Providencia*	*Chromobacter violaceum*
Oxidase	−	−	−	−	+	+	−	−	−	−	−	−	−	−(W+)
Christensen's urea	+	+	+	−	−	−(+)	−(+)	−(+)	+,−	+,−	+	+	−	−(L+)
Simmon's citrate	−	+,−	−	−	+(−)	+(−)	+	+	+	+,−	−	+	−	+(L)
Motility −25°C	+	+	+	−	+	+	+	+	+	−	+	+	+	+
Motility −36°C	−	−	−	−	+	+	+	+	+	−	+	+	+	+
Arginine dihydrolase	−	−	−	−	−	+	−	−(+)	+	−	−	−	−	+
Lysine decarboxylase	−	−	−	−	+	−(+)	+	+(−)	−	+,−	−	−	−	−
Ornithine decarboxylase	+	+	−	−	+	−	+	+(−)	+	−	+	−	−	−
Phenylalanine deaminase	−	−	−	−	−	−	−	+,−	−	−	+	+	+	−
Melibiose −25°C	−	+,−	+	−	−	−	−	+	−	+	−	−	−	−
Raffinose −25°C	−	+,−	−	−	−	−	−	+(−)	−	+	−	−	−	−
Rhamnose −25°C	−	+,−	+	−	−	−	−	+	+	+	−	+(−)	−	−

Note: Tests performed at 36°C unless otherwise noted ( ) = minority of reactions; W = weak; L = late.

[a] *Y. frederiksenii, Y. intermedia, Y. kristensenii.*

48 hr) in diameter and are pale pink or peach colored. *Y. enterocolitica* colonies are more easily recognized when a dissecting microscope is used with oblique lighting.[2,19,22] This is especially recommended for examination of MacConkey agar.

E. Selection of Suspect Colonies

1. Biochemical Screening

At least three lactose-negative colonies should be selected from each selective agar medium. Although lactose is usually not fermented by *Y. enterocolitica* on these media, occasional organisms can be encountered that rapidly ferment this carbohydrate. For this reason, several lactose-positive colonies should also be selected when they closely resemble *Y. enterocolitica* colonies in size and morphology.

Each suspect colony should be screened by inoculating triple sugar iron agar (TSI), two semisolid tubes of motility media, and a urea agar slant. All media are incubated at 25°C except one of the motility media which is incubated at 36°C. The TSI reaction at 24 hr is an acid slant, acid butt, with no gas and no H_2S. Reversion of the slant may occur at 25°C because of very rapid fermentation of sucrose followed by oxidative degradation of peptones. *Y. enterocolitica* is nonmotile at 36°C but is motile at 25°C; it is urea-positive.

2. Presumptive Identification

Isolates resembling *Y. enterocolitica* in the screening media should be tested with the biochemicals listed in Table 1. These are minimal tests which separate *Yersinia* species from other closely related bacteria.

3. Definitive Identification

Isolates that have been presumptively identified as a *Yersinia* sp. should be further tested according to the schema presented by Brenner (Chapter 1), as outlined by Bottone (Chapter 3), and elsewhere.[23-25]

IV. CONCLUSION

Although the frequency of recovery of *Y. enterocolitica* from clinical specimens has dramatically increased in recent years, there is still a need for improvement in the methods for rapid detection and recovery of this bacterium. This is especially true in the U.S. where timely clinical diagnosis is prevented by the need for cold enrichment of specimens that may contain other than "classic" serotypes.

Advances in the processing of environmental and food specimens are needed, as well as in methods for the selective recovery of *Y. enterocolitica* of specific serotype from specimens containing species which are ubiquitously present in the environment. The development of such methodologies is a key to the ultimate understanding of the epidemiology and ecology of this bacterium.

REFERENCES

1. **Bottone, E. J.,** *Yersinia enterocolitica:* a panoramic view of a charismatic microorganism, *CRC Crit. Rev. Microbiol.,* 5, 211, 1977.
2. **Toma, S.,** Survey on the incidence of *Yersinia enterocolitica* in the province of Ontario, *Can. J. Public Health,* 64, 477, 1973.
3. **Morris, G. K. and Feeley, J. C.,** *Yersinia enterocolitica:* a review of its role in food hygiene, *Bull. W. H. O.,* 54, 79, 1976.
4. **Feeley, J. C., Lee, W. H., and Morris, G. K.,** *Yersinia enterocolitica,* in *Compendium of Methods for the Microbiological Examination of Foods,* Speck, M. L., Ed., American Public Health Association, Washington, D. C., 1976.
5. **Highsmith, A. K., Feeley, J. C., and Morris, G. K.,** Isolation of *Yersinia enterocolitica* from water, in *Bacterial Indicators/Health Hazards Associated with Water,* Hoadley, A. W. and Dutka, B. J., Eds., American Society for Testing and Materials, Philadelphia, 1977.
6. **Morris, G. K., Feeley, J. C., Martin, W. T., and Wells, J. G.,** Isolation and identification of *Yersinia enterocolitica, Public Health Lab.,* 35, 217, 1977.
7. **Patterson, J. S. and Cook, R.,** A method for recovery of *Pasteurella pseudotuberculosis* from feces, *J. Pathol. Bacteriol.,* 85, 241, 1963.
8. **Tsubokura, M., Orsuki, K., and Itagaki, K.,** Studies on *Yersinia enterocolitica.* I. Isolation of *Yersinia enterocolitica* from swine, *Jpn. J. Vet. Sci.,* 35, 419, 1973.
9. **Niléhn, B.,** Studies on *Yersinia enterocolitica* with special reference to bacterial diagnosis and occurrence in human acute enteric disease, *Acta Pathol. Microbiol. Scand.,* Suppl. 206, 5, 1969.
10. **Cary, S. G. and Blair, E. B.,** New transport medium for shipment of clinical specimens. I. Fecal specimens, *J. Bacteriol.,* 88, 96, 1964.
11. **Sorger, S., Pai, C. H., Lafleur, L., Lackman, L., Hammerberg, O., and Marks, M. I.** Is Cold Enrichment Necessary for Isolation of *Yersinia enterocolitica* (YE) from Stool in a Clinical Laboratory, Abstr. No. 124, Intersci. Conf. Antimicrobial Agents Chemotherapy, American Society for Microbiology, Washington, D.C., 1978.
12. **Highsmith, A. K., Feeley, J. C., Skaliy, P., Wells, J. G., and Wood, B. T.,** Isolation of *Yersinia enterocolitica* from well water and growth in distilled water, *Appl. Environ. Microbiol.,* 745, 1977.
13. **Black, R. E., Jackson, R. J., Tsai, T., Medvesky, M., Shayegani, M., Feeley, J. C., MacLeod, K. I. E., and Waklee, A. M.,** Epidemic *Yersinia enterocolitica* infection due to contaminated chocolate milk, *N. Engl. J. Med.,* 298, 76, 1978.
14. **Schiemann, D. A. and Toma, S.,** Isolation of *Yersinia enterocolitica* from raw milk, *Appl. Environ. Microbiol.,* 35, 54, 1978.
15. **Lee, W. H.,** Two tween media for the isolation of *Yersinia enterocolitica, Appl. Environ. Microbiol.,* 35, 215, 1977.
16. **Saari, T. N. and Quan, T. J.,** Waterborne *Yersinia enterocolitica* in Colorado, Abstr. No. C119, Annu. Meet. American Society Microbiology, Washington, D.C., 1976.
17. **Dudley, V. M. and Shotts, E. B.,** A medium for isolating *Yersinia enterocolitica, J. Clin. Microbiol.,* Aug. 1979.

18. **Hanna, M. O., Stewart, J. C., Carpenter, Z. L., and Vanderzant, C.**, Development of *Yersinia enterocolitica*-like organisms in pure and mixed cultures on different bismuth sulfite agars, *J. Food Prot.*, 40, 676, 1977.
19. **Wauters, G.**, Improved methods for the isolation and the recognition of *Yersinia enterocolitica*, *Contrib. Microbiol. Immunol.*, 2, 68, 1973.
20. **Wetzler, T. F.**, Pseudotuberculosis, in *Diagnostic Procedures for Bacterial, Mycotic and Parasitic Infections*, Bodily, H. L., Updyke, E. L., and Mason, J. O., Eds., American Public Health Association, Washington, D.C., 1970.
21. **Mollaret, H. H. and Thal, E.**, Yersinia, in *Bergey's Manual of Determinative Bacteriology*, 8th ed., Buchanan, R. E. and Gibbons, N. E., Eds., William & Wilkins, Baltimore, 1974.
22. **Highsmith, A. K., Feeley, J. C., and Morris, G. K.**, *Yersinia enterocolitica:* a review of the bacterium and recommended laboratory methodology, *Health Lab. Sci.*, 14, 253, 1977.
23. **Brenner, D. J.**, Classification of *Yersinia enterocolitica*, in *Yersinia enterocolitica: A Multifaceted Human Pathogen*, Bottone, E. J., Ed., CRC Press, Boca Raton, Fla., 1980.
24. **Brenner, D. J., Steigerwalt, A. G., Falcao, D. P., Weaver, R. E., and Fanning, G. R.**, Characterization of *Yersinia enterocolitica* and *Yersinia pseudotuberculosis* by deoxyribonucleic acid hybridization and by biochemical reactions, *Int. J. Syst. Bacteriol.*, 26, 180, 1976.
25. **Hawkins, T. M. and Brenner, D. J.**, Isolation and Identification of *Yersinia*, Center for Disease Control Manual, Atlanta, 1978.

Chapter 3

YERSINIA ENTEROCOLITICA: AN APPROACH TO LABORATORY IDENTIFICATION WITH REFERENCE TO DEOXYRIBONUCLEIC ACID HYBRIDIZATION GROUPS AND BIOCHEMICAL CHARACTERISTICS

Edward J. Bottone

TABLE OF CONTENTS

I.	Introduction	18
II.	Microscopic Morphology	18
III.	Cultural Characteristics	18
IV.	Biochemical Characteristics	20
	A. Typical *Y. enterocolitica*	20
	B. Rhamnose-Fermenting Strains	25
	C. Sucrose-Negative Strains	27
Acknowledgment		28
References		28

I. INTRODUCTION

The laboratory identification of the Gram-negative bacterium, *Yersinia enterocolitica*, proceeds through two distinct but interwoven avenues — namely recognition of suspect colonies and awareness of several biochemical characteristics, some temperature-dependent, necessary for ascription of an isolate to the genus *Yersinia*. Having achieved these initial parameters, of equal importance is the recognition of subtleties regarding biochemical characteristics of suspect isolates which diverge from prototype *Y. enterocolitica*. In this regard, unlike *Y. pseudotuberculosis* which is biochemically homogeneous (see Chapter 4), *Y. enterocolitica*-like isolates, especially rhamnose-fermenting varieties, display a wide range of biochemical heterogeneity which may mask identification as *Yersinia*.

The design of this chapter is to present to the reader an approach to the laboratory identification of *Y. enterocolitica* which utilizes a general schema based on the four DNA-DNA homology groups of Brenner et al.[1] which correlates with biochemical activity. Briefly, as shown in Table 1, Brenner et al., in their significant contribution, showed that *Y. enterocolitica* strains formed three and possibly four DNA relatedness groups which could be defined biochemically. Group one corresponds to typical *Y. enterocolitica* both indole-positive and -negative strains; members of the second group differed from typical strains in being rhamnose-positive (but raffinose-, melibiose-, and α-methyl glucoside-negative); group three strains were positive for all four characteristics; the fourth hybridization group was comprised of sucrose-negative isolates. The nomenclature to be proposed for members of groups two, three, and four (as proposed by Brenner[2] and Brenner, Bercovier, Ursing, Alonso, Steigerwalt, Fanning, Carter and Mollaret) is also included.

The ensuing discussion will deal with the morphological, cultural, and biochemical attributes of typical *Y. enterocolitica* and *Y. enterocolitica*-like species.

II. MICROSCOPIC MORPHOLOGY

Microscopically, *Y. enterocolitica* may present as small *Pasteurella*-like coccobacilli when observed in direct smears of clinical material (Figure 1) or after 24 hr growth on blood agar at either 25 or 37°C. Bipolar staining may be evidenced in smears but this staining tendency is not considered distinctive of this species.[3] *Enterobacteriaceae*-like bacilli ranging from 1 to 3.5 μm in length and from 0.5 to 1.3 μm in width[4] may be encountered in smears prepared from more selective "enteric" media. Irrespective of nutrient base, pleomorphic filament forms with vacuolations may also be observed in smears from older cultures incubated especially at 37°C. Capsules may be visualized in clinical exudates (Figure 1).[5]

Y. enterocolitica is motile at 25°C by peritrichous flagella[3] which impart to the microorganism a tumbling, spinning motion analogous to that observed with *Listeria monocytogenes*. At strict 37°C incubation temperatures, *Y. enterocolitica* is nonmotile, or at best a sluggish motility may be observed with a few cells. Bissett,[6] however, has reported motility at 35°C while Brenner[35] has encountered strains that lack motility at 25°C.

III. CULTURAL CHARACTERISTICS

Procedures for the recovery of *Y. enterocolitica* from primary specimens are described in Chapter 2 and are highlighted in Chapter 17. The text of these chapters deals primarily with the cultivation of typical *Y. enterocolitica* of which colony characteristics on a

Table 1
CORRELATION BETWEEN DEOXYRIBONUCLEIC ACID HYBRIDIZATION GROUPS AND BIOCHEMICAL REACTIONS OF *Y. ENTEROCOLITICA*[a]

Species	Representative serotype	DNA homology group	Biochemical characteristics
Y. enterocolitica	O:3, O:5, 27 O:8, O:9	I	Typical
Y. frederiksenii[a]	O:16	II	Rhamnose-positive, raffinose, melibiose α-methyl glucoside negative
Y. intermedia[a]	O:17, NAG	III	Rhamnose, raffinose melibiose, α-methyl glucoside positive
Y. kristensenii[a]	O:11, 0:12	IV	Typical — except for negative sucrose fermentation

[a] After Brenner.[1]

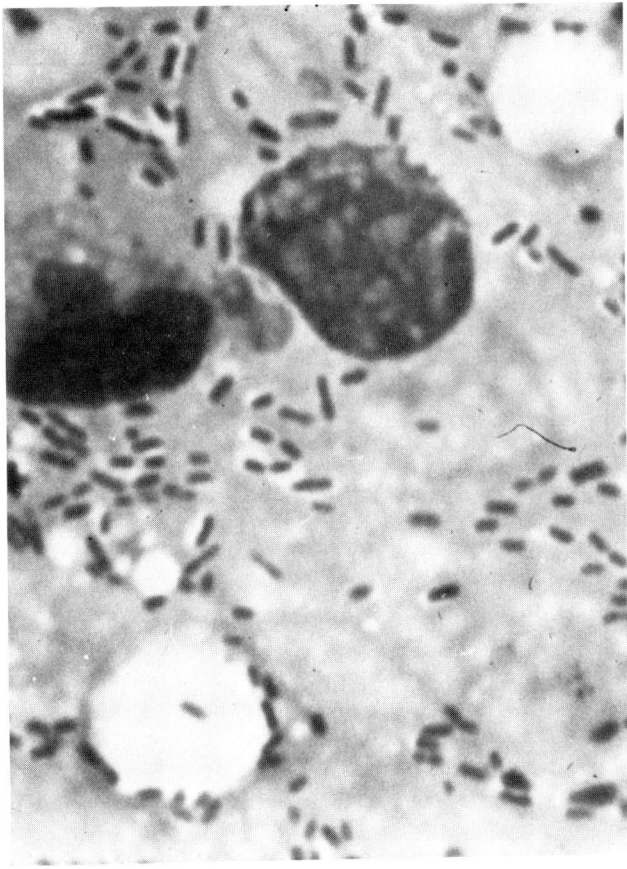

FIGURE 1. *Y. enterocolitica;* smear of peritoneal exudate of experimentally infected white mouse showing bacillary and cocco-bacillary forms with rounded ends (Magnification × 1000.)

variety of routinely used media are presented in Table 2. It is imperative that attempts to isolate *Y. enterocolitica* be accompanied by (1) awareness of the diversity of "color" reactions rendered by this microorganism on the more commonly utilized "enteric" media and (2) the inability of some biochemically and serologically atypical isolates to initiate growth, or at best grow sparsely, on Hektoen-enteric (H-E), xylose lysine deoxycholate (XLD), or *Salmonella-Shigella* (SS) agar.[5-10] In our experience, cultural attempts at isolation of *Y. enterocolitica* as well as *Y. enterocolitica*-like strains should include MacConkey or Endo agar in addition to other "enteric" media. It is also advisable to inoculate duplicate sets of media and incubate one set at 25 to 29°C.[3,4]

Yersinia species generally do not ferment lactose, (although a lactose-raffinose plasmid has been reported) but, with the exception of Brenner's group 4 strains, utilize sucrose readily. Thus, colonies of strains growing on H-E, XLD, or EMB, each of which contains sucrose, will mimic those of typical "coliforms" and hence often be disregarded. Additionally, colonies on these and other isolation media are pinpoint after 24 hr incubation (Table 2) and frequently overlooked.

IV. BIOCHEMICAL CHARACTERISTICS

A. Typical *Y. enterocolitica*

The biochemical characterization of typical *Y. enterocolitica* should proceed without great difficulty. Those strains most frequently encountered in human disease present with relatively stable biochemical (and serological) characters. As noted in several chapters herein, serotypes O:3 and O:9 comprise the majority of human isolates on a global basis, while in the U.S. serotypes O:5,27, and especially O:8 predominate. Although serotype O:3 is considered rare in the U.S., two serotype O:3 isolates were encountered in the survey by Bissett[6] and it is beginning to appear that serotype O:3 is emerging as a causative agent of yersiniosis in the U.S.[11-14] and is being encountered from environmental samples in New York state (Chapter 16).

Biochemically, the salient constellation of characteristics suggesting that an isolated glucose-fermenting bacterium could be *Y. enterocolitica* or even *Y. pseudotuberculosis* is denoted in Table 3 and includes negative reactions for oxidase, phenylalanine deaminase, lysine decarboxylase, and arginine dihydrolase. Ornithine is decarboxylated by *Y. enterocolitica*, which is one of the criteria enabling distinction from *Y. pseudotuberculosis* (Chapter 5). Positive reactions also include urease and the temperature-related expression of β-galactosidase, acetylmethylcarbinol (Voges-Proskauer test), and motility at 25°C but not at 37°C. Indole production is variable for *Y. enterocolitica* but negative in *Y. pseudotuberculosis*. Exceptions to some of these characteristics do occur and are discussed below with special precautions for choice of substrates necessary to elicit the distinguishing features of *Y. enterocolitica*.

The biochemical identification of a suspect *Y. enterocolitica* as alluded to should be carried out at both 25 and 37°C, as many of the indicated results are manifested only at the lower incubation temperature. This facet of laboratory identification becomes even more critical when dealing with the rhamnose-fermenting varieties which display a greater range of temperature-dependent features.[7-9,15] Some of the more important phenotypic traits affected by temperature or substrates utilized include the following.

Indole production — Strain and serotype variability exist. For instance, serotype O:3 is constantly indole-negative whereas serotypes O:8 and O:9 are indole producers. The capability or lack thereof of producing indole from tryptophane is of prime importance in distinguishing among *Y. enterocolitica* biotypes. It is stressed that the degree of positivity will vary with incubation temperature, source of tryptophane, and reagent used for detection. Although exceptions have been reported by Niléhn,[4] in general,

Table 2
COLONY CHARACTERISTICS OF *Y. ENTEROCOLITICA* SEROTYPE 0:3, 0:8, AND 0:9 ON COMMONLY USED MEDIA

Agar medium	Colony characteristics				
	24 hr		48 hr		Comment
	22°C(mm)	37°C(mm)	22°C(mm)	37°C(mm)	
Blood[+]	Pinpoint	0.05	1.0—2.5	2.0—3.5	Smooth irregular contours
Deoxycholate (DC)	Pinpoint[a]	0.05—1	1.5—2	2—2.5	Colorless
Endo	0.5—1	0.05—1.5	1.5—2	1.5—2.0	Colorless
Eosin methylene blue (EMB)	Pinpoint	Pinpoint	0.01—0.5	0.1—0.5	Lavender (24 hr) metallic sheen (48 hr)
Hektoen-enteric (H-E)[b]	Pinpoint	Pinpoint	2.0	2.0	Salmon colored
MacConkey	Pinpoint	1.5—2	1—1.5	2.0	Pinkish hue
Salmonella-Shigella (SS)[b]	Pinpoint—0.05	Pinpoint—0.05	0.5—2.0	1.0—2.0	Colorless
Xylose-lysine deoxycholate (XLD)	Pinpoint	Pinpoint	0.5—2.5	1.0—1.5	Yellow; irregular edges

Note: [+] 5% Sheep blood-trypticase soy agar.

[a] Barely perceptible.
[b] Rhamnose-positive strains failed to grow on H-E and SS agars.

Table 4
BIOTYPES OF *YERSINIA ENTEROCOLITICA* AS PROPOSED BY NILÉHN,[4] WAUTERS,[3] AND BERCOVIER[27]

Tests	Nilehn					Wauters					Bercovier
	1	2	3	4	5	1	2	3	4	5	Indole
Lecithinase[a]	\multicolumn{5}{c}{Not tested}	+	−	−	−	−	
Salicin (acid)	+	−	−	−	−	\multicolumn{5}{c}{Not included}	
Esculin hydrolysis	+	−	−	−	−	\multicolumn{5}{c}{Not included}	
Indole production[a]	+	+	−	−	−	+	+(L)	−	−	−	
Xylose (acid)[a]	+	+	+	−	−	+	+	+	−	−	
Lactose (acid, OF medium)[a]	+	+	+	−	−	+	+	+	−	−	
Nitrate reduction	+	+	+	+	−	+	+	+	+	−	
Trehalose (acid)	+	+	+	+	−	+	+	+	+	−	
Sorbitol (acid)	+	+	+	+	−	\multicolumn{5}{c}{Not included}	
Ornithine decarboxylase[a]	+	+	+	+	−	+	+	+	+	−	
Voges-Proskauer[a]	+	+	+	+	−	\multicolumn{5}{c}{Not included}	
β-Galactosidase[a]	+	+	+	+	−	+	+	+	+	−	
Sucrose[a]	+	+	+	+	−	\multicolumn{5}{c}{Not included}	
Sorbose	+	+	+	+	−	\multicolumn{5}{c}{Not included}	

Bercovier indole/xylose dichotomy:
- Indole: + → Biotypes 1,2 ; − (via Xylose)
- Xylose: + → Biotype 3 ; − → Biotype 4

Note: OF, oxidation-fermentation; +, positive; −, negative; L, late.

[a] Tests performed at 25°C.

absence serves to distinguish group 4 strains (see below) and *Y. pseudotuberculosis*. Such isolates are dealt with separately in Sections IV. B and IV. C.
3. Substrates irregularly acidified which include lactose, esculin, and salicin.
4. Substrates not fermented by *Y. enterocolitica* include adonitol, dulcitol, inulin, and raffinose (except group 3 strains and rare group 1 strains).

Recognizing the biochemical variability that exists among *Y. enterocolitica* isolates from diverse sources (human, animal, environment), Niléhn[4] and Wauters[3] translated such heterogeneity into a format for grouping (biotyping) *Y. enterocolitica* which is reproduced in Table 4. Bercovier,[27] using a Guttman scale analysis, simplified the biotyping procedure on the basis of indole production and xylose fermentation. This schema has been appended to the chart for comparison.

Through the detailed and meticulously carried out studies of Mollaret,[28] Wauters,[29] and Bercovier[27] a well-constructed relationship has evolved between the ecological niche of *Y. enterocolitica* and *Y. enterocolitica*-like strains and their sero-biochemical profile. Thus, strains well-adapted in human or porcine hosts, for example serotypes O:3 and O:9, are biochemically homogeneous as are strains isolated from the chinchilla (O:1) and from rabbit (O:2). The exact natural reservoir for serogroups O:5, 27, and O:8 which are isolated frequently in the U.S. remains to be fully established. Mollaret[28] has proposed the pig, monkey, carnivores, and man as hosts. The biochemical characteristics of these homogeneous *Y. enterocolitica* are shown in Table 5. These strains produce the "classic" syndromes such an enteritis, mesenteric lymphadenitis and/or terminal ileitis which may mimic appendicitis and, rarely, septicemia.

The second sero-biochemical group of *Y. enterocolitica* is comprised of strains derived mainly from environmental sources which display an extremely variable range of

Table 5
BIOCHEMICAL CHARACTERISTICS OF THE MORE FREQUENTLY ENCOUNTERED SEROTYPES OF *YERSINIA ENTEROCOLITICA* CAUSING HUMAN INFECTIONS[3,29]

Characteristics	*Y. enterocolitica* serotype				
	O:1,2	O:3	O:5,27	O:8	O:9
Indole[a]	0	0	(+)	+	+
Lecithinase[a]	0	0	0	+	0
Glucose	+	+	+	+	+
Mannitol	+	+	+	+	+
Sorbitol	+	+	+	+	+
Cellobiose	+	+	+	+	+
Trehalose	+	+	+	+	+
Galactose	+	+	+	+	+
Xylose	+	0	+	+	+
Maltose[a]	+	+	+	+	+
Lactose	0	0	0	0	0
Rhamnose	0	0	0	0	0
Raffinose	0	0	0	0	0
Adonitol	0	0	0	0	0
Dulcitol	0	0	0	0	0
Salicin	0	0(+)	0	0	0
Esculin hydrolysis	0	0	0	0	0
β-Galactosidase (ONPG)[b]	+	+	+	+	+
Citrate (Simmons)	0	0	0	0	0
Arginine dihydrolase	0	0	0	0	0
Lysine decarboxylase	0	0	0	0	0
Ornithine decarboxylase	+	+	+	+	+
Nitrate reduction[a]	+	+	+	+	+

Note: 0 = negative; + = positive; (+) = late positive.

[a] Temperature-dependent.
[b] ONPG reaction for β-galactosidase reported negative for South African isolates for some strains isolated in Europe and from rabbits.[21]

biochemical and serological heterogeneity. Such isolates may be exemplified by the rhamnose-fermenting and to a lesser degree the sucrose-negative isolates. The former all fall predominately into biogroup 1 whereas Wauters[3,29] has suggested that perhaps a new biogroup (biogroup 6) could be established to accommodate sucrose-negative isolates. *Y. enterocolitica* and *Y. enterocolitica*-like strains comprising this group are apparently not as invasive as "typical" isolates and have not been associated with the more serious forms of yersiniosis.[28,30]

B. Rhamnose-Fermenting Strains

Rhamnose-fermenting *Y. enterocolitica*-like isolates, until recently, presented a particular diagnostic problem for microbiologists, working in the hospital setting. Reasons for this inherent confusion stem from perpetuating in textbooks that the lack of rhamnose fermentation is a feature used to distinguish *Y. enterocolitica* from the closely related *Y. pseudotuberculosis* which readily utilizes this carbohydrate. Additionally, as noted above, the role of rhamnose-fermenting species in human pathology is only

Table 6
REPRESENTATIVE BIOCHEMICAL PATTERNS REPORTED FOR RHAMNOSE FERMENTING *Y. ENTEROCOLITICA*-LIKE ISOLATES AT 25°C

Author	Rhamnose	Raffinose	Melibiose	Lactose	α-Methyl glucoside	Simmons citrate	V-P	Indole
Bottone et al.[7]	+	+	+	0(+)	ND	+	+	+
Chester and Stotzky[9]	+	+	+	(+)		+	+	+(37°C)
Brenner et al[1]	+	0	0	0	0	0	0	+
	+	+	+	+	+	+	0	+
	+	+	+	+	+	0	NG	+
	+	0	+	+	+	+	NG	+
	+	+	+	+	+	0	NG	+
	+	+	+	0	+	0	NG	+
	+	0	+	+	0	+	0	+
	+	+	+	(+)	NG	+	0	+
Hanna et al.[22]	+	0	0	(+)	NG	0	+	+
	+	+	+	(+)	NG	+	+	+
Highsmith et al.[23]	+	+	0	0	NG	NG	0	+
Dabernat et al.[15]	+	0	0	(+)	NG	NG	0	+
	+	0	0	(+)	NG	NG	+	0
	+	+	+	(+)	—	—	+	+
	+	+	+,0	(+)	+,0	+	+	+
Lassen[33]	+	0	0	+	NG	NG	+	+
Saari and Quan[34]	+	+	+	NG	NG	(+)	NG	NG
Bercovier[27]	+	+	+	0(+)	+	V(+)	+	+
	+	0	0	0(+)	0	V	+	+
Alonso et al.[31]	+	0,+	NG	0	NG	V	+	+

Note: ND = note done; NG = not given; 0 = negative; + = positive; (+) = late positive; V-P = Voges-Proskauer.

recently emerging[5,7,8,15,30] while most of the previously reported cases of *Y. enterocolitica* infection have documented the well-established syndromes of gastroenteritis, mesenteric lymphadenitis, terminal ileitis, and septicemia as being caused by the biochemically typical, rhamnose-negative strains.

Rhamnose-fermenting strains comprise Brenner's DNA homology groups 2 and 3. Biochemically, while variability exists with regard to overall characteristics, there appears within these groups a constancy relative to the results obtained with raffinose, melibiose, and α-methyl glucoside. This relationship was noted first by Brenner,[1] and subsequently confirmed by Alonso et al.[31] and Bercovier et al.[21]

Depicted in Table 6 are the partial biochemical profiles obtained by several investigators reporting the recovery of rhamnose-fermenting isolates. It will be noted that a degree of variability exists for the Voges-Proskauer and indole reactions. That is, little agreement exists between the presence or absence of acetylmethylcarbinol or indole production and the capability of an isolate to utilize rhamnose. When viewed, however, in terms of the reactions rendered by these strains with raffinose, melibiose, and α-methylglucoside, rhamnose-fermenting *Y. enterocolitica*-like isolates may be divided into the two groups.[2,21,31] Thus with the minor exceptions (perhaps related to incubation temperatures) noted in Table 6, a relationship does exist between rhamnose fermentation and the capability or not to attack raffinose, melibiose, and α-methylglucoside which is confirmed by the genetic delineation of Brenner et al.[1]

Table 7
BIOCHEMICAL CHARACTERISTICS[a] OF RHAMNOSE FERMENTING Y. ENTEROCOLITICA-LIKE ISOLATES ENABLING DIFFERENTIATION FROM TYPICAL Y. ENTEROCOLITICA AND FROM Y. PSEUDOTUBERCULOSIS

Test	Species			
	Rhamnose Positive		Y. entero-colitica	Y. pseudo-tuberculosis
	group 2[a]	group 3[a]		
Simmons' citrate[b]	V	V	0	0
Indole	V	V	V	0
Voges-Proskauer	V	V	+[c]	0
Lysine decarboxylase	0	0	0	0
Ornithine decarboxylase	+	+	+	0
Arginine dihydrolase	0	0	0	0
Sucrose	+	+	+[c]	0
Rhamnose	+	+	0	+
Raffinose	0	+	0	0
Melibiose	0	+	0	+
α-Methylglucoside	0	+	0	0
Sorbitol	+	+	+	0

Note: 0 = negative; + = positive; V = variable.

[a] According to Brenner,[1] group 2 = *Y. frederiksenii*; group 3 = *Y. intermedia*.
[b] 25°C.
[c] Except sucrose-negative isolates.

The biochemical characteristics of the rhamnose-fermenting group are markedly affected by incubation temperature. In addition to the temperature-related features noted with typical *Y. enterocolitica* which apply to these yersiniae as well, expression of rhamnose, raffinose, melibiose, and sodium citrate utilization is pronounced at 25°C and sparse or absent at 37°C.[7,9] Susceptibility to antibiotics among rhamnose-fermenting strains is also temperature-dependent. Increased in vitro resistance to ampicillin, carbenicillin, cephalothin, chloramphenicol, gentamicin, kanamycin, and streptomycin has been observed after growth at 22°C as compared to 37°C.[8-10,32]

Rhamnose-fermenting *Y. enterocolitica*-like isolates must be distinguished from *Y. pseudotuberculosis* with which they share surface antigens and hence agglutinate with *Y. pseudotuberculosis* antiserum.[31] Characteristics present among rhamnose-fermenting *Y. enterocolitica*-like isolates enabling their differentiation from typical *Y. enterocolitica* as well as from *Y. pseudotuberculosis* are presented in Table 7. Chester et al.[10] have also shown that group 3 rhamnose-fermenting isolates produce succinic acid which is lacking among typical *Y. enterocolitica* and *Y. pseudotuberculosis*.

C. Sucrose-Negative Strains

Sucrose-negative isolates of *Y. enterocolitica* have only infrequently been recovered from human sources[6,8] although reports of their occurrence in animals have appeared.[29] Such isolates differ from typical *Y. enterocolitica* by their absence of sucrose fermentation and acetylmethyl-carbinol production. Growth on agar media is accompanied by a strong musty[29] or potato or cabbage-like odor[8] which signals the in vitro presence of sucrose-negative *Yersinia* species.

In our laboratory four sucrose-negative isolates (stool, blood, eye) have been examined which conformed to the above and previously reported[3,6] descriptions. One of

Table 8
DIFFERENTIAL CHARACTERISTICS OF SUCROSE-NEGATIVE Y. ENTEROCOLITICA

	Y. enterocolitica		
Test	Sucrose-negative	Typical isolates	Y. pseudotuberculosis
Sucrose	0	+	0
Voges-Proskauer 25°C	0	+,0	0
Ornithine decarboxylase	+	+	0
Rhamnose	0	0	+
Salicin	0	0	+
Sorbitol	+	+	0
Lactose	0,+	0,+	0
Lecithinase	(+),0	+,0	UNK
Indole	$+^w$	+,0	0
Yellow pigment	+,0	UNK	UNK

Note: + = positive; (+) = delayed; $+^w$ = weak positive; 0 = negative; UNK = unknown.

these isolates was unusual in yielding a lactose-positive variant which rapidly (18 hr) degraded this disaccharide in both peptone agar and in O-F medium. Additionally, this same isolate produced a nondiffusible yellow pigment on blood, chocolate, and trypticase soy agars after several days incubation at both 25 and 37°C. Characteristics of sucrose-negative yersiniae derived from our studies as well as those of Wauters[29] and Bissett[6] enabling differentiation from typical *Y. enterocolitica* and *Y. pseudotuberculosis* (sucrose-negative) are detailed in Table 8. Sucrose-negative isolates have been mainly serotype O:11, O:12, or nonagglutinable,[8,29] and appear to be intermediary in their human disease producing capability as evidenced in at least one instance by the antibody response engendered by its presence in a symptomatic patient.[8]

It is emphasized that the sucrose-negative isolates discussed herein correspond to Brenner's DNA group 4 and should be distinguished from Biogroup 5, the hare strains, which belong to DNA group 1. The hare strains are suc⁻ (some are sucrose⁺) ornithine negative, and generally biochemically inactive (Table 4).

ACKNOWLEDGMENT

The author wishes to express his gratitude to Dr. Don J. Brenner for critically reviewing this manuscript.

REFERENCES

1. **Brenner, D. J., Steigerwalt, A. G., Falcao, D. P., Weaver, R. E., and Fanning, G. R.**, Characterization of *Yersinia enterocolitica* and *Yersinia pseudotuberculosis* by deoxyribonucleic acid hybridization and by biochemical reactions, *Int. J. Syst. Bacteriol.*, 26, 180, 1976.
2. **Brenner, D. J.**, Speciation in *Yersinia*, in *Contributions to Microbiology and Immunology*, Vol. 5, Karger, Basel, 1979, 33.
3. **Wauters, G.**, Contribution a l'etude de *Yersinia enterocolitica;* these d'agrege Universite Catholique de Louvain, Vander Louvain, 1970.
4. **Niléhn, B.**, Studies on *Yersinia enterocolitica* with special reference to bacterial diagnosis and occurrence in human acute enteric disease, *Acta Pathol. Microbiol. Scand. Suppl.*, 206, 1, 1969.

5. Bottone, E. J., *Yersinia enterocolitica:* a panoramic view of a charismatic microorganism, *Crit. Rev. Microbiol.,* 5, 211, 1977.
6. Bissett, M. L., *Yersinia enterocolitica* isolates from humans in California, 1968—1975, *J. Clin. Microbiol.,* 4, 137, 1976.
7. Bottone, E. J., Chester, B., Malowany, M. S., and Allerhand, J., Unusual *Yersinia enterocolitica* isolates not associated with mesenteric lymphadenitis, *Appl. Microbiol.,* 5, 858, 1974.
8. Bottone, E. J. and Robin, T., *Yersinia enterocolitica:* recovery and characterization of two unusual isolates from a case of acute enteritis, *J. Clin. Microbiol.,* 5, 341, 1977.
9. Chester, B. and Stotzky, G., Temperature-dependent cultural and biochemical characteristics of rhamnose-positive *Yersinia enterocolitica, J. Clin. Microbiol.,* 3, 119, 1976.
10. Chester, B., Stotzky, G., Bottone, E. J., Malowany, M. S., and Allerhand, J., *Yersinia enterocolitica:* biochemical, serological, and gas-liquid chromatographic characterization of rhamnose, raffinose, melibiose, and citrate utilizing strains, *J. Clin. Microbiol.,* 6, 461, 1977.
11. Thirumoorthi, M. C. and Dajani, A. S., *Yersinia enterocolitica* osteomyelitis in a child, *Am. J. Dis. Child.,* 132, 578, 1978.
12. Sierra, M. F., personal communication, Kings County Hospital, Brooklyn, N. Y., 1978.
13. Berman, M., personal communication, North Shore University Hospital, Long Island, N.Y., 1977.
14. Cohen, M., personal communication, St. Francis Hospital, Long Island, N.Y., 1977.
15. Dabernat, H. J., Bauriaud, R., Lemozy, J., Lefevre, J. C., and Lareng, M. B., *Yersinia enterocolitica* fermentant le rhamnose. A propos de 15 souches isolees chez des enfants, *Med. Mal. Infect.,* 5, 158, 1978.
16. Kahn, B., A rapid microtechnique for the detection of indole in primary cultures, *Can. J. Med. Technol.,* 30, 145, 1968.
17. Rustigan, R. and Stuart, C., Decomposition of urea by *Proteus, Proc. Soc. Exp. Biol. Med.,* 47, 108, 1941.
18. Van Noyen, R., Isebaert, A., and Vandepitte, J., Sur un biotype urease negatif de *Yersinia enterocolitica, Ann. Inst. Pasteur Paris,* 117, 658, 1969.
19. LeMinor, L., Coynault, C., and Guiso, N., Discordances entre la positivite du test a l'ONPG et la presence d' une B-galactosidase chez les *Enterobacteriaceae* et autres bacilles gram — a metabolisme fermentatif, *Ann. Microbiol. Paris,* 128B, 35, 1977.
20. Hugh, R. and Leifson, E., The toxonomic significance of fermentative versus oxidative metabolism of carbohydrates by various gram-negative bacteria, *J. Bacteriol.,* 66, 24, 1953.
21. Bercovier, H., Alonso, J. M., Bentaiba, N., Brault, J., and Mollaret, H. H., Contribution a la definition et a la taxonomie de *Yersinia enterocolitica,* in *Contributions to Microbiology and Immunology,* Vol. 5, S. Karger, Basel, 1979, 12.
22. Hanna, M. O., Zink, D. L., Carpenter, Z. L., and Vanderzant, C., *Yersinia enterocolitica*-like organisms from vacuum packaged beef and lamb, *J. Food Sci.,* 41, 1254, 1976.
23. Highsmith, A. K., Feeley, J. C., Skaliy, P., Wells, J. G., and Wood, B. T., Isolation of *Yersinia enterocolitica* from well water and growth in distilled water, *Appl. Environ. Microbiol.,* 34, 745, 1977.
24. Caprioli, T., Drapeau, A. J., and Kasatiya, S., *Yersinia enterocolitica:* serotypes and biotypes isolated from humans and the environment in Quebec, Canada, *J. Clin. Microbiol.,* 8, 7, 1978.
25. Cornelis, G., Bennett, P. M., and Grinsted, J., Properties of pGC1, a lac plasmid originating in *Yersinia enterocolitica* 842, *J. Bacteriol.,* 127, 1058, 1976.
26. Cornelis, G., Luke, R. K. J., and Richmond, M. H., Fermentation of raffinose by lactose-fermenting strains of *Yersinia enterocolitica* and by sucrose-fermenting strains of *Escherichia coli, J. Clin. Microbiol.,* 7, 180, 1978.
27. Bercovier, H., Contribution a l etude epidemiologique des infections a *Yersinia enterocolitica.* I. Interet du traitment automatise d'une banque de donnees, *Med. Mal. Infect.,* 6, 425, 1976.
28. Mollaret, H. H., Contribution a l'etude epidemiologique des infections a *Yersinia enterocolitica.* III. Bilan provisoire des connaissances, *Med. Mal. Infect.,* 6, 442, 1976.
29. Wauters, G., Correlation between ecology, biochemical behaviour, and antigenic properties of *Yersinia enterocolitica,* in, *Contribution to Microbiology and Immunology,* Vol. 2, Winblad, S., Ed., S. Karger, Basel, 1973, 38.
30. Bottone, E. J., Atypical *Yersinia enterocolitica:* clinical and epidemiological parameters, *J. Clin. Microbiol.,* 7, 562, 1978.
31. Alonso, J. M., Bejot, J., Bercovier, H., and Mollaret, H. H., Sur un groupe de souches de *Yersinia enterocolitica* fermentant le rhamnose, *Med. Mal. Infect.,* 10, 490, 1975.
32. Dabernat, H. J., Antimicrobial susceptibility of rhamnose-positive *Yersinia enterocolitica, Ann. Microbiol. Paris,* 129A, 503, 1978.
33. Lassen, J., *Yersinia enterocolitica* in drinking water, *Scand. J. Infect. Dis.,* 4, 125, 1972.
34. Saari, T. H. and Quan T. J., *Waterborne Yersinia enterocolitica in Colorado,* Abstr. Annu. Meet. American Society for Microbiology, Washington, D.C., 1975, 45.
35. Brenner, D. J., personal communication, 1977.

Chapter 4

MICROBIOLOGICAL ASPECTS OF *YERSINIA PSEUDOTUBERCULOSIS*

Marjorie L. Bissett

TABLE OF CONTENTS

I.	Introduction	32
II.	Morphology	32
III.	Cultural Characteristics	32
IV.	Isolation from Clinical Material	33
V.	Biochemical Characteristics	34
VI.	Serologic Typing of *Yersinia pseudotuberculosis*	35
VII.	Differentiation of *Y. pseudotuberculosis* from Other Organisms	36
	References	39

I. INTRODUCTION

The first visualization of *Yersinia pseudotuberculosis* was made by Malassez and Vignal[1] in 1883. Investigators, in the years since that time, have studied the properties of the organism and sought methods to distinguish it from closely related members of the genus *Pasteurella* and other *Yersinia* (*Yersinia pestis* and *Yersinia enterocolitica*) as well as from other members of the family *Enterobacteriaceae*. The purpose of this chapter is to present the results of these investigations, define the microbiology of *Y. pseudotuberculosis*, and indicate those properties that are most useful in differentiating it from other organisms with which it may be confused.

II. MORPHOLOGY

Y. pseudotuberculosis is a pleomorphic Gram-negative coccobacillus whose morphology varies from coccoid (measuring 0.8 μm in width and from 0.7 to 1.7 μm in length) to rod-shaped with rounded ends (measuring from 0.8 μm in width and from 1.5 to 6.0 μm in length). The coccoid forms are usually arranged singly, while rod forms may occur singly, in groups, or in short chains. Filamentous forms may occur. The form of the coccobacillus depends upon the condition of culture, i.e., media, atmospheric environment, and length of time in culture. Staining may be irregular with barred and granular types present [2] and vacuolization may be apparent.[3] The organism may show bipolar staining, but this is not a constant characteristic. It is nonspore-forming and nonencapsulated. Some investigators have reported a capsular-like material variously described as a "slime layer",[4] a "clear envelope",[5] or a "viscous layer",[6] but no true capsule has been demonstrated by electron microscopy[7] or by scanning electron microscopic studies.[8]

Y. pseudotuberculosis is characteristically nonmotile at 37°C and motile at 25°C. Preston and Maitland[9] have determined that the change to nonmotility occurred in a narrow range of temperature around 30°C. These investigators reported that *Y. pseudotuberculosis* had from one to six peritrichous flagella and found that more flagella per organism could be observed if flagellar stains were performed from broth cultures. Gunnison et al.[4] reported the observation of strains with as many as 16 flagella. Knapp,[10] using flagellar stains and electron microscopic studies, reported that, as a rule, motile organisms possessed from three to six peritrichous flagella which occurred optimally in a temperature range of 20 to 30°C. Motility of *Y. pseudotuberculosis* at 22°C serves as a characteristic that aids in the differentiation of this organism from the closely related *Y. pestis* organism, which is never motile.

III. CULTURAL CHARACTERISTICS

Y. pseudotuberculosis is a facultative anaerobe that grows well on ordinary media, i.e., nutrient agar, tryptic soy agar, blood agar, and chocolate agar. After 24-hr incubation on blood agar, the colonies are white, opaque, circular, and convex with an entire edge. The surface of the colonies is smooth and glistening. At 37°C colonies measure about 0.5 to 1 mm in diameter; at 25°C they are slightly smaller, the majority being 0.5 mm in diameter. At 48 hr the colonies range in size from 1 to 3 mm in diameter, the center of the colony being slightly raised and the surface glistening but tending to appear dry. With further incubation the colonies begin to dissociate into the rough (R) form and appear flatter with a definite raised center; the surface becomes dry and the consistency more granular. A transparent flat "skirt" may develop around some of the colonies.

On MacConkey agar the colonies are smaller in size with a transparent periphery surrounding an opaque pale lavender raised center. The edge may be entire or undulate

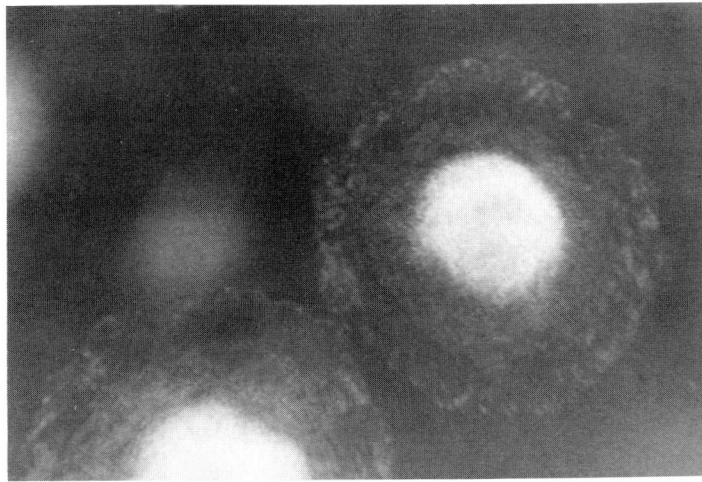

FIGURE 1. Colonies of *Y. pseudotuberculosis* on MacConkey agar after incubation for 24 hr at 35°C as viewed with a dissecting microscope. Note raised central portion with transparent to granular flat periphery. (Magnification × 18.)

and the surface often granular-appearing in the central portion with a glistening to granular transparent flat periphery (Figure 1). On *Salmonella-Shigella* (SS), eosin methylene blue (EMB), and xylose-lysine-deoxycholate (XLD) agars the colonies develop more slowly. They are pale pink on oil and EMB. On XLD the center of the colonies are yellow with a pink transparent periphery. Growth is definitely inhibited on bismuth sulfite agar. Optimal growth of *Y. pseudotuberculosis* occurs at pH 6.0 to 8.0. Dissociation to the rough form occurs more frequently under acidification and more rapidly at 37°C.

IV. ISOLATION FROM CLINICAL MATERIAL

Although *Y. pseudotuberculosis* grows well on ordinary media after primary isolation, initial isolation from human or animal specimens may require special handling to be successful. The presence of *Y. pseudotuberculosis* may be overlooked if cultures are incubated only at 37°C. Some strains require incubation of the primary culture at 22 to 25°C for successful isolation of the organisms.[10-13] The work of Burrows and Gillett[14] offers one explanation for this observation. They showed that *Y. pseudotuberculosis* requires additional nutritional factors for growth at 37°C over those needed for growth at 28°C. Because some strains have been isolated only when cultures have been incubated under anaerobic conditions, Knapp[7,10,15] recommends that cultures for primary isolation be incubated both aerobically and anaerobically.

Isolation from specimens where the causative organism may be expected to occur in pure culture, i.e., blood, lymph nodes, or biopsy material from liver or spleen, may be handled by the procedures usually recommended for these materials as long as incubation temperature and atmospheric conditions are considered.

Isolation from blood may be attempted using any one of the multipurpose broths currently recommended for blood culture work, i.e., trypticase soy broth, brain heart infusion broth, etc. Anaerobic and aerobic culture should be performed and incubated at 22 and 35°C. Because of the slow growth of the organism in primary cultures, it is recommended that blood cultures be incubated for at least 10 days before being reported as negative.[10,13]

Recommendations for isolation from lymph nodes or other tissues vary. Several investigators[11,16,17] have indicated the difficulties of isolating the organism from lymph nodes. In 117 cases of mesenteric lymphadenitis due to *Y. pseudotuberculosis*, Knapp[16] isolated the organism from lymph nodes in only 13 cases and from the blood in 2 cases. The balance of the cases were diagnosed on the basis of histologic examinations and serologic tests. In reviewing the methods used by those investigators who have reported success in isolating *Y. pseudotuberculosis* from lymph nodes, the procedure most often used was inoculation of a suspension of ground tissue into a supportive liquid medium such as blood digest broth,[13] trypticase soy yeast broth,[18] or glucose tryptone broth[19] and onto several solid media such as blood agar, chocolate agar, and MacConkey agar.[13,18-20] Thioglycolate broth has also been used.[18] Simultaneously inoculated liquid and solid media are incubated at 22 and 35°C.

Primary isolation from specimen material when mixed flora are expected presents further problems. Sources of specimen material, such as peritoneal fluid, appendices, and feces require selective media and special techniques. Selective media used frequently in isolation of other *Enterobacteriaceae* such as MacConkey, SS, EMB, and Endo agars have been used. Specialized media have been developed for isolation of yersiniae from specimens with mixed flora.[21-23] These media contain antibiotics and substances to inhibit growth of other organisms. Use of the specialized media has been successful in the isolation of *Y. pseudotuberculosis* from animal feces, but insufficient information is available concerning their use in isolation from human specimens.

Paterson and Cook[22] have recommended use of a cold enrichment technique for isolation of *Y. pseudotuberculosis* from animal fecal material. The technique consists of holding a 10% suspension of fecal material in phosphate buffer solution (pH 7.6) at 3 to 4°C for 7 to 28 days. Timofeeva et al.[24] have found this technique to be very useful for isolation of *Y. pseudotuberculosis* from feces of humans with the scarlatina fever form of pseudotuberculosis reported in Russia. Isolation from human feces is, however, a rarely reported event in the rest of the world. The first isolate of *Y. pseudotuberculosis* from human feces was reported by Daniels[12] in 1961. The only other reports in the European, British, or Western literature were an unpublished observation by Kampelmacher (cited by Knapp[15]) and an isolation from a 10-year-old girl reported by Zaremba and Borowski.[25] Our laboratory has received for confirmation two fecal isolates of *Y. pseudotuberculosis*; one of these was isolated only on EMB agar after 72 hr incubation at room temperature. In all isolation attempts, the use of a dissecting microscope to scan cultures for developing colonies on solid media is strongly recommended.

It is apparent that we have not yet achieved a method for the consistent isolation of *Y. pseudotuberculosis* from specimen material. The majority of cases of *Y. pseudotuberculosis* reported in the world literature have been confirmed by histologic and serologic tests rather than by culture. Until a highly selective medium for these organisms has been developed and evaluated, their isolation will continue to depend on the knowledge and acuity of the microbiologist in utilizing those techniques which are available and in recognizing the organism when it does appear on solid substrates.

V. BIOCHEMICAL CHARACTERISTICS

Y. pseudotuberculosis strains, unlike *Y. enterocolitica*, are exceptionally constant in their biochemical characteristics with only rare individual strain variation. Like other members of the *Yersinia* group, *Y. pseudotuberculosis* strains are catalase-positive, oxidase-negative, do not produce phenylalanine deaminase, are methyl-red-positive, and

Table 1
CHARACTERISTICS OF TYPICAL
Y. PSEUDOTUBERCULOSIS

Test	Result	Test	Result
Motility			
20—25°C	+	Adonitol	v
35°C	−	alpha-Methyl glucoside	−
Urease	+	Arabinose	+
Catalase	+	Cellobiose	−
Oxidase	−	Dulcitol	−
Nitrate reductase	+	Erythritol	−
Gelatinase	−	Glucose	+
Citrate (Simmon's)	v	Glycerol	v
Indole	−	Inositol	−
Voges Proskauer	−	Lactose	−
Phenylalanine deaminase	−	Maltose	+
beta-Galactosidase	+	Mannitol	+
Methyl red	+	Melibiose	+
Esculin hydrolysis	+	Raffinose	−
Lysine decarboxylase	−	Rhamnose	+
Ornithine decarboxylase	−	Salicin	+
Arginine dihydrolase	−	Sorbitol	−
Triple sugar iron agar (TSI)	K/A	Sucrose	−
Kligers' iron agar (KIA)	K/A	Trehalose	+
Lysine iron agar (LIA)	K/A	Xylose	+
H₂S (TSI)	−		

Note: +, positive; −, negative; v, variable; K, alkaline; A, acid.

do not liquefy gelatin. Neither lysine nor ornithine decarboxylase nor arginine dihydrolase are produced. No growth occurs on Simmon's citrate medium when incubated at 37°C, but delayed growth occurs with some strains when incubated at 18 to 27°C.[26] Characteristically, *Y. pseudotuberculosis* strains attack carbohydrates with acid production but without gas formation (Table 1). Acid is produced from glucose, xylose, maltose, mannitol, salicin, trehalose, rhamnose, melibiose, and arabinose. Lactose, sucrose, raffinose, adonitol, dulcitol, inositol, sorbitol, cellobiose, erythritol, and α-methyl glucoside are not utilized. The action on raffinose and sorbitol may be variable. Very rare strains may produce acid from dulcitol or sucrose and other rare strains may fail to produce acid from adonitol, melibiose, or salicin. The organism produces urease, reduces nitrate to nitrite, produces β-galactosidase, and gives an alkaline slant with acid butt with no H₂S production in triple sugar iron agar (TSI). Weak hydrogen sulfide production has been observed in SIM medium after 72 hr incubation at room temperature[19] and Devignat and Boivin[5] reported that some strains produced H₂S slowly in Difco lead acetate agar.

VI. SEROLOGIC TYPING OF *YERSINIA PSEUDOTUBERCULOSIS*

Y. pseudotuberculosis has been separated into six groups, Types I through VI, on the basis of type-specific thermostable somatic O antigens[27-30] (Table 2). Subtypes labeled A and B have been determined for Types I, II, IV, and V.

In studies of the complex thermolabile flagellar (H) antigens, five different H-antigens (a to e) have been discerned.[27,28,31] H-antigen a is found in all strains with the exception of some strains of Type IV. The rough antigen (R) is common to all types and is shared with

Table 2
THE ANTIGENIC SCHEME OF *Y. PSEUDOTUBERCULOSIS*

O-types	O-subtypes	Rough antigens "R"	O-antigens (thermostable)	H-antigens (thermolabile)
I	A	(1)	2,3	a,c
	B	(1)	2,4	a,c
II	A	(1)	5,6	a,d
	B	(1)	5,7	a,d
III		(1)	8	a
IV	A	(1)	9,11	b;a,b
	B	(1)	9,12	a,b,c
V	A	(1)	10,14	a;a,e(b)
	B	(1)	10,15	a
VI		(1)	13	a

From Thal, E. and Knapp, W., *Ser. Immunobiol. Stand.*, 15, 219, 1971. With permission.

all strains of *Y. pestis* that have been tested. The O-antigen serotyping factor 5 of Types IIA and IIB is shared with *Salmonella* group B[27] and the serotyping factor 9 of types IVA and IVB is shared with *Salmonella* group D.[32] Therefore, to establish specificity for serologic diagnosis (Chapter 9), the sera of a patient reacting with *Y. pseudotuberculosis* Type II or Type IV antigen must be absorbed with salmonellae of the corresponding group. Type VI strains have also been shown to share common antigens with *E. coli* O group 55.[33]

There is no cross-agglutination between the *Y. pseudotuberculosis* typing antisera and the vast majority of *Y. enterocolitica* organisms. However, Alonso et al.[34] found that 15 of 53 strains of rhamnose-positive *Y. enterocolitica* were agglutinated by the anti-*Y. pseudotuberculosis* — 13 by *Y. pseudotuberculosis* Type II antiserum, 1 by Type I, and 1 by Type V antiserum.

Although cross-agglutination does not occur between the *Y. pseudotuberculosis* antisera used in typing and *Y. pestis*, these two species do share other antigens. Further information regarding these commonly shared antigens and chemical analysis of the antigen structure of *Y. pseudotuberculosis* may be obtained by consulting the referenced articles.[6,35-38]

The determination of the serotype of suspected *Y. pseudotuberculosis* isolates aids in the confirmation of their identification as well as being useful in epidemiologic investigation. Cross-agglutination does not occur between the six types. Typing is accomplished by use of O agglutinating antisera prepared in rabbits hyperimmunized with strains of the described types. To prevent spontaneous agglutination which occurs when rough strains of the organism are used, the isolated strain to be typed must be in the smooth form before it is used as an antigen. Presently antisera for typing *Y. pseudotuberculosis* are not commercially available. Therefore, isolated strains of suspected *Y. pseudotuberculosis* should be forwarded to a reference laboratory for this service.

VII. DIFFERENTIATION OF *Y. PSEUDOTUBERCULOSIS* FROM OTHER ORGANISMS

Differentiation of *Yersinia* species from other related species, such as *Pasteurella*, is relatively simple because the cultural and biochemical properties of the *Pasteurella* species differ considerably from those of the *Yersinia*. The more difficult problem is to differentiate the *Yersinia* from those members of the *Enterobacteriaceae* that are usually considered as part of the normal intestinal flora. Such a differentiation depends upon the cultural, biochemical, and serologic properties of the suspect organism. Sonnenwirth[39] has stated that "any lactose-negative, Gram-negative, fermentative rod which is oxidase-negative, does not produce H_2S in TSI, is motile only at room temperature, and is urease-positive on Christensen urease agar in 3 to 24 hr, but phenylalanine-negative, should be suspected of being *Y. enterocolitica* or *Y. pseudotuberculosis* and its further identification should be pursued."

Once the organism has been determined to be a member of the *Yersinia* genus, it then becomes necessary to specifically confirm the isolate as *Y. pseudotuberculosis*, *Y. enterocolitica*, or *Y. pestis*. *Y. pestis* is usually a concern in this differential diagnosis only in patients from areas of the world where sylvatic plague is endemic, among these being the southwestern and western U.S.

In Table 3 are listed the test reactions that are of particular significance in the definitive determination of these species. In light of our present knowledge of these organisms, some of the tests used for differentiation, i.e., motility at 20 to 25°C, require special techniques or interpretation. Motility at 25°C is difficult to demonstrate with some strains of *Y. pseudotuberculosis*.[7,13,41] Girard[41] recommends the use of flagellar stains, while Knapp[7] and Mair[13] recommend the use of U tubes[7] or Craigie tubes[13] with strains where demonstration of motility is difficult. Hanging drop methods are not recommended.

There have been reports of rare strains of *Y. pseudotuberculosis* that do not produce urease[24,40] and a report of one strain of *Y. pestis* that produced urease.[41] However, authorities in this field agree that production of urease by *Y. pseudotuberculosis* and its nonproduction by *Y. pestis* constitutes a major differential test in the identification of these organisms.[7,26,40-42]

Fermentation of the carbohydrates listed in Table 3 is helpful in the specific identification of the three species. Tests using *Y. pestis* and *Y. pseudotuberculosis* bacteriophage are useful in confirmation of these organisms. The specificity of tests with these two bacteriophage are temperature-dependent[49,50] and require experience for their interpretation. In sylvatic plague endemic areas both bacteriophage (*Y. pseudotuberculosis* and *Y. pestis*) should be used in tests of suspicious isolates. An additional aid in the rapid and specific identification of *Y. pestis* is the fluorescent antibody test using *Y. pestis* Fraction I antiserum. This test is highly specific for *Y. pestis*. Very rare atypical strains of *Y. pseudotuberculosis* may give a false-positive fluorescence due to presence of an antigen that cross-reacts with the Fraction I conjugate.[51] Animal pathogenicity tests are infrequently used. Studies have shown that *Y. pestis* is pathogenic for the guinea pig and the white rat while *Y. pseudotuberculosis* is not pathogenic for the white rat unless the strain is a toxin-producing Type III *Y. pseudotuberculosis*.[28] *Y. enterocolitica* does not cause pathology in either one of these laboratory animals.

Differentiation of *Y. pseudotuberculosis* from *Y. enterocolitica* strains is usually made by the evaluation of reactions obtained for ornithine decarboxylase, fermentation of sucrose, rhamnose, melibiose, cellobiose, and sorbitol. Serologic typing results, the action of bacteriophage, and animal pathogenicity studies are further confirmatory tests that are utilized in the specific identification.

Differentiation of *Y. pseudotuberculosis* and *Y. pestis* can usually be accomplished by testing for motility at 20 to 25°C, for production of urease, and for acid production from rhamnose and melibiose. In some instances, application of bacteriophage utilizing both *Y. pestis* and *Y. pseudotuberculosis* phage and use of *Y. pestis* Fraction I conjugate in fluorescent antibody tests are necessary. Animal pathogenicity studies can be of further assistance in specific identification.

Table 3
DIFFERENTIAL CHARACTERISTICS OF Y. PSEUDOTUBERCULOSIS, Y. PESTIS, AND Y. ENTEROCOLITICA

	Y. pseudotuberculosis	Y. pestis	Y. enterocolitica
Motility[a]			
20°C	+	−	+
35°C	−	−	−
Indole	−	−	v
Ornithine decarboxylase	−	−	+
Urea[b]	+	−	+
Sucrose[c]	−	−	+
Rhamnose[d]	+	−	v
Melibiose[e]	+	−	−
Sorbitol[f]	v	−	+
Inositol	−	−	v
Cellobiose	−	−	+
Pseudotuberculosis-phage[g]			
20°C	+	+	−
37°C	+	+	−
Pestis-phage[g]			
20°C	−	+	−
37°C	+	+	−
FA with conjugated Fraction I antiserum[h]	−	+	−
Animal pathogenicity[i]			
Guinea pigs	+	+	−
White rats	−	+	−

Note: +, positive; −, negative; v, variable.

[a] Motility: Some strains of *Y. pseudotuberculosis* require special techniques such as use of "U" tubes[7] or Craigie tubes[13] to demonstrate motility.
[b] Urea: There have been reports of rare strains of *Y. pseudotuberculosis* that do not produce urease.[24,40] There has been one report of a strain of *Y. pestis* that produces urease.[41]
[c] Sucrose: Rare strains of *Y. pseudotuberculosis* may ferment sucrose.[3,7] Sucrose-negative strains of *Y. enterocolitica* have been reported.[43,44]
[d] Rhamnose: Rare strains of *Y. pestis* may show delayed fermentation of rhamnose.[5] Rhamnose-positive strains of *Y. enterocolitica* have been reported and reviewed by Bottone[45,46] and Alonso et al.[34]
[e] Melibiose: Occasional strains of *Y. pseudotuberculosis* may not ferment melibiose.[47] Rhamnose-positive strains of *Y. enterocolitica* may also ferment melibiose.[44,46,48]
[f] Sorbitol: Some strains of rhamnose-positive *Y. enterocolitica* do not ferment sorbitol.[34] *Y. pseudotuberculosis* strains usually do not ferment this substance.
[g] Bacteriophage testing: The specificity of tests with *Y. pseudotuberculosis* and *Y. pestis* bacteriophage is temperature-dependent.[49,50] In sylvatic plague endemic areas both *Y. pseudotuberculosis* and *Y. pestis* bacteriophage should be applied to suspicious cultures.
[h] Fluorescent antibody tests with Fraction I antiserum: This antiserum is specific for the Fraction I antigen of *Y. pestis*. Very rare atypical strains of *Y. pseudotuberculosis* may possess an antigen that crosses with conjugated antiserum to *Y. pestis* Fraction I.[51]
[i] Virulence in animals: *Y. pseudotuberculosis* strains are not pathogenic for the white rat unless they are toxin-producing strains of Type III.[28]

The three species of *Yersinia* under discussion here are obviously closely related and, thus, exhibit many similar biological and biochemical characteristics. Their differentiation and specific identification depend on a few tests, some readily available in the usual clinical microbiology laboratory and others available only at reference laboratories.

Individual, rarely encountered strains of each species may show atypical results with any one of the tests considered to be necessary for specific identification. However, the likelihood of these strains giving aberrant results in all of the tests used for specific identification is very remote. Thus, confirmation of the identification of a strain is dependent on the recognition of the pattern of results considered specific for that species and not on any single characteristic.

REFERENCES

1. **Malassez, L. and Vignal, W.**, Tuberculose zooglóeique, *Arch. Phys.*, 2, 369, 1883.
2. **Wilson, G. S. and Miles, A. A.**, *Principles of Bacteriology and Immunity*, Vol. 1, 6th ed., Williams & Wilkins, Baltimore, 1975, 998.
3. **Moss, E. S. and Battle, J. D.**, Human infection with *Pasteurella pseudotuberculosis rodentium* of Pfeiffer, *Am. J. Clin. Pathol.*, 11, 677, 1941.
4. **Gunnison, J. B., Shevky, M. C., Zion, V. K., and Abbott, M. J.**, Lysis of *Pasteurella pseudotuberculosis* by bacteriophage, *J. Infect. Dis.*, 88, 187, 1951.
5. **Devignat, R. and Boivin, A.**, Comportement biologique et biochimique de *P. pestis* et de *P. pseudotuberculosis.*, *Bull. W. H. O.*, 10, 463, 1954.
6. **Meyer, K. F.**, *Pasteurella* and *Francisella*, in *Bacterial and Mycotic Infections of Man*, 4th ed., Dubos, R. J. and Hirsch, J. G., Eds., J. B. Lippincott, Philadelphia, 1965, 659.
7. **Knapp, W.**, *Pasteurella pseudotuberculosis* unter besonderer Berücksichtigung ihrer humanmedizinischen Bedeutung, *Ergeb. Mikrobiol.*, 32, 196, 1959.
8. **Chen, T. H. and Elberg, S. S.**, Scanning electron microscopic study of virulent *Yersinia pestis* and *Yersinia pseudotuberculosis* Type 1, *Infect. Immun.*, 15, 972, 1977.
9. **Preston, N. W. and Maitland, H. B.**, The influence of temperature on the motility of *Pasteurella pseudotuberculosis*, *J. Gen. Microbiol.*, 7, 117, 1952.
10. **Knapp, W.**, Die Laboratoriumdiagnose von Infektionen mit *Pasteurella pseudotuberculosis*, *Ärztl. Labor.*, 6, 197, 1960.
11. **Mair, N. S., Mair, H. J., Stirk, E. M., and Corson, J. G.**, Three cases of acute mesenteric lymphadenitis due to *Pasteurella pseudotuberculosis*, *J. Clin. Pathol.*, 13, 432, 1960.
12. **Daniels, J. J. H. M.**, Enteral infection with *Pasteurella pseudotuberculosis*, *Br. Med. J.*, 2, 997, 1961.
13. **Mair, N. S.**, The laboratory diagnosis of infection with *Pasteurella pseudotuberculosis*, in *Recent Advances in Clinical Pathology*, Series 5, Dyke, S. S., Ed., Churchill Ltd., London, 1968, 35.
14. **Burrows, T. W. and Gillett, W. A.**, The nutritional requirements of some *Pasteurella* species, *J. Gen. Microbiol.*, 45, 333, 1966.
15. **Knapp, W.**, Klinisch-bakteriologische und epidemiologische Befunde bei der Pseudotuberkulose des Menschen, *Arch. Hyg. Bakteriol.*, 147, 369, 1963.
16. **Knapp, W.**, Mesenteric adenitis due to *Pasteurella pseudotuberculosis* in young people, *N Engl. J. Med.*, 259, 776, 1958.
17. **Mollaret, H. H.**, L'adénite mésentérique aiguë á "*Pasteurella pseudotuberculosis*". (Bacille de Malassez et Vignal): A propos de 30 observations. I. Etude clinique sérologique et bacteriologique, *Presse Méd.*, 68, 1375, 1960.
18. **Saari, T. N. and Triplett, D. A.**, *Yersinia pseudotuberculosis* mesenteric adenitis, *J. Pediatr.*, 85, 656, 1974.
19. **Hewstone, A. E. and Campbell, P. E.**, Mesenteric lymphadenitis due to *Pasteurella pseudotuberculosis*, *Aust. Paediatr. J.*, 6, 129, 1970.
20. **Szita, J. and Svidró, A.**, Bacteriologic diagnosis of *Yersinia pseudotuberculosis*, *Acta Microbiol. Acad. Sci. Hung.*, 18, 87, 1971.
21. **Morris, E. J.**, Selective media for some *Pasteurella* species, *J. Gen. Microbiol.*, 19, 305, 1958.
22. **Paterson, J. S. and Cook, R.**, A method for recovery of *Pasteurella pseudotuberculosis* from faeces, *J. Pathol. Bacteriol.*, 85, 241, 1963.
23. **Knisely, R. F., Swaney, L. M., and Friedlander, H.**, Selective media for the isolation of *Pasteurella pestis*, *J. Bacteriol.*, 88, 491, 1964.
24. **Timofeeva, L., Mironova, L., and Golovačeva, V.**, Les Souches de "*Pasteurella pseudotuberculosis*" isolees en Siberie et en Extreme-Orient, *Symp. Ser. Immunobiol. Stand.*, 9, 219, 1968.
25. **Zaremba, M. and Borowski, J.**, *Yersinia pseudotuberculosis* in man in Poland, in *Contributions to Microbiology and Immunology*, Vol. 2, Winblad, S., Ed., S. Karger, Basel, 1973, 217.
26. **Mollaret, H. H.**, Contribution a l'Etude des Caractères biochimiques de *Pasteurella pseudotuberculosis* (Bacille de Malassez et Vignal), *Ann. Inst. Pasteur Paris*, 100, 685, 1961.

27. **Schütze, H.,** Studies on *B. pestis* antigens. II. The antigenic relationship of *B. pestis* and *B. pseudotuberculosis rodentium, Br. J. Exp. Pathol.,* 13, 289, 1932.
28. **Thal, E.,** Untersuchungen über *Pasteurella pseudotuberculosis,* Ph.D. thesis, Veterinär Högskola, Lund, Sweden, 1954.
29. **Tsubokura, M., Itagaki, K., and Kawamura, K.,** Studies on *Yersinia (Pasteurella) pseudotuberculosis.* II. A new type of *Y. pseudotuberculosis* Type VI, and subdivision of Type V strains, *Jpn. J. Vet. Sci.,* 33, 137, 1971.
30. **Thal, E. and Knapp, W.,** A revised antigenic scheme of *"Yersinia pseudotuberculosis", Symp. Ser. Immunobiol. Stand.,* 15, 219, 1971.
31. **Thal, E.,** Weitere Untersuchungen über die thermolabilen Antigene der *Yersinia pseudotuberculosis* (syn. *Pasteurella pseudotuberculosis), Zentralbl. Bakteriol. I Abt. Orig.,* 200, 56, 1966.
32. **Knapp, W.,** Die diagnostische Bedeutung der antigenen Beziehungen zwischen *Pasteurella pseudotuberculosis* und der Salmonella-Gruppe, *Zentralbl. Bakteriol. I Abt. Orig.,* 164, 57, 1955.
33. **Mair, N. S. and Fox, E.,** An antigenic relationship between *Yersinia pseudotuberculosis* Type 6 and *Escherichia coli* O-group 55, in *Contributions to Microbiology and Immunology,* Vol. 2, Winblad, S., Ed., S. Karger, Basel, 1973, 180.
34. **Alonso, J. M., Bejot, J., Bercovier, H., and Mollaret, H. H.,** Sur un groupe de souches de *Yersinia enterocolitica* fermentant le rhamnos. Interet diagnostic et particularities ecologiques, *Med. Malad. Infect.,* 5, 470, 1975.
35. **Davies, D. A. L.,** The smooth and rough somatic antigens of *Pasteurella pseudotuberculosis, J. Gen. Microbiol.,* 18, 118, 1958.
36. **Burrows, T. W. and Bacon, G. A.,** V and W antigens in strains of *Pasteurella pseudotuberculosis, Br. J. Exp. Pathol.,* 41, 38, 1960.
37. **Crumpton, M. J. and Davies, D. A. L.,** An antigenic analysis of *Pasteurella pestis* by diffusion of antigens and antibodies in agar, *Proc. R. Soc. B,* 145, 109, 1965.
38. **Brubaker, R. R.,** The genus *Yersinia:* biochemistry and genetics of virulence, *Curr. Top. Microbiol. Immunol.,* 57, 111, 1972.
39. **Sonnenwirth, A. C.,** Yersinia, in *Manual of Clinical Microbiology,* 2nd ed., Lennette, E. H., Spaulding, E. H., and Truant, J. P., Eds., American Society for Microbiology, Washington, D. C., 1974, chap. 19.
40. **Fauconnier, J.,** La Decomposition de l'Urée en milieu synthétique de Ferguson par *Past. pseudotuberculosis, Ann. Inst. Pasteur,* 79, 104, 1950.
41. **Girard, G.,** Méthodes permettant de différencier *P. pestis* de *P. pseudotuberculosis.* Possibilité d'uniformiser ces méthodes, *Bull. W. H. O.,* 9, 645, 1953.
42. **Thal, E. and Chen, T. H.,** Two simple tests for the differentiation of plague and pseudotuberculosis bacilli, *J. Bacteriol.,* 69, 103, 1955.
43. **Bissett, M. L.,** *Yersinia enterocolitica* isolates from humans in California, 1968—1975, *J. Clin. Microbiol.,* 4, 137, 1976.
44. **Bottone, E. J. and Robin, T.,** *Yersinia enterocolitica:* recovery and characterization of two unusual isolates from a case of acute enteritis, *J. Clin. Microbiol.,* 5, 341, 1977.
45. **Bottone, E. J., Chester, B., Malowany, M. S., and Allerhand, J.,** Unusual *Yersinia enterocolitica* isolates not associated with mesenteric lymphadenitis, *Appl. Microbiol.,* 27, 858, 1974.
46. **Bottone, E. J.,** *Yersinia enterocolitica:* A panoramic view of a charismatic microorganism, *CRC Crit. Rev. Clin. Lab. Sci.,* 5, 211, 1977.
47. **Tsubokura, M., Otsuki, K., Fukuda, T., Kubota, M., Imamura, M., Itagaki, K., Yamaoka, K., Wakatsuki, M.,** Studies on *Yersinia pseudotuberculosis.* IV. Isolation of *Y. pseudotuberculosis* from healthy swine, *Jpn. J. Vet. Sci.,* 38, 549, 1976.
48. **Chester, B. and Stotzky, G.,** Temperature-dependent cultural and biochemical characteristics of rhamnose-positive *Yersinia enterocolitica, J. Clin. Microbiol.,* 3, 119, 1976.
49. **Gunnison, J. B., Larson, A., and Lasarus, A. S.,** Rapid differentiation between *Pasteurella pestis* and *Pasteurella pseudotuberculosis* by action of bacteriophage, *J. Infect. Dis.,* 88, 254, 1951.
50. **Cavanaugh, D. C. and Quan, S. F.,** Rapid identification of *Pasteurella pestis* using specific bacteriophage lyophilized on strips of filter paper. A preliminary report, *Am. J. Clin. Pathol.,* 23, 619, 1953.
51. **Quan, S. F., Knapp, W., Goldenberg, M. I., Hudson, B. W., Lawton, W. D., Chen, T. H., and Kartman, L.,** Isolation of a strain of *Pasteurella pseudotuberculosis* from Alaska identified as *Pasteurella pestis:* an immunofluorescent false positive, *Am. J. Trop. Med. Hyg.,* 14, 424, 1965.

Chapter 5

ANTIGENS OF *YERSINIA ENTEROCOLITICA*

Georges Wauters

TABLE OF CONTENTS

I.	Introduction	42
II.	O-Antigens	42
	A. Study of the O-Antigen	42
	B. The O-Antigen Scheme	43
	1. Previously Described O-Factors	43
	2. New O-Factors	45
	C. Cross-Reacting Antigens Between *Y. enterocolitica* and Other Gram-Negative Bacilli	46
	D. Biochemical Nature of the O-Antigen	48
III.	K-Antigens	48
IV.	H-Antigens	49
V.	Serological Typing of *Y. enterocolitica* in Diagnostic Bacteriology	50
References		52

I. INTRODUCTION

After a first study on the serological behavior of *Yersinia enterocolitica* by Knapp and Thal[1] in 1963, a provisional antigenic scheme was worked out by Winblad[2] in 1967. At that time, only a limited number of strains were known and this investigator was able to describe eight antigenic O-factors. In 1969 a ninth factor was mentioned by Niléhn.[3] Ahvonen et al.[4] were the first to detect a cross-reaction between the latter serotype 0:9 and the *Brucella* antigens.

From then on, it was clear that the strains of *Y. enterocolitica* involved in human diseases belonged mainly to three different serotypes. Types O:3 and O:9 are spread over different regions in the world, while type O:8 is limited to North America.

Later on the antigenic scheme was extended to 17 O-factors by Wauters et al.[5] in 1971 and even further to 34 O-factors by Wauters et al.[6] in 1972. These authors also started the study of flagellar and capsular antigens.

In 1973, Knapp and Thal[7] proposed a simplified antigenic scheme that contained only six serological groups. These authors excluded from the species *Y. enterocolitica* some strains belonging to the biotypes 1 and 5 described by Niléhn[3] and Wauters.[8] The corresponding O factors were, therefore, excluded and the scheme was reduced to six main antigenic O-groups.

However, Brenner et al.[9] have demonstrated by deoxyribonucleic acid hybridization that there is a high degree of homology between all these strains. Even if new species were created within this group of microorganisms, it seems justified that *Y. enterocolitica* and *Y. enterocolitica*-like strains fit into one single antigenic scheme. The same situation exists among the species belonging to the *Klebsiella* and the *Salmonella* groups.

II. O-ANTIGENS

Y. enterocolitica, like the other *Enterobacteriaceae*, possess lipopolysaccharide determinants corresponding to the O-antigen which are heat-stable and alcohol resistant.

A. Study of the O-Antigen

Specific O-antisera are prepared in rabbits by means of a bacterial antigen, heated for 60 min at 120°C or for 2.50 hr at 100°C. The rabbits are injected intravenously four times with increasing amounts, i.e., 0.5, 1, 2, and 4 mℓ, of a bacterial suspension containing 2.10^9 organisms per milliliter.

The study of the O-antigens of *Y. enterocolitica* is obstructed by the marked tendency of this organism to change to a rough (R) form. This happens at a much higher frequency than in other *Enterobacteriaceae* and this phenomenon is enhanced by growth at 37°C. As in *Escherichia coli* and in *Salmonella*, there is a gradual transition between the pure S-(Smooth) form and the complete R-form. A great number of strains, now available in several collections, are more or less in a state of "roughness".

The agglutination test in a 0.3% auramine solution, as recommended by LeMinor,[10] can be used to distinguish the S- and R-forms of *Y. enterocolitica*. The S-colonies cannot always be easily recognized by their aspect in indirect light. In order to select S-forms, we often used a medium containing bile salts such as *Salmonella-Shigella* (SS) H, MacConkey (T) or (H) deoxycholate-citrate-lactose (DCL) agar. On such media, the S-colonies are mostly larger than those of the R-variants, which are smaller or do not even grow at all. However, this rule does not apply always for strains belonging to biotype 1 and to the rhamnose-positive strains.

Intermediate R-forms contain, apart from the specific O-antigen, an R-antigen responsible for nonspecific cross-reactions between strains of unrelated O-groups and

Table 1
CORRELATION BETWEEN AURAMINE-AGGLUTINABILITY AND CROSS-REACTIONS IN LIVING STRAINS BELONGING TO DIFFERENT O-GROUPS OF *Y. ENTEROCOLITICA*

Living Strains			O-sera								
No.[a]	O-factor	Auramine	O:8	O:15	O:21	O:4	O:6	O:9	O:3	O:5	O:16
161	O:8	++[b]	+++	+	++	−	−	−	−	−	−
614	O:15	++	+	+++	+	−	−	−	−	−	−
1110	O:21	++	−	−	+++	−	−	−	−	−	−
96	O:4	+	−	−	−	+++	−	−	−	−	−
102	O:6	+	−	−	−	−	+++	−	−	−	−
383	O:9	+	−	−	−	−	−	+++	−	−	−
134	O:3	−	−	−	−	−	−	−	+++	−	−
124	O:5	−	−	−	−	−	−	−	−	+++	−
867	O:16	−	−	−	−	−	−	−	−	−	+++

[a] Numbering according to the collection of the Institut Pasteur, Paris.
[b] + to +++ = moderate to strong reaction on slide-agglutination.

even with other Gram-negative bacteria. This especially happens with heated bacteria, while the nonheated antigen still gives specific agglutination in the O-serum. These cross-reactions occurring with the heated antigen were reported by Winblad[11] in 1973 who ascribed it to a "common Gram-negative antigen".

The difficulties many workers met in typing *Y. enterocolitica* strains, particularly when using a heated antigen, can be explained, according to our experience, by the frequent presence of a partial R-antigen that is exposed by heating. Most antisera, except those prepared with perfectly smooth strains, contain a few R-agglutinins besides the specific O-agglutinins. The R-agglutinins can be removed after absorption by a rough unrelated *Y. enterocolitica* strain or even by a rough form of *E. coli*.

Although nonspecific reactions mainly occur in heated strains, it must be emphasized that the use of sera prepared with intermediate R-strains can lead to some cross-reactions even with living bacteria.

The study of the O-antigens of *Y. enterocolitica* must take into account the following points:

1. The strains used for preparing the O-antisera must be carefully selected for smooth form. The auramine agglutinability gives useful information, but does not warrant that the strain lacks any rough antigen.
2. The agglutinations with the O-antisera are carried out, if possible, with nonheated strains in order to avoid nonspecific reactions.
3. When typing requires the use of a heated antigen, for instance, with strains possessing a K-antigen, the O-serum should be absorbed, if necessary, for R agglutinins by means of a heated rough strain of *E. coli*.

The correlation between the auramine-agglutinability and the aspecific cross-reactions in heated and nonheated strains is illustrated in Tables 1 to 3.

B. The O-Antigen Scheme
1. Previously Described O-Factors

The eight O-antigenic factors recorded by Winblad[2] appeared in the reference strains in the following order. 1,2,3 — 2,3 — 3 — 4 — 5 — 6 — 7,8 — 8.

Table 2
CORRELATION BETWEEN AURAMINE-AGGLUTINABILITY AND CROSS-REACTIONS IN HEATED STRAINS BELONGING TO DIFFERENT O-GROUPS OF *Y. ENTEROCOLITICA*

Heated strains			O-sera								
No.	O-factor	Auramine	O:8	O:15	O:21	O:4	O:6	O:9	O:3	O:5	O:16
161	O:8	++	+++	++	+++	++	++	+	++	++	++
614	O:15	+++	+++	+++	++	++	++	+	++	++	++
1110	O:21	++	+++	+	+++	++	−	−	++	+	++
96	O:4	++	+++	+++	−	+++	+	−	+	+	+
102	O:6	++	+++	+++	−	++	+++	+	−	+	−
383	O:9	++	++	+	−	−	−	+++	−	−	−
134	O:3	+	+	++	−	−	−	−	+++	−	−
124	O:5	−	++	++	−	−	−	−	−	+++	−
867	O:16	−	++	−	−	−	−	−	−	−	++

Table 3
CORRELATION BETWEEN AURAMINE-AGGLUTINABILITY AND CROSS-REACTIONS IN HEATED STRAINS BELONGING TO DIFFERENT O-GROUPS OF *Y. ENTEROCOLITICA*. O-SERA ARE ABSORBED BY A BOILED ROUGH STRAIN OF *ESCHERICHIA COLI*

Heated strains			O-sera (absorbed by a rough *E. coli*)								
No.	O-factor	Auramine	O:8	O:15	O:21	O:4	O:6	O:9	O:3	O:5	O:16
161	O:8	++	+++	+	++	−	−	−	−	−	−
614	O:15	+++	+	+++	−	+	−	−	−	−	−
1110	O:21	++	+	−	+++	−	−	−	−	−	+
96	O:4	++	−	+	−	+++	−	−	−	−	−
102	O:6	++	−	−	−	−	+++	−	−	−	−
383	O:9	++	−	−	−	−	−	+++	−	−	−
134	O:3	+	−	−	−	−	−	−	+++	−	−
124	O:5	−	−	−	−	−	−	−	−	+++	−
867	O:16	−	−	−	−	−	−	−	−	−	++

Extending this antigenic scheme, Wauters et al.[5,6] brought the number of O-factors up to 34 and their work gave the opportunity to revise some formulas previously established.

1. The first two groups contain a common factor called factor 2. It appears that group 2 possess a distinct antigen, named 2b while number 2a was used for the common antigen. Furthermore, the growth temperature has a major effect on the antigenic pattern of these groups. Factors 1 and 2 are dominant at 22°C while factor 3 prevails above 28°C. Thus, the formulas can be written in the following way:

   22°C: 1, 2a, (3)   28°C: (1, 2a), 3
   22°C: 2a, 2b, (3)   28°C: (2a), 2b, 3

2. The strains agglutinated by serum O:5 belong to two different biotypes. The strains of biotype 1 only have the factor O:5. The strains of biotype 2 (rarely 3) possess the supplementary factor 0:27.

Table 4
LIST OF THE O-FACTORS OF *Y. ENTEROCOLITICA* DESCRIBED UP TO 1972, WITH THE REFERENCE STRAINS

O-factors	N^r strains[a]	Biochemical group[b]	Origin	Country
O:1,2a,3	135	Biot. 3	Chinchilla	Netherlands
O:2a,2b,3	178	Biot. 5	Hare	Great Britain
O:3	134	Biot. 4	Man	Sweden
O:4,32	96	Biot. 1	Chinchilla	Denmark
O:4,33	1476	Biot. 1	Water	Norway
O:5	124	Biot. 1	Cow	France
O:5,27	885	Biot. 2	Dog	Great Britain
O:6,30	102	Biot. 1	Man	Denmark
O:6,31	1477	Biot. 1	Water	Norway
O:7,8	106	Biot. 1	Guinea pig	Denmark
O:8	161	Biot. 1	Man	U.S.
O:9	383	Biot. 2	Man	Belgium
O:10	500	Biot. 1	Ice cream	France
O:11,23	105	Sacch.−	Man	Denmark
O:11,24	841	Sacch.−	Man	U.S.
O:12,25	490	Sacch.−	Hare	France
O:12,26	103	Sacch.−	Sheep	Denmark
O:13,7	553	Biot. 1	Man	Belgium
O:14	480	Biot. 1	Man	Belgium
O:15	614	Biot. 4	Man	Netherlands
O:16	1475	Sacch.−	Water	Norway
O:16,29	867	Rha.+	Man	Belgium
O:17	955	Rha.+, Mel.+	Water	Norway
O:18	846	Biot. 1	Man	U.S.
O:19,8	842	Biot. 1	Man	U.S.
O:20	845	Biot. 1	Man	U.S.
O:21	1110	Biot. 1	Man	U.S.
O:22	1367	Biot. 1	Man	Czechoslovakia
O:28	1474	Sacch.−	Water	Norway

[a] Numbering according to the collection of the Institut Pasteur, Paris.
[b] Biochemical grouping based on the biotypes 1 to 5 according to Wauters[8] and on further biochemical groups (rhamnose +, rhamnose and melibiose +, saccharose −) according to Brenner et al.[9]

3. Minor accessory factors numbered from O:29 to O:33 were found, respectively, in groups O:16, O:6, and O:4. Presumably, more such factors will be detected when a larger number of strains have been investigated. In groups O:11 and O:12 as well, most of the strains seem to have additional factors, some of which were numbered O:23 to O:26.
4. Factor O:34, described in 1972, appeared later on to be identical to the O:10 factor and should, therefore, be excluded from the antigenic scheme.

Table 4 shows the list of the strains with their origin, used as a reference for the first 33 O-factors.

2. New O-Factors

Since the description of 33 O-antigenic factors up to 1972, it appeared that a great number of strains isolated in recent years could not be agglutinated by the available sera. These "new serotypes" mostly belong to the strains originating from the environment.

Table 5
LIST OF THE NEW O-FACTORS OF *Y. ENTEROCOLITICA* WITH THE REFERENCE STRAINS

O-factors	N^r strains	Biochemical group	Isolated by or received from	Origin	Country
O:35	3842	Rha.+	Wauters W342	Pig	Belgium
O:36	2222	Rha.+, Mel.+	Zen-Yoji Te 17	Oyster	Japan
O:37	7224	Rha.+, Mel.+	Van Landuyt	Man	Belgium
O:38	7175	Rha.+	Wauters W466	Pig	Belgium
O:39	7142	Rha.+	Vandepitte 13/12/78	Man	Belgium
O:40	2677	Biot. 1	Quan 722614	Man	U.S.
O:41,42	2223	Biot. 1	Zen-Yoji Te 18	Man	Japan
O:41,43	3235	Biot. 1	Toma 164	Man	Canada
O:44	7146	Rha.+	Vandepitte 13/22/78	Man	Belgium
O:44,45	7210	Rha.+	Wauters W630	Man	Belgium
O:46	7230	Sacch.−	Mair 20203/70	Mouse	Great Britain
O:47	7184	Biot. 3	Wauters W546	Man	Belgium
O:48	3960	Rha.+, Mel.+	Toma 224	Man	Canada
O:49,51	7231	Biot. 1	Lassen 546/71		Norway
O:50,51	7229	Sacch.−	Aldova Z7 ST2	Microtus	Czechoslovakia
O:52	7209	Sacch.−	Wauters W629	Man	Belgium
O:52,53	2842	Rha.+, Mel.+	Toma 120	Water	Canada
O:52,(54)	2835	Rha.+, Mel.+	Toma 112	Beaver	Canada
O:55		Rha.+	Van Pee 6-3/78 (WE111/78)	Water	Belgium
O:56		Biot. 3	Toma 822	Piglet	Canada
O:57		Rha.+, Mel.+	Toma 500	Man	Canada

Several fit into biotype 1 or may be classified as *Y. enterocolitica*-like strains, i.e., rhamnose-positive and sucrose-negative strains. We carried out the antigenic study of a limited number of such strains. This allowed us to recognize a number of new O-factors. A description of the strains used for this study is given in Table 5.

In most of these strains the O-antigen does not cross-react with the formerly known O-groups or with the recently investigated strains. However, a few strains exhibit some cross-agglutination among each other. In such cases, the partial antigens were demonstrated by cross-absorption of the corresponding sera. The titers of the sera prepared against the new O-factors, before and after absorption, are given in Table 6.

C. Cross-Reacting Antigens Between *Y. enterocolitica* and Other Gram-Negative Bacilli

Ahvonen et al.[4] have drawn the attention on the strong cross-reaction existing between *Y. enterocolitica* type 9 and different *Brucella* species. Akkermans and Hill[12] and Bockemühl and Roth[13] have stressed the importance of this common antigen in the serological detection of brucellosis of cattle. Corbel and Cullen[14] proposed a Rose Bengal plate test in order to differentiate the serological response of cattle to the two organisms. Similarly, brucellosis and yersiniosis in humans may be misdiagnosed by serology. Ahvonen and Sivers[16] stated in 1969 that serum of patients suffering from *Y. enterocolitica* type O:9 infection often presents equal agglutination titers against the two bacteria.

In 1971 Hurvell et al.[15] investigated the cross-reacting antigens and antibodies in different *Brucella* species and in *Y. enterocolitica* type O:9. According to these authors, there is no qualitative difference in the cross-reaction between *Y. enterocolitica* on the one hand and, respectively, *B. melitensis*, *B. abortus*, *B. suis*, and *B. neotomae* on the

Table 6

TITERS OF O-SERA AGAINST *Y. ENTEROCOLITICA* (FACTORS O:35 TO O:57) BEFORE AND AFTER CROSS-ABSORPTION

O-sera	O-factor	3842	2222	7224	7175	7142	2677	2223	3235	7146	7210	7230	7184	3960	7231	7229	7209	2842	2835	E111	T822	T500
3842	O:35	3200[a]																				
2222	O:36		12800																			
7224	O:37			12800																		
7175	O:38				6400																	
7142	O:39					1600																
2677	O:40						3200															
2223	O:41,42							12800	3200													
abs. 3235	O:42							800	0													
3235	O:41,43							400	3200													
abs. 2223	O:43							0	1600													
7146	O:44									6400	6400											
abs. 7210	—									0	0											
7210	O:44,45									800	3200											
abs. 7146	O:45									0	800											
7230	O:46											12800										
7184	O:47												3200									
3960	O:48													6400								
7231	O:49,51														25600	800						
abs. 2229	O:49														6400	0						
7229	O:50,51														200	1600						
abs. 7231	O:50														0	1600						
7209	O:52																800	400	800			
abs. 2842	—																0	0	0			
abs. 2835	—																0	0	0			
2842	O:52,53																1600	3200	6400			
abs. 7209	O:53																0	400	0			
abs. 2835	O:53																0	800	0			
2835	O:52,(54)[b]																3200	1600	3200			
abs. 7209	O(54)																0	0	0			
abs. 2842	O(54)																0	0	200			
E. 111/78	O:55																			3200		
T. 822	O:56																				400	
T. 500	O:57																					800

[a] Titers are expressed as the reciprocal of the highest dilution giving an agglutination.
[b] Brackets indicate a very weak and irregular factor.

other hand. There is, however, a quantitative difference, *B. abortus* being responsible for the strongest cross-reaction. Furthermore, the cross-reacting antigens seem to contribute to only a smaller part of the total antigenic spectrum of *Y. enterocolitica* than they do in *Brucella*. According to Hurvell,[17,18] with hot phenol-water extracted polysaccharides the cross-reactivity is unaffected by heat or by lipolytic or proteolytic enzymes, suggesting the carbohydrate nature of these determinants. Hurvell[19,20] was also able to distinguish the specific antibodies of each species by means of immunodiffusion, immunoelectrophoresis, and electroimmunoassay. Carlsson et al.[21] applied the ELISA method for this purpose.

Y. enterocolitica O:9 is also related to the O:30 factor of *Salmonella urbana*, as stated by Hurvell[22] in 1973.

On the contrary, while *Brucella* and *Vibrio cholerae* share a common antigen, reported by Feeley,[23] *Y. enterocolitica* type O:9 and *Vibrio cholerae* do not exhibit any cross-reaction.

A cross-reaction between *Y. enterocolitica* O:12, 25, and the antigen O:47 of *S. bergen* has been reported by Wauters et al.[5]

The O:46 factor carried by *Y. enterocolitica* IP 7230 (Mair 20203) strongly cross-reacts with *Y. pseudotuberculosis* of serogroup IV. The latter organism is also agglutinated by a serum prepared against *Salmonella* O:9.

Morganella (*Proteus*) *Morganii* O:43 and O:44 are related to *Y. enterocolitica*, respectively, O:9 and O:17, as reported by Velin and Rauss.[24]

In 1975 Alonso et al.,[25] examining several rhamnose-positive strains, noticed that some of them were agglutinated by sera against *Y. pseudotuberculosis* belonging to serogroups O:1, O:2, and O:5.

D. Biochemical Nature of the O-Antigen

In 1973, Rische et al.[26] and Beer and Seltman[27] studied the polysaccharide components of *Y. enterocolitica*. In 11 strains investigated, the following elements were always present: 2-keto-3-deoxyoctonate (KDO), two heptoses, glucosamine, galactosamine, glucose, and galactose. Rhamnose and 6-deoxyhexoses were found in some strains.

In 1975, Wartenberg et al.[28] carried on the study of the polysaccharides of *Y. enterocolitica* and confirmed the findings of the former authors, except that galactose and galactosamine were not always present in all the strains investigated. Apart from the basic components mentioned above, the strains have additional residues such as galactose, mannose, rhamnose, fucose, galactosamine, and two different 6-deoxyhexoses. These latter components allowed the authors to recognize seven different chemotypes among the six serogroups analyzed.

Acker and Wartenberg[29] were able, in 1976, to compare the ultrastructure of the lipopolysaccharide of *Y. enterocolitica*, *Salmonella typhimurium*, and *Escherichia coli* by electron microscopy. Strand-like LPS structure appears to be common to the three species. The strands containing longitudinal fibrils of helicoidal structure are thought to be responsible for a "double track" profile of the outer membrane. In 1977 the study of the arrangement of the LPS on the outer membrane of *Y. enterocolitica* was carried on by Acker.[30] This author, using an antiserum against the S-form of the organism, demonstrated through electron microscopy that an outer layer consisted of the sidechains of the specific O-polysaccharide. The absence of this layer in a deep rough mutant made it possible to discern the strand-like structures as contiguous strands of the LPS.

III. K-ANTIGENS

Because of the presence of an envelope antigen, the living cells of some *Y. enterocolitica* strains cannot be agglutinated by the homologous O sera. This was first

Table 7
K-ANTIGENS OF *Y. ENTEROCOLITICA*, TITERS OF THE CORRESPONDING K-ANTISERA[a] AND REFERENCE STRAINS

N[r] strains[b]	K-antigen	O-antigen	Titer K-ser	Strain received from or isolated by	Origin	Country
551	K:1	O:10	3200	Wauters W51	Man	Belgium
129	K:2	O:8	3200	Mair 9625	Coypu	Great Britain
—	K:3	O:56	800	Toma 822	Piglet	Canada
—	K:4	O:57	3200	Toma 500	Man	Canada
7142	K:5	O:39	1600	Vandepitte 13/12/78	Man	Belgium
7209	K:6	O:52	1600	Wauters W629	Man	Belgium

[a] Titers are expressed as the reciprocal of the highest dilution giving a positive reaction. There are no cross-reactions between the different K-antigens.
[b] Numbering of the strains according to the collection of the Institute Pasteur, Paris.

observed by Wauters et al.[5] in strain IP 551 belonging to serogroup O:10 and possessing a K-antigen called K:1.

Recently, Aleksic et al.[31] have demonstrated with the electron microscope that the K:1 antigen of strain IP 551 consisted of fimbriae. These authors have also studied the immunologic properties of the fimbriae-antigen. It was found to be relatively thermostable, as it is only destroyed completely after steaming for at least 30 min at 120°C. The O inagglutinable strains still recover their O-agglutinability by heating at 100°C.

K-antisera are prepared in the following way: the strain is cultured at 37°C in order to avoid the presence of flagellar antigen. Rabbits are immunized by means of unheated, formalized antigen. The serum contains both O- and K-agglutinins. The O-agglutinins are removed by absorption with a strain belonging to the same O-group without K-antigen, if such a strain is available. If not, absorption of the O-agglutinins may be carried out by means of the homologous strain heated for 1 hr at 120°C. The pure K-antiserum (antifimbriae serum) should be used for typing of the living bacteria.

Presently, we have been able to detect six different K-antigens. Some of them were found associated with only one O-factor. Others, like the K:1 antigen, occur in strains belonging to different O-groups. Whether the antigens K:2 to K:6 are fimbriae or not has not yet been investigated.

Most strains bearing an envelope antigen were isolated from the environment or belong to biochemical groups known for their ubiquitous character, like biotype 1 and the rhamnose-positive and the sucrose-negative strains. They are particularly frequent among the rhamnose-positive water strains, and we have met several more strains belonging to this group and having a K-antigen which is not yet serologically identified.

At this present time, strains usually involved in human diseases do not seem to carry envelope antigens. Strain IP 129 with antigen K:2 belong to serogroup O:8, like many North American isolates. However, this strain is biochemically different from the human O:8 strains.

The K-antigens are listed in Table 7.

IV. H-ANTIGENS

The motility of *Y. enterocolitica* varies according to their biotypes. Strains of biotype 1 are usually very motile, while those belonging to biotype 4 and still more to biotype 5 display a rather weak motility. Moreover, the growth temperature has an effect upon flagella formation and at 37°C most strains completely lack flagellar antigens.

Therefore, we prefer the term "unheated antigen", rather than "OH-antigen" when speaking about living or nonheated bacterial suspensions. Indeed, these antigenic suspensions often do not contain any H-antigen, particularly when the strains are not previously trained for motility.

The study of the H-antigens of *Y. enterocolitica* must be done on strains cultured at 25°C and beforehand trained on a soft agar medium in order to enhance their motility.

H-antisera are prepared in rabbits by means of formalin-killed bacteria. The OH-sera are further absorbed by the homologous heated antigen in order to discard the O-agglutinins. The titer of H-antisera is often very high, reaching 1/25,000 and more.

Wauters et al.[5,6] investigated the H-antigens of the strains used for the O-antigen scheme. They described 19 different H-factors which were designated by small letters.

Many H-patterns are very complex, mainly in strains of biotype 1 where up to 5 different factors often occur in one single strain.

The strains of *Y. enterocolitica* and *Y. enterocolitica*-like organisms behave in three different ways according to their flagellar antigens:

1. The strains of biotypes 2, 3, 4, and 5 only share the factors a, b, and c in various combinations.
2. As a rule, strains of biotype 1 have a complex pattern composed by the factors from a to k. Moreover, their H-formulas are highly diversified even in the strains of the same O group.
3. The H-patterns of the *Y. enterocolitica*-like strains, i.e., rhamnose-positive and sucrose-negative strains, share no common antigen with the typical *Y. enterocolitica* strains. This could be of concern in taxonomic problems.

The flagellar antigens of the strains possessing the new O-factors described in this chapter have not yet been investigated. Table 8 reports the H-antigens as described up to 1972 by Wauters et al.[5,6]

At the present time, the great variety of the H-antigens confers a limited interest in typifying the H-antigen. Indeed, especially in biotype 1, different H-patterns are scattered over an unlimited number of individual strains. Perhaps it will be possible to simplify the H-antigen scheme by using methods other than agglutination. A specific motility inhibition by H-antisera was proposed by Le Minor and Pigache[32] for determining the H-antigens of *Serratia marcescens*. This approach might facilitate the typing of the H-antigens of *Y. enterocolitica* and improve its usefulness. However, there is no evidence that this technique can be applied to slightly motile strains.

V. SEROLOGICAL TYPING OF *Y. ENTEROCOLITICA* IN DIAGNOSTIC BACTERIOLOGY

There is a fair correlation between the antigenic pattern, the biochemical features, and the ecological behavior of some *Y. enterocolitica* strains, as it was stated by Wauters.[33] The host-bounded strains, particularly those involved in human diseases, fit into a small numbers of serotypes, corresponding to well-defined biotypes. The typing of the antigens of such strains may, therefore, help lead to their identification.

On the other hand, some environmental strains may become pathogenic for man as opportunistic microorganisms. This is clearly illustrated by the observation made by Dabernat et al.,[34] who isolated, over a period of several months, 15 strains of a rhamnose-positive *Y. enterocolitica* in hospitalized children. Most of the strains were recovered

Table 8
H-ANTIGENS OF
Y. ENTEROCOLITICA[a]

N[r] strains[b]	O-factors	H-factors
64	O:1,2a,3	H:a,b,c
178	O:2a,2b,3	H:b,c
175	O:3	H:a,b,c
211	O:9	H:a,b
614	O:15	H:(b),c
96	O:4,32	H:b,e,f,i
123	O:5	H:b,c,d,e,i
102	O:6,30	H:a,b,d,g,i
106	O:7,8	H:d,e,f,g,h
161	O:8	H:b,e,f,i
551	O:10	H:b,(f),k
105	O:11,23	H:l
490	O:12,25	H:o
553	O:13,7	H:n
480	O:14	H:m
867	O:16	H:p
955	O:17	H:q
841	O:11,24	H:r
1474	O:28	H:s
1494	O:12,26	H:t

[a] According to data of Wauters et al.[5,6]
[b] Numbering of the strains according to the collection of the Institute Pasteur, Paris.

from blood cultures and almost all belonged to serotype O:14. In such a case, too, serotyping can help to assess the similarity between the strains and allows us to look for a common nosocomial origin.

However, it now appears that many O-factors, even those present in human pathogenic strains, may occur in strains of various origin and belong to different biochemical groups. For example, factor O:8 is not only found in human strains, but also in a few strains originating both from animals and from the environment. The latter, however, are biochemically different from the typical human O:8 strains. Furthermore, factor O:3 seemed to be quite specific for the human biotype 4 strains. Nevertheless, it has been detected as the single O-factor present in some waterstrains of biotype 1.

Therefore, it must be emphasized that serotyping of *Y. enterocolitica* does not constitute a tool for classification of the strains by themselves. Antigenic study should be performed only after proper biotyping of the strains. Only in these conditions will serological typing prove to be helpful both in identification of the specific pathogenic strains and in contributing to epidemiological investigation.

For routine work, only the O-typing should be made by the diagnostic laboratories on the strains currently encountered in human diseases, i.e., serotypes O:3, O:8, O:9, and perhaps O:5, 27. The other strains occasionally found in man, animals, food, and the environment should be submitted to a reference center for *Yersinia*. Several such laboratories are already working in different areas.

Particular care must be taken in keeping this organism in a smooth form whenever an antigenic typing has to be carried out. In connection with this, routine techniques should avoid any incubation at 37°C, as very few passages at this temperature may favor the occurrence of intermediate rough forms. Moreover, when cross-reactions appear in several unrelated O-sera, it must be kept in mind that nonspecific reactions may be due to R agglutinins, especially when using antigens prepared by heating.

Typing of the H antigens does not yet have a practical value for diagnostic purpose in routine laboratories.

REFERENCES

1. **Knapp, W. and Thal, E.**, Untersuchungen über die kulturellbiochemischen, serologischen, tierexperimentellen und immunologischen Eigenschaften einer vörlaufig *"Pasteurella* X" benannten Bakterienart., *Zbl. Bakt. I Abt. Orig.,* 190, 472, 1963.
2. **Winblad, S.**, Studies on serological typing of *Yersinia enterocolitica, Acta Pathol. Microbiol. Scand.,* Suppl. 187, 115, 1967.
3. **Niléhn, B.**, Studies on *Yersinia enterocolitica, Acta Pathol. Microbiol. Scand.,* Suppl. 206, 10, 1969.
4. **Ahvonen, P., Jansson, E., and Aho, K.**, Marked cross-agglutination between *Brucellae* and a subtype of *Yersinia enterocolitica, Acta Pathol. Microbiol. Scand.,* 75, 291, 1969.
5. **Wauters, G., Le Minor, L., and Chalon, A. M.**, Antigènes somatiques et flagellaires des *Yersinia enterocolitica, Ann. Inst. Pasteur (Paris),* 120, 631, 1971.
6. **Wauters, G., Le Minor, L., Chalon, A. M., and Lassen, J.**, Supplément au schéma antigénique de *Yersinia enterocolitica, Ann. Inst. Pasteur (Paris),* 122, 951, 1972.
7. **Knapp, W. and Thal, E.**, Die biochemische Charakterisierung von *Yersinia enterocolitica* (syn. *"Pasteurella* X") als Grundlage eines vereinfachten O-Antigenschemas, *Zentralbl. Bakteriol. Hyg. I Abt. Orig. A.,* 223, 88, 1973.
8. **Wauters, G.**, *Contribution à l'étude de Yersinia enterocolitica,* Vander, Louvain, 1970, 58.
9. **Brenner, D. J., Steigerwalt, A. G., Falcao, D. P., Weaver, R. E., and Fanning, G. R.**, Characterization of *Y. enterocolitica* and *Y. pseudotuberculosis* by deoxyribonucleic acid hybridization and by biochemical reactions, *Int. J. Syst. Bact.,* 26, 180, 1976.
10. **Le Minor, L.**, Le Diagnostic de Laboratoire des Bacilles à Gram Négatif. Entérobactéries Vol. 1, Ed. de la Tourelle, St-Mandé, 1972, 35.
11. **Winblad, S.**, Studies on the O-serotypes of *Yersinia enterocolitica, Proc. Int. Symp. Yersinia, Pasteurella and Francisella,* Winblad, S., Ed., S. Karger, Basel, 1973, 27.
12. **Akkermans, J. P. W. M. and Hill, W. K. W.**, *Yersinia enterocolitica* serotype 9 infection as a factor interfering with serodiagnosis of *Brucella* infections in swine, *Neth. J. Vet. Sci.,* 5, 73, 1972.
13. **Bockemühl, J. and Roth, J.**, *Brucella*-titer bei subklinischen Infektionen mit *Yersinia enterocolitica* Serotyp 0:9 in einem Schweine-Zuchtbetrieb, *Zentralbl. Bakteriol. Hyg. I Abt. Orig. A,* 240, 86, 1978.
14. **Corbel, M. J. and Cullen, G. A.**, Differentiation of the serological response to *Yersinia enterocolitica* and *Brucella abortus* in cattle, *J. Hyg.,* 68, 519, 1970.
15. **Hurvell, B., Ahvonen, P. and Thal, E.**, Serological cross-reactions between different *Brucella* species and *Yersinia enterocolitica, Acta Vet. Scand.,* 12, 86, 1971.
16. **Ahvonen, P. and Sievers, K.**, *Yersinia enterocolitica* infection associated with *Brucella* agglutinins. Clinical features of 24 patients, *Acta Med. Scand.,* 185, 121, 1969.
17. **Hurvell, B.**, Serological cross-reactions between different *Brucella* species and *Yersinia enterocolitica. Acta Pathol. Microbiol. Scand. Sect. B,* 81, 105, 1973.
18. **Hurvell, B.**, Serological cross-reactions between different *Brucella* species and *Yersinia enterocolitica, Acta Pathol. Microbiol. Scand. Sect. B,* 81, 113, 1973.
19. **Hurvell, B.**, Serological cross-reactions between different *Brucella* species and *Yersinia enterocolitica.* Immunodiffusion and immunoelectrophoresis, *Acta Vet. Scand.,* 13, 472, 1972.
20. **Hurvell, B.**, Differentiating of cross-reacting antibodies against *Brucella abortus* and *Yersinia enterocolitica* by electroimmuno assay, *Acta Vet. Scand.,* 16, 318, 1975.
21. **Carlsson, H. E., Hurvell, B., and Lindberg, A. A.**, Enzyme linked immunosorbent assay (ELISA) for titration of antibodies against *Brucella abortus* and *Yersinia enterocolitica, Acta Pathol. microbiol. Scand. Sect. C,* 84, 168, 1976.
22. **Hurvell, B.**, Serological cross-reactions between different *Brucella* species and *Yersinia enterocolitica, Acta Vet. Scand.,* 14, 1, 1973.
23. **Feeley, J. C.**, Somatic O antigen relationship of *Brucella* and *Vibrio cholerae, J. Bacteriol.,* 99, 645, 1969.
24. **Velin, D. and Rauss, K.**, Antigenic relationship between *Morganella morganii* and *Yersinia enterocolitica, Acta Microbiol. Acad. Sci. Hung.,* 23, 83, 1976.

25. **Alonso, J. M., Bejot, J., Bercovier, H., and Mollaret, H. H.**, Sur un groupe de souches de *Yersinia enterocolitica* fermentant le rhamnose, *Méd. Mal. Infect.*, 5, 490, 1975.
26. **Rische, H., Beer, W., Seltman, G., Thal, E., and Horn, G.**, Die Zusammensetzung der Lipopolysaccharide von *Yersinia enterocolitica* und *Yersinia pseudotuberculosis* und die Empfindlichkeit gegenüber Bakteriophagen, *Proc. Int. Symp. Yersinia, Pasteurella and Francisella*, Winblad, S., Ed., S. Karger, Basel, 1973, 23.
27. **Beer, W. and Seltman, G.**, Zuckersammensetzung der Lipopolysaccharide einiger *Yersinia enterocolitica*-Stämme, *Z. Allg. Mikrobiol.*, 13, 167, 1973.
28. **Wartenberg, K., Lysy, J., and Knapp, W.**, On the sugar content of the lipopolysaccharides of the various strains known as *Yersinia enterocolitica*, *Zentralbl. Bakteriol. Hyg. I Abt. Orig. A*, 230, 361, 1975.
29. **Acker, G. and Wartenberg, K.**, Ultrastructure of lipopolysaccharides of *Yersinia enterocolitica, Salmonella typhimurium* and *Escherichia coli, Zentralbl. Bakteriol. Hyg. I Abt. Orig. A*, 235, 439, 1976.
30. **Acker, G.**, The arrangement of lipopolysaccharides on the outer membrane of *Yersinia enterocolitica*: an electron microscopic study, *Zentralbl. Bakteriol. I Abt. Orig. A.*, 237, 504, 1977.
31. **Aleksic, R., Rohde, R., Müller, G., and Wohlers, B.**, Examination of the envelope antigen K1 in *Yersinia enterocolitica* which was identified as fimbriae, *Zentralbl. Bakteriol. Hyg. I Abt. Orig. A*, 234, 513, 1976.
32. **Le Minor, L. and Pigache, F.**, Etude antigénique de souches de *Serratia marcescens* isolées en France, *Ann. Microbiol. (Paris)*, 128B, 207, 1977.
33. **Wauters, G.**, Correlation between ecology, biochemical behaviour and antigenic properties of *Yersinia enterocolitica., Proc. Int. Symp. Yersinia, Pasteurella and Francisella*, Winblad, S., Ed., S. Karger, Basel, 1973, 38.
34. **Dabernat, H. J., Bauriaud, R., Lemozy, J., Lefèvre, J. C., and Lareng, M. B.**, *Yersinia enterocolitica* fermentant le rhamnose. A propos de 15 souches isolées chez des enfants, *Méd. Mal. Infect.*, 8, 158, 1978.

Chapter 6

ANTIBIOTIC RESISTANCE IN *YERSINIA ENTEROCOLITICA*

Guy Cornelis

TABLE OF CONTENTS

I.	Chromosomal Resistance Toward the β-Lactam Antibiotics............. 57	
	A.	β-Lactamase Production 57
	B.	β-Lactamases from Strain W222 57
	C.	β-Lactamases from Strain H66 58
	D.	β-Lactamases from Other Strains of Serotypes O:3 and O:9 59
	E.	Role of β-Lactamases A and B in Antibiotic Resistance of Strains Belonging to Serotypes O:3 and O:9 59
	F.	Cellular Location of the β-Lactamases in *Y. enterocolitica* 61
	G.	β-Lactamases from *Y. enterocolitica* Strains that Belong to Serotype O:5b 63
	H.	Inducibility of β-Lactamase B 63
	I.	Mutants of *Y. enterocolitica* IP97 that Synthesize Type B β-Lactamase Constitutively 63
	J.	Correlation Between Serological and Biochemical Groupings of *Y. enterocolitica* with the β-Lactamases of the Strains 65
II.	Resistance Plasmids in *Y. enterocolitica* 70	
Acknowledgments 70		
References 70		

I. CHROMOSOMAL RESISTANCE TOWARD THE β-LACTAM ANTIBIOTICS

A. β-Lactamase Production

In the early papers, several authors reported that antibiotics that belong to the β-lactam group have little or no activity against *Yersinia enterocolitica*.[1-6] As shown in Table 1, there is indeed a striking difference between *Y. enterocolitica* and *Y. pseudotuberculosis* in their sensitivity to β-lactam antibiotics.

Members of serological groups of *Y. enterocolitica* are resistant to benzylpenicillin and cephalothin and most of the groups also to ampicillin and carbenicillin. The minimal inhibitory concentration is highest with carbenicillin. With respect to sensitivity to β-lactam antibiotics, strains belonging to the different serological and biochemical groups of *Y. enterocolitica* appear homogeneous. However, considerable differences exist between the various groups. In parallel with this resistance, *Y. enterocolitica* produces fair amounts of β-lactamase.[6,7]

Since all the *Y. enterocolitica* strains that have been tested (more than 400) produce significant levels of β-lactamase and since it is exceptional, so far, to find plasmid-coded resistance toward other antibiotics in this species (see Section II), one can assume that β-lactamase production is chromosomally mediated. This is further supported by the fact that strain W277OI (a nalidixic acid-resistant derivative of W277 which is a typical strain of serotype O:9) does not contain any plasmid DNA while its derivative carrying the *lac* plasmid pGCI does.[8]

B. β-Lactamases from Strain W222

β-Lactamase from strain W222 (a representative of serotype O:3) has been purified by ion exchange chromatography.[9] Surprisingly, this strain turned out to produce two different β-lactamases instead of a single enzyme as was expected (Figure 1). The two β-lactamases have been called A and B. As shown in Table 2, enzyme A hydrolyzes a wide variety of penicillins and cephalosporins, including carbenicillin, while enzyme B is mainly a cephalosporinase. Of the substrates tested, cephaloridine was the best for enzyme A while cephalosporin C was the best for enzyme B. Treatment with para-chloromercuribenzoate (5×10^{-4} M), that reacts with thiol groups, causes more than 99% inhibition of lactamase A but less than 15% inhibition of lactamase B.[9] Conversely, cloxacillin and carbenicillin are strong inhibitors of B while A is less affected (see Table 3).[9,10] The molecular weights of A and B were estimated by gel filtration on Sephadex® G-100. As shown in Figure 2, lactamase A has a mol wt of 20,000 daltons and B has a mol wt of 34,000 daltons. In an attempt to determine whether strain W222 produced an inducible β-lactamase, 6-amino-penicillanic acid (6-APA) was added to the culture. The quotient (β-lactamase activity of cultures grown with 6-APA)/(β-lactamase activity of cultures grown without 6-APA) was 1.9 when cephalosporin C was used as a substrate, but only 1.4 when cephaloridine was used as a substrate. The most probable explanation is that β-lactamase B is inducible while β-lactamase A is not.

C. β-Lactamase from Strain H66

Nearly all the strains of *Y. enterocolitica* that belong to serotype O:3 are highly resistant to carbenicillin (MIC = 256 to 512 μg/mℓ, see Table 1). However, strain H66, isolated in Belgium by Greven-Brauns and given to us by Professor Vandepitte, is exceptional among those of serotype O:3. It is inhibited by 0.5 μg/mℓ of carbenicillin, but nevertheless it is still resistant to cephalothin.

Table 1
MEAN MINIMUM INHIBITORY CONCENTRATIONS OF β-LACTAM ANTIBIOTICS

Minimum inhibitory concentration (μg/mℓ)

Organism	6APA	Oxacillin	Methicillin	Cloxacillin	Carbenicillin	Cefamandole	Ampicillin	Benzylpenicillin	Cephaloridine	Cephalexin	Cefoxitin	Cephalothin	Cephalosporin C
Y. pseudotuberculosis	32	32	32	16	0.5	0.5	0.25	0.5	1	2	0.5	0.5	4
Y. enterocolitica group O:1	128	128	512	128	512	4	64	128	16	32	64	256	1024
Y. enterocolitica group O:2													
Strain A14700	128	64	256	64	256	1	32	32	4	8	4	32	512
Strain F96-1	64	32	256	32	128	2	32	16	2	8	1	16	16
Y. enterocolitica group O:3	128	128	512	64	512	4	64	128	8	32	4	128	512
Strain H66	32	32	32	32	0.5	0.5	8	32	4	8	4	32	512
Y. enterocolitica group O:5b (except indole(−) strains)	64	64	128	64	2	2	32—64	256	32	64	64	512	1024
Y. enterocolitica group O:9	128	64	256	64	256	2	64	128	32	64	64	512	1024
Strain W23801	32	—	16	—	2	0.25	2	8	2	—	8	32	1024
Y. enterocolitica group O:16	64	128	256	128	256	2	128	256	16	64	32	256	1024

From Cornelis, G., J. Gen. Microbiol., 91, 391, 1975. With permission.

Table 2
SUBSTRATE PROFILES OF β-LACTAMASE FROM Y. ENTEROCOLITICA, STRAIN W222[a]

Substrate	Crude extract	β-Lactamase A		β-Lactamase B	
		Rate of hydrolysis	K_m (μM)	Rate of hydrolysis	K_m (μM)
Penicillin G	100	100	39	100	17
Ampicillin	43	77	—	<1	—
Carbenicillin	21	25	—	<1	—
Cephaloridine	248	300	175	277	—
Cephalosporin C	88	35	—	605	—
Cephalothin	—	134	—	214	—
Oxacillin	—	16	—	<1	—
Cloxacillin	—	2	—	<1	—
Methicillin	—	8	—	<1	—
Cephalexin	—	5	—	8	—
Cefoxitin	—	<1	—	<1	—
Formylcefamandole	—	317	—	40	—
Cefamandole	—	127	—	<1	—
6-APA	—	13	—	—	—

[a] The rates of hydrolysis were determined at 30°C with substrate concentrations of 2 mg/mℓ (about 5 mM). With 6-APA the rate was measured spectrophotometrically (Waley, 1974) at pH 6.8. With other substrates measurements were made in the pH stat at pH 6.5. The rates given are relative to an arbitrary value of 100 for benzylpenicillin.

From Cornelis, G. and Abraham, E. P., *J. Gen. Microbiol.*, 87, 273, 1975. With permission.

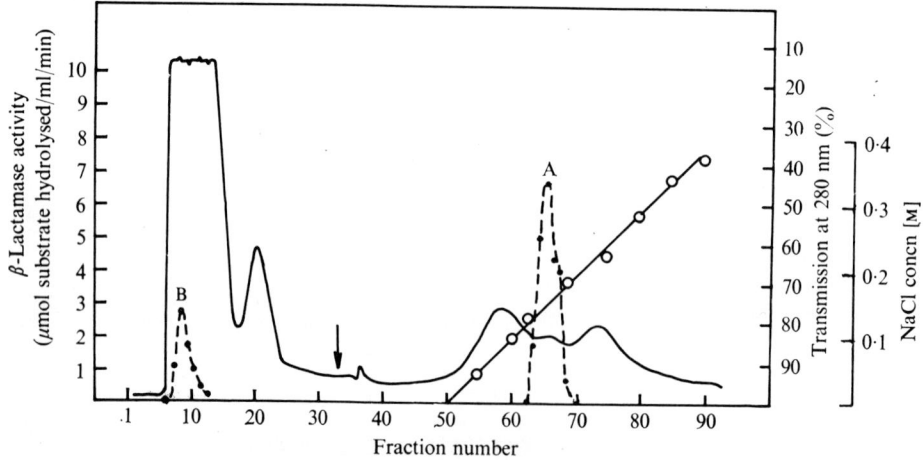

FIGURE 1. Chromatographic separation of β-lactamases A and B from *Y. enterocolitica*, strain W222. Crude extract was chromatographed on a column of Sephadex CM_{50} in sodium phosphate buffer (0.025 M) pH 7.3 at 4°C. The flow rate was 26 mℓ/hr and fractions were collected every 15 min. NaCl gradient (0 to 0.5M) was applied after the collection of 33 fractions. Cephalosporin C was used as a substrate in the assay of enzyme B, and Cephaloridine in the assay of enzyme A. ●——●, β-lactamase activity; ○——○, NaCl concentration; —, light transmission at 280 nm. (From Cornelis, G. and Abraham, E. P., *J. Gen. Microbiol.*, 87, 273, 1975. With permission.)

Table 3
**INHIBITION OF *Y. ENTEROCOLITICA*
β-LACTAMASES BY POORLY HYDROLYZED
β-LACTAM ANTIBIOTICS**[a]

Inhibitors	Inhibition (%)		
	β-Lactamase A (W222)	β-Lactamase B (W222)	Crude extract (M771)
Cloxacillin	28	>99	>99
Methicillin	64	>99	NT
Carbenicillin	NT	>99	>99
Cefoxitin	22	93	>99
Cefamandole	6	18	NT
Ampicillin	NT	64	77

Note: NT = not tested.

[a] The activity of enzyme A was determined with cephaloridine as a substrate. Cephalosporin C was the substrate used for enzyme B and for the crude extract of strain M771 serotype O:5b. The inhibitor concentration was 10^{-4} *M*. Preparations of enzyme A (W222) and B (W222) were described previously.

From Cornelis, G. and Abraham, E. P., *J. Gen. Microbiol.*, 87, 273, 1975. With permission.

Considering that a "normal" strain of serotype O:3 synthesizes two β-lactamases, it was tempting to believe that an "abnormal" strain, such as H66, synthesized only one of these two β-lactamases. This hypothesis turned out to be the correct one: all the β-lactamase activity contained in a crude extract of H66 emerged as a single peak from a column of Sephadex® CM-50 and had the substrate profile of β-lactamase B from strain W222.[9] Moreover, the β-lactamase synthesized by strain H66 was clearly inducible (see Table 4) like enzyme B from strain W222.

D. β-Lactamases from Other Strains of Serotypes O:3 and O:9

In order to show that the production of β-lactamases A and B was the rule among the strains of serotypes O:3 and O:9, we analyzed five different strains among which was strain IP 134 (serotype O:3), isolated from a patient with terminal ileitis.[11] Clearly, all these strains synthesized two β-lactamases similar to A and B.[9] The fact that the vast majority of the strains synthesized both lactamases implies that the genes encoding these two β-lactamases must be on the chromosome, at least at this stage of the evolution. This assumption is reinforced by the fact that neither A nor B has the same isoelectric focusing pattern as any plasmid-coded β-lactamase described so far.[12] It must be stressed however that the production of two different chromosomal β-lactamases is very exceptional among the *Enterobacteriaceae* (for review see Sykes and Matthew[13]). β-Lactamase B resembles the inducible chromosomal lactamases from *Enterobacter*,[14] *Proteus*,[15] and *Pseudomonas aeruginosa*,[16,17] whereas β-lactamase A, seems of a much less ubiquitous type.

E. Role of β-Lactamases A and B in Antibiotic Resistance of Strains Belonging to Serotypes O:3 and O:9

We have seen that strain H66 (serotype O:3), sensitive to carbenicillin and ampicillin,

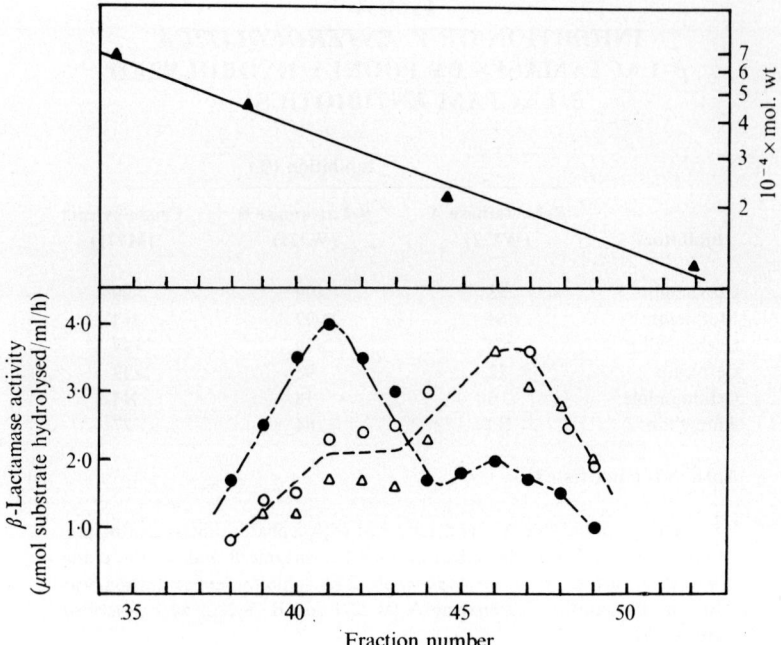

FIGURE 2. Estimation of the molecular weights of β-lactamases A and B. Gel filtration of the crude extract was carried out on a column of Sephadex® G_{100} at 4°C in sodium phosphate buffer, 0.025 M, pH 7.3; the flow rate was 35 mℓ/hr and 8.6-mℓ fractions were collected. The reference substances (▲) were bovine serum albumin (mol wt 67,000); ovalbumin (mol wt 45,000); soybean trypsin inhibitor (mol wt 21,500); cytochrome C (mol wt 12,500). The dextran blue peak appeared in fraction 27 and the DNP-lysine peak in fraction 71. ●, rate of hydrolysis with cephalosporin C as substrate; ○, rate with cephaloridine as substrate; △, rate with formylcefamandole as substrate. (From Cornelis, G. and Abraham, E. P., *J. Gen. Microbiol.*, 87, 273, 1975. With permission.)

Table 4
INDUCTION OF β-LACTAMASE PRODUCTION IN *Y. ENTEROCOLITICA* STRAIN H66 BY 6-APA[a]

6-APA (μg/mℓ)	β-Lactamase activity (relative)
0	100
25	260
50	407
100	584
250	840
500	815

[a] 6-APA was added 4.5 hr after inoculation and the cultures harvested after 7 hr. Enzyme activities are related to an arbitrary value of 100 for the culture grown without 6-APA.

From Cornelis, G. and Abraham, E. P., *J. Gen. Microbiol.*, 87, 273, 1975. With permission.

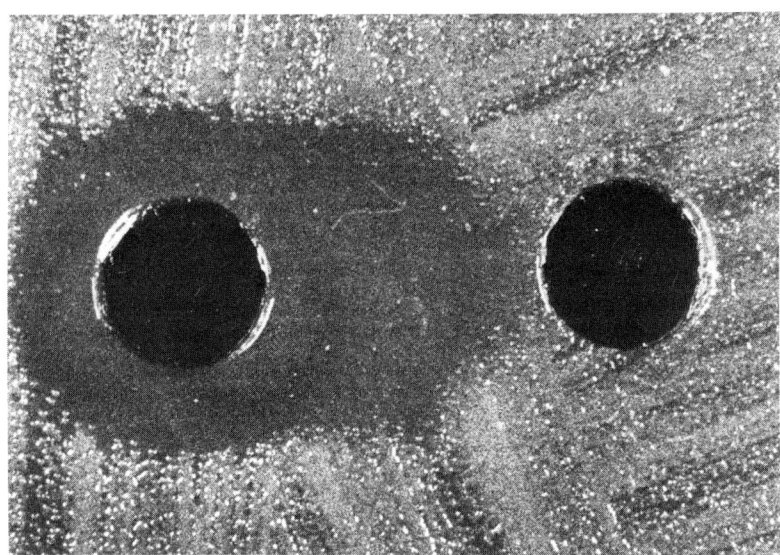

FIGURE 3. Synergy between carbenicillin and cephalosporin C. The plate was seeded with *Y. enterocolitica* strain W239 (serotype O:3). Left: carbenicillin (1 mg/mℓ; right: cephalosporin C (4 mg/m). Each hole received 50-µℓ antibiotic solution. (From Cornelis, G., *J. Gen. Microbiol.*, 91, 391, 1975. With permission.)

can be considered as a naturally occurring mutant deficient in β-lactamase A production. This clearly suggests that β-lactamase A plays a major role in resistance toward carbenicillin and ampicillin (see Table 1), while β-lactamase B contributes to resistance to the cephalosporins. This is confirmed by the effect of competitive inhibitors of β-lactamase B such as cloxacillin and carbenicillin. As shown in Figure 3, there is a strong synergy between carbenicillin and cephalosporin C. Similarly, there is also a synergy between cloxacillin and the cephalosporins, but no synergy can be detected between cloxacillin and carbenicillin. This shows that β-lactamase B plays a role in resistance toward cephalosporin C, but does not protect the cell from carbenicillin.

Although cephaloridine is a very good substrate for both β-lactamases A and B, the MIC values of this antibiotic are rather low (see Table 1). This can be explained by the fact that *Y. enterocolitica* has no permeability barrier to cephaloridine.[10] Richmond and Sykes showed indeed that a β-lactam antibiotic with good permeability greatly reduces the protective effect of a given β-lactamase, while the efficiency of a β-lactamase is enhanced towards a β-lactam with poor permeability (generally exhibited by carbenicillin).[18]

F. Cellular Location of the β-Lactamases in *Y. enterocolitica*

β-Lactamases A and B from *Y. enterocolitica* are cell-bound and are released only when the cells are subjected to some form of lysis. This is not a peculiarity of *Y. enterocolitica*, but a common phenomenon in β-lactamase-producing Gram-negative bacteria.

There is now wide agreement among people working in the field of β-lactamases that these enzymes are located in the periplasmic space of Gram-negative bacteria. This space is located between the bilayered inner cytoplasmic membrane and the outer envelope.[18] Periplasmic enzymes can be released without complete cell disruption by procedures

Table 5
SELECTIVE RELEASE (%) OF β-LACTAMASES BY OSMOTIC SHOCK AND SPHEROPLASTS FORMATION

Treatment	Strain	% release of β-lactamase	
		A	B
Spheroplast formation	W238(O:9)	6.3	15
Osmotic shock	W238	12	80
	W222(O:3)	18	84

such as spheroplast formation or osmotic shock. Spheroplasts are made by treatment with a combination of lysozyme and EDTA. Osmotic shock is achieved by suspending bacterial cells first in Tris buffer containing EDTA and 20% sucrose and then immediately in distilled water.[19] The EDTA and Tris treatment presumably causes some damage in the outer envelope and the sudden transition to distilled water creates a rapid flow of water into the cytoplasm. This expands the inner membrane against the peptidoglycan, forcing out the periplasmic components. Smith and Wyatt[20] observed that in *E. coli*, some β-lactamases could be released by osmotic shock while some could not. According to these authors, the criterion which governs the retention or release of β-lactamases from Gram-negative bacteria during osmotic shock is the molecular weight of the enzyme: those with a mol wt of about 20,000 are released, while those of mol wt 30,000 or more are not. In their hands, however, all the β-lactamases were liberated by spheroplast formation, thus confirming the periplasmic location of these enzymes. *Y. enterocolitica* W238 (serotype O:9) and W222 (serotype O:3) were subjected to osmotic shock and spheroplast formation. Release of β-lactamases A and B was inferred from assay of β-lactamase recovered from supernatants and sonicated pellets, using cephalosporin C and cephaloridine as substrates.

A suspension containing 3.6×10^{10} cells of *Y. enterocolitica* W238 per milliliter was converted to spheroplasts by treatment with lysozyme (1 mg/mℓ). The conversion was monitored by the lysis of the cells contained in a small aliquot when diluted in distilled water. As shown in Table 5, neither β-lactamase A nor β-lactamase B was significantly released by this treatment. At first sight, this would suggest that both β-lactamases in *Y. enterocolitica* are not periplasmic enzymes. However, if β-lactamase A were cytoplasmic, i.e., located behind the target it protects, it would be difficult to understand the great protection that it affords the organism. The absence of release by spheroplast formation could be understood if the β-lactamases from *Y. enterocolitica* are periplasmic, but also bound to some surface component.

The osmotic shock gave different results. As shown in Table 5, this procedure significantly releases β-lactamase B while β-lactamase A remains mostly cell-bound. Since A has a mol wt of 20,000 and B has a mol wt of 34,000, this observation is in contradiction with the results of Wyatt and Smith with *E. coli*. Indeed, in *Y. enterocolitica*, the osmotic shock selectively releases the enzyme with high molecular weight and not the other one. This experiment demonstrates at least that β-lactamase B is a periplasmic enzyme, though it is not released by spheroplast formation. This tends to suggest that the release by spheroplast formation might, in this case, not be an essential requirement for periplasmic enzymes. The fact that A is not released by osmotic shock would then mean either that it is not periplasmic, or it is periplasmic, but nevertheless quite firmly bound to some surface component.

G. β-Lactamases from *Y. enterocolitica* Strains that Belong to Serotype O:5b

The strains carrying the O antigens n° 5 and 27 (serotype O:5b) are biochemically identical to the strains of serotype O:9.[4] Surprisingly, these strains are sensitive to carbenicillin but still resistant to most of the other β-lactam antibiotics (see Table 1).[10] As would be expected, they synthesize only one β-lactamase with a substrate profile and an isoelectric focusing pattern identical to those of β-lactamase B from serotype O:3 and O:9.[10,12] Among the strains of serotype O:5b, two exceptional strains are indole-negative. Interestingly enough, these strains have also a different behavior toward β-lactam antibiotics.[10]

H. Inducibility of β-Lactamase B

Since the strains of serotype O:5b synthesize only β-lactamase type B, they represent the ideal group to investigate the induction of this enzyme.

We recently analyzed 20 strains of serotype O:5b.[21] These strains, originating from Western Europe, Canada, U.S., and Japan, were kindly supplied by Professor G. Wauters. In order to investigate the inducibility of β-lactamase B, the strains were grown at 30°C, in 10 mℓ tryptic soy broth contained in 100 mℓ conical flasks shaken at 150 rpm. When the O.D. (660 nm) reached 0.3, 250 μg/mℓ of 6-amino-penicillanic acid (6-APA) was added to the culture. Under these standardized conditions, the induction ratio (amount of β-lactamase produced in the presence of inducer/amount of β-lactamase produced without inducer) was higher than eight for five strains and higher than three for nine other strains. For the last six strains, the induction ratio ranged only from 1.5 to 2.5. Although none of these induction ratios is very high, this experiment clearly supports the idea that β-lactamase B is an inducible enzyme in the majority of the strains. Strain IP 97 (serotype O:5b), isolated in 1960 from a chinchilla in Denmark, showed the highest induction ratio under these conditions. This strain was therefore selected for a study of the induction phenomenon. Cultures were grown at 30°C in 200 mℓ tryptic soy broth contained in 500-mℓ conical flasks with four baffles, shaken at 150 rpm. Various β-lactam antibiotics were added (at a concentration = MIC × 4) when the O.D. reached 0.3 and the culture was stopped when the O.D. reached 1.0. The β-lactamase activity in the crude extracts was assayed with cephalosporin C as a substrate. Since carbenicillin is a strong inhibitor of β-lactamase B, the crude extracts of the cultures induced with this agent were treated before the assay of β-lactamase B by a penicillinase from *Bacillus cereus* (Becton-Dickinson) that has no detectable activity against cephalosporin C. Surprisingly, as shown in Table 6, carbenicillin was by far the best inducer, at least under these conditions. Cephalothin and ampicillin, on the other hand, did not significantly induce β-lactamase B synthesis. 6-APA and carbenicillin at various concentrations were subsequently added as inducers when the O.D. reached 0.275. Under these conditions, 250 μg/mℓ of 6-APA was found to be the most effective concentration (induction ratio = 33). Carbenicillin used at a concentration of 50 μg/mℓ gave a still higher induction ratio (58).

Figure 4 shows the kinetics of the induction of β-lactamase B in the presence of 250 μg/mℓ of 6-APA. The lag between the addition of the inducer and the response is shorter than 30 min and the specific activity (units of enzyme per milligram of protein) keeps increasing for 8 hr. If the inducer is withdrawn, the specific activity decreases rapidly, but even after 24 hr it is still 3 times higher than in the uninduced control. The induction in the presence of 15 μg/mℓ carbenicillin is similar. The specific activity also increases very rapidly and the maximum is reached 6 hr after the addition of the inducer.

I. Mutants of *Y. enterocolitica* IP97 that Synthesize Type B β-Lactamase Constitutively

By serial subcultures in presence of increasing concentrations of carbenicillin, Flett et

Table 6
INDUCTION OF β-LACTAMASE B SYNTHESIS IN Y. ENTEROCOLITICA, STRAIN IP 97 (SEROTYPE O:5b) BY VARIOUS β-LACTAM ANTIBIOTICS

Inducer	Concentration (μg/mℓ)	Specific activity (units/mg of proteins)	Induction ratio
None	—	0.08	—
Benzylpenicillin	1000	1.14	14
6-Aminopenicillanic acid	250	0.71	9
Ampicillin	250	0.10	1
Carbenicillin	16	2.33	29
Cephalosporin C	4000	0.39	5
Cefamandole	8	0.45	6
Cephaloridine	125	0.52	7
Cephalexin	250	0.71	9
Cephalothin	2000	0.10	1
Cefoxitin	250	0.59	7

Note: 20 mℓ overnight culture was inoculated into 180 mℓ of fresh tryptic soy broth contained in 500-mℓ conical flasks with four baffles and incubated at 30°C with shaking at 150 rpm. When the O.D. reached 0.3, an inducer was added at the given final concentration. When the O.D. reached 1.0, the cultures were centrifuged and the specific activity were determined.[23]

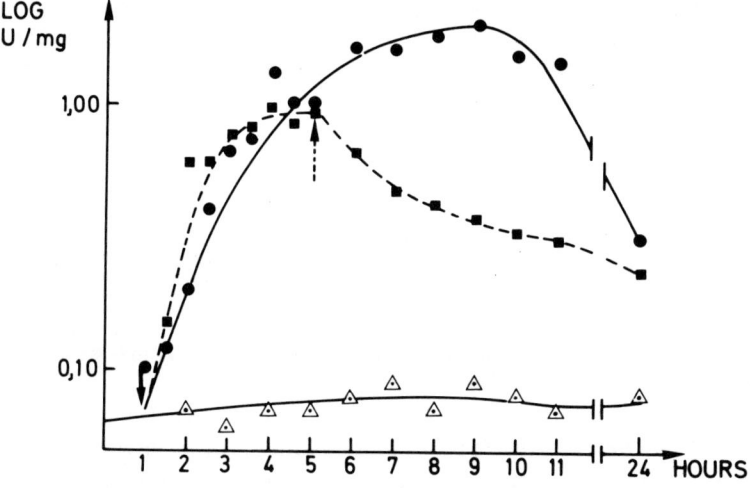

FIGURE 4. Kinetics of the induction of β-lactamase B formation in *Y. enterocolitica*, strain IP 97 (serotype O:5b). An overnight culture was diluted 10 times into fresh tryptic soy broth (3 × 180 mℓ) contained in 500-mℓ conical flasks with four baffles. The cultures were incubated at 30°C with shaking (150 rpm). When the O.D. reached 0.275, 6-APA was added to two cultures (■ and ●) at a final concentration of 250 μg/mℓ. The time of addition is indicated by an arrow. A third culture was not treated with 6-APA (△). At the time indicated by the dotted arrow, one of the 6-APA-treated cultures was centrifuged and resuspended in 6 APA free medium (■).[23]

al.[22] have isolated a mutant of *Pseudomonas aeruginosa* in which the chromosomal β-lactamase (Sabath and Abraham enzyme) is constitutive.

By subculturing *Y. enterocolitica* IP97 30 times in the presence of carbenicillin (concentration increasing from 4 to 70 μg/mℓ), four mutants in which the induction ratio is about one (against 30 for the parent strain) have been isolated.[23] Repeated subcultures of the same strain in the presence of cefamandole and cefoxitin selected mutants highly resistant to these molecules as well as to cephaloridine. In these mutants, the β-lactamase is still inducible, albeit poorly. The specific activity (units of enzyme per milligram protein) is however 50 times higher than that of the fully induced wild-type parental strain. Moreover, as shown in Table 7, these mutants also developed a permeability barrier against cephaloridine.

Treatment with 6-APA also selected mutants with a low inducibility (Table 7). Clearly, the selection procedure used in these experiments (repeated subcultures) must favor the emergence of multiple mutants and it is therefore difficult to draw conclusions from the analysis of the results. Nevertheless, these experiments clearly show that constitutive mutants can be isolated in *Y. enterocolitica* as has been shown with other inducible enzymes. Carbenicillin seems to be the sole agent used that selects for such mutants. Surprisingly, the results indicate that β-lactamase B, present in large quantities, can somehow protect the cell against carbenicillin, though this agent is a powerful inhibitor of the enzyme.

This experiment also shows that mutants highly resistant to cefoxitin and cefamandole could easily emerge during a treatment with these drugs. The mutants we observed synthesize a β-lactamase more efficient at hydrolyzing cephaloridine and cefamandole than the β-lactamase of the parental strain, but still with an unaltered isoelectric focusing pattern. Moreover, these two mutants have developed a permeability barrier against various cephalosporins (see Table 7).

J. Correlation Between Serological and Biochemical Groupings of *Y. enterocolitica* with the β-Lactamases of the Strains

Isoelectric focusing allows one to determine and to compare the isoelectric points of β-lactamases with great accuracy and extreme sensitivity. Visual comparison of β-lactamases by examining the pattern of enzymatically active bands has shown that chromosomally mediated β-lactamases are specific for genus, species, and subspecies.[24] The crude extracts of the 37 strains listed in Table 8 were submitted to this analysis.[12] As shown in Figure 5, the β-lactamase isoelectric focusing patterns of all the strains examined consisted of several bands. Some of these focused close together and were regarded as a group of bands representing a single β-lactamase. In isoelectric focusing, an apparently single β-lactamase commonly focuses as a group of bands, often a main band accompanied by satellites; the reason for this appearance is unknown.[25]

When β-lactamases from more than one strain of a certain serotype were focused, the β-lactamase patterns for each of these strains were identical (Figure 5, serotypes O:1, O:2, O:3, O:5, O:8, O:9)

1. Serological groups O:1, O:2, O:3, and O:9 — All the strains belonging to these groups had the same β-lactamase isoelectric focusing patterns, with the exception of strain H66 (see below). The pI 8.7 band corresponded to the broad-spectrum β-lactamase A. Enzyme B was represented by bands with pI 5.3 and 5.7. As expected, strain H66 produced no type A β-lactamase (Figure 5, strain number 37). In common with other strains of serotypes O:1, O:2, O:3, and O:9 strain H66 gave a weak β-lactamase band at pI 6.7.

Table 7
CHARACTERISTICS OF MUTANT STRAINS SELECTED BY 30 REPEATED SUBCULTURES OF Y. ENTEROCOLITICA STRAIN IP 97 (SEROTYPE O:5b) IN THE PRESENCE OF INCREASING CONCENTRATIONS OF VARIOUS β-LACTAM ANTIBIOTICS

Strain	Selected by serial subcultures on	β-lactamase B production after induction (% of the wild type)	Induction ratio[+] of β-lactamase	Substrate profile of β-lactamase	Crypticity factor[++] determined with cephaloridine
IP 97	—	100	20	N	1.0
9704	Carbenicillin	614	1.3	N	1.0
9708	Cefamandole	5857	3.3	M	5.0
9710	Cefoxitin	5900	4.2	M	3.4
9714	6-APA	786	4.9	N	10.0

Note: N = typical of enzyme B.[9,10] M = modified; the modification is an increased hydrolytic activity against cephaloridine and cefamandole. + = 6-APA was used as an inducer. ++ = The crypticity factor is the ratio between the enzyme activity on cephaloridine found in intact and disrupted cells; if higher than 1.0, this value reflects an accessibility barrier that restricts the access of cephaloridine.[23]

Table 8
STRAINS OF *Y. ENTEROCOLITICA* FROM WHICH INTRACELLULAR β-LACTAMASES WERE ANALYSED BY ISOELECTRIC FOCUSING

Series no.	Source no.[a]	Serological group[b]	Biochemical group[c]	Species of isolation	Worker	Country
1	IP92 [NCTC10462]	1	3	Chinchilla	Frederiksen	Denmark
2	IP110	1	3	Chinchilla	Akkermans	Netherlands
3	IP8	2	5	Hare	Lucas	France
4	IP178	2	5	Hare	McDiarmid	U.K.
5	F96-1	2	5	Hare	Fievez	Belgium
6	A14700	2	5	Hare	Aldova	Czechoslovakia
7	W222	3	4	Human	Wauters	Belgium
8	W223	3	4	Human	Wauters	Belgium
9	W224	3	4	Pig	Wauters	Belgium
10	W230	3	4	Pig	Wauters	Belgium
11	W237	3	4	Pig	Wauters	Belgium
12	W239	3	4	Human	Wauters	Belgium
13	IP134 ["Winblad"]	3	4	Human	Winblad	Sweden
14	IP97	5,27	2	Chinchilla	Frederiksen	Denmark
15	IP130	5,27	2	Monkey	Mair	U.K.
16	IP199	5,27	2	Chinchilla	County Vet.	U.S.A.
17	IP1607	5,27	2	Human	Zen-Yogi	Japan
18	M771	5,27	2	Human	Moinet	Belgium
19	IP636[NCTC10598]	8	1	Human	Sonnenwirth	U.S.A.
20	IP1453	8	1	Human	Sonnenwirth	U.S.A.
21	IP1628	8	1	Human	Weaver	U.S.A.
22	IP2026	8	1	Human	Weaver	U.S.A.
23	IP410	9	2	Human	Ahvonen	Finland
24	W227	9	2	Human	Wauters	Belgium
25	W228	9	2	Human	Wauters	Belgium
26	W238	9	2	Human	Wauters	Belgium
27	W244	9	2	Human	Wauters	Belgium
28	IP123	5	1	Cow	Vallee	France

Table 8 (continued)
STRAINS OF *Y. ENTEROCOLITICA* FROM WHICH INTRACELLULAR β-LACTAMASES WERE ANALYSED BY ISOELECTRIC FOCUSING

Series no.	Source no.[a]	Sero-logical group[b]	Bio-chemical group[c]	Species of isolation	Origin Worker	Country
29	IP102	6	1	Human	Frederiksen	Denmark
30	AZ1523	16[d]	1	Mouse	Aldova	Czechoslovakia
31	IP867	16[d]	2	Human	Graux	Belgium
32	IP955	17	1	Water	Lassen	Norway
33	IP841	11,24		Human	Weaver	U.S.
34	IP103	12,26		Sheep	Frederiksen	Denmark
35	V.2. 73(RY$_3$)	9	2	Human	Vandepitte	Belgium
36	V.2. 73(R$^-$)	9	2	Human	Derived from V.2. 73 (RY$_3$)	
37	H66	3	4	Human	Greven-Brauns	Belgium

[a] Strains designated IP were received from the collection of Prof. H. H. Mollaret, Institute Pasteur, Paris.
[b] Classified as described by Winblad[30,31] and Wauters et al.[29,40]
[c] Classified according to the criteria proposed by Niléhn[28] slightly modified by Wauters.[4]
[d] As suggested by the different biotypes, these two strains of serological group O:16 exhibit different biochemical behavior. These strains do not fit exactly in any of the described biotypes.

From Matthew, M., Cornelis, G., and Wauters, G., *J. Gen. Microbiol.*, 102, 55, 1977. With permission.

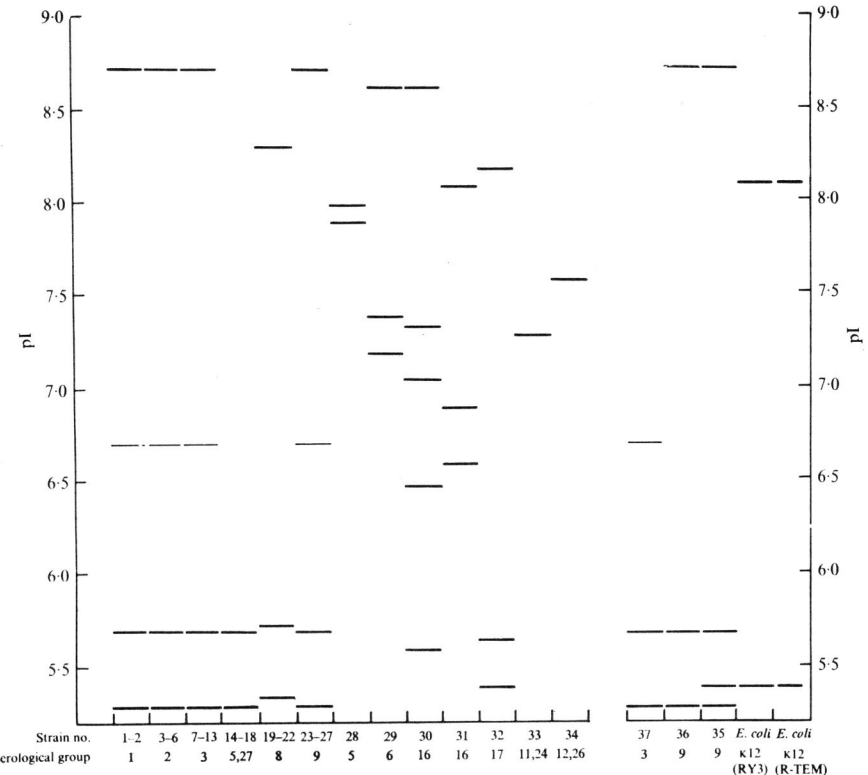

FIGURE 5. Analytical isoelectric focusing patterns of intracellular β-lactamases produced by *Y. enterocolitica*. Particulars of the strains are given in Table 8. Plasmid RY3 was transferred from *Y. enterocolitica* V.2.73(RY3) into *E. coli* K12. *E. coli* K12(R-TEM) is a reference strain that produces the TEM-I plasmid-mediated β-lactamase. (From Matthew, M., Cornelis, G., and Wauters, G., *J. Gen. Microbiol.*, 102, 55, 1977. With permission.)

2. Serological group O:5b — As already mentioned, strains of this group, except those that were indole (−), produced only the type B β-lactamase.
3. Other serological groups — None of the β-lactamases produced by strains of the other serological groups studied were identical with the type A and type B enzymes, although strains from serotype O:8 and O:17 gave a doublet of bands with pI values close to those of the B β-lactamase.

With the exception of strain IP841 (serotype O:11, O:24) and strain IP103 (serotype O:12, O:26), all strains apparently produced more than one β-lactamase. Strains of different serotypes produced different β-lactamases. The patterns of the A and B β-lactamases from *Y. enterocolitica* strains of serotypes O:1, O:2, O:3, O:9 (and enzyme B for group O:5b) clearly suggest that these strains may have a close taxonomic relationship in spite of the biochemical differences between them. Knapp and Thal[26] questioned whether strains of serological group O:2, usually isolated from hares, should be included in the species *Y. enterocolitica*. Results suggest that they should.

Strains belonging to serotypes O:1, O:2, O:3, O:9, and O:5b are all regularly encountered in specific hosts and all have an established pathogenicity. Apart from the β-lactamase patterns, strains of serological groups O:1, O:2, and O:3 have other features in common. Most of the strains are lysogenic:[27] many phages liberated by group O:1 and

O:3 strains plaque on group O:2 strains and many phages liberated by group O:2 strains plaque on strains of groups O:1 and O:3, but not on representatives of other serological groups.[28] In addition, all these strains possess the O:3 antigen.[29-31] Isoelectric focusing of β-lactamases from *Y. enterocolitica* thus tends to group together these pathogenic strains despite their wide differences in geographical origin.

II. RESISTANCE PLASMIDS IN *Y. ENTEROCOLITICA*

The recipient ability of *Y. pestis* and *Y. pseudotuberculosis* for R plasmids was demonstrated very early by Ginoza and Matney.[32] Afterwards, several authors showed that *Y. enterocolitica* could also act as a recipient for R plasmids.[33,34] As might be expected, R factors were subsequently detected in *Y. enterocolitica* strains isolated in 1970.[6,35]

These three plasmids were conjugative and of the fi(+) type. Plasmid RY2 was later on assigned to the FII incompatibility group.[36] Plasmid RY3, isolated from a strain of serotype O:3, conferred an increased resistance to ampicillin and other antibiotics.[6] This suggested that RY3 encoded a third β-lactamase for its *Y. enterocolitica* (O:9) host strain. This was later confirmed by isoelectric focusing: apart from A and B, the host strain synthesized the very widely distributed TEM-I plasmid-mediated β-lactamase[12] (Figure 5). Later, six R$^+$ *Y. enterocolitica* strains were isolated in Japan. These plasmids were also conjugative, but belonged to the incompatibility group N.[37] The very limited number of reports of R$^+$ strains seems to reflect an extremely low proportion of R$^+$ strains among the strains isolated so far. This low proportion is presumably the consequence of the very low recipient ability observed with *Y. enterocolitica* strains,[38] since plasmids from several incompatibility groups (including the very widely distributed PI group)[39] can be maintained in this host. Colson and Cornelis showed that a DNA restriction and modification system contributes to the low recipient ability of *Y. enterocolita* for plasmids transmitted from other species.[38]

ACKNOWLEDGMENTS

A great deal of the work reported here was realized in collaboration with Professor E.P. Abraham (University of Oxford), Professor G. Wauters (University of Louvain), and Doctor M. Matthew (Glaxo Research Ltd) and I wish to express my deep gratitude for their help, contribution, valuable suggestions, and discussions.

I also wish to thank E. P. Abraham and M. Matthew for their critical reading of this manuscript.

REFERENCES

1. **Niléhn, B.,** Studies on *Yersinia enterocolitica*. Characterization of 28 strains from human and animal sources, *Acta Pathol. Microbiol. Scand.*, 69, 83, 1967.
2. **Martinevski, I. L. and Stogova, A. G.,** Antibiotic sensitivity in *Yersinia, Antibiotiki*, 14, 61, 1969.
3. **Van Noyen, R., Vandepitte, J., and Isebaert, A.,** *Yersinia enterocolitica*: een belangrijk oorzaak van darmontsteking, *Ned. Tijdschr. Geneeskd.*, 114, 188, 1970.
4. **Wauters, G.,** Contribution a l'étude de *Yersinia enterocolitica*. Ph.D. thesis, University of Louvain, Vander, Louvain, Belgium, 1970, 118.
5. **Zen-Yoji, H. and Maruyama, T.,** The first successful isolations and identifications of *Yersinia enterocolitica* from human cases in Japan, *Jpn. J. Microbiol.*, 16, 493, 1972.
6. **Cornelis, G., Wauters, G., and Vanderhaeghe, H.,** Presence de β-lactamase chez *Yersinia enterocolitica, Ann. Microbiol. (Paris)*, 124B, 139, 1973.
7. **Mishankin, B. N., Ryzhko, I. V., and Grigorian, E. G.,** Study of penicillinase activity of *Pasteurella pseudotuberculosis* and *Pasteurella X, Antibiotiki*, 18, 621, 1973.
8. **Cornelis, G., Bennett, P. M., and Grinsted, J.,** Properties of pGC1, a *lac* plasmid originating in *Yersinia enterocolitica* 842, *J. Bacteriol.*, 127, 1058, 1976.

9. **Cornelis, G. and Abraham, E. P.**, β-lactamases from *Yersinia enterocolitica, J. Gen. Microbiol.,* 87, 273, 1975.
10. **Cornelis, G.**, Distribution of β-lactamases A and B in some groups of *Yersinia enterocolitica* and their role in resistance, *J. Gen. Microbiol.,* 91, 391, 1975.
11. **Winblad, S., Niléhn, B., and Sternby, N. H.**, *Yersinia enterocolitica (Pasteurella X)* in human enteric infections, *Br. Med. J.,* 2, 1363, 1966.
12. **Matthew, M., Cornelis, G., and Wauters, G.**, Correlation of serological and biochemical grouping of *Yersinia enterocolitica* with the β-lactamases of the strains, *J. Gen. Microbiol.,* 102, 55, 1977.
13. **Sykes, R. B. and Matthew, M.**, The β-lactamases of Gram-negative bacteria and their role in resistance to β-lactam antibiotics, *J. Antimicr. Chemother.,* 2, 115, 1976.
14. **Hennessey, T. D.**, Inducible β-lactamase in *Enterobacter, J. Gen. Microbiol.,* 49, 277, 1967.
15. **Ayliffe, G. A. J.**, Cephalosporinase and penicillinase activity of Gram-negative bacteria, *J. Gen. Microbiol.,* 40, 119, 1965.
16. **Sabath, L. D., Jago, M., and Abraham, E. P.**, Cephalosporinase and penicillinase activities of a β-lactamase from *Pseudomonas pyocyaneae, Biochem. J.,* 96, 739, 1965.
17. **McPhail, M. and Furth, A.**, Purificiation and properties of an inducible β-lactamase from *Pseudomonas aeruginosa* NCTC 8203, *Biochem. Soc. Trans.,* 1, 1260, 1973.
18. **Richmond, M. H. and Sykes, R. B.**, The β-lactamases of Gram-negative bacteria and their possible physiological role, in *Advances in Microbial Physiology,* Vol. 9, Rose, A. H. and Wilkinson, J. F., Eds., Academic Press, New York, 1973, 69.
19. **Heppel, L. A.**, Selective release of enzymes from bacteria, *Science,* 156, 1451, 1967.
20. **Smith, J. T. and Wyatt, J. M.**, Relation of R factor and chromosomal β-lactamase with the periplasmic space, *J. Bacteriol.,* 117, 931, 1974.
21. **Vandermeeren, M. C., Carlier, M. B., and Cornelis, G.,** unpublished experiments, 1977.
22. **Flett, F., Curtis, N., and Richmond, M. H.**, Mutant of *Pseudomonas aeruginosa* 185 that synthesize type 1d β-lactamase constitutively, *J. Bacteriol.,* 127, 1585, 1976.
23. **Carlier, M. B. and Cornelis, G.**, unpublished experiments, 1978.
24. **Matthew, M. and Harris, A.**, Identification of β-lactamases by isoelectric focusing: correlation with bacterial taxonomy, *J. Gen. Microbiol.,* 94, 56, 1976.
25. **Matthew, M., Harris, A. M., Marshall, M. J., and Ross, G. W.** The use of analytical isoelectric focusing for detection and identification of β-lactamases, *J. Gen. Microbiol.,* 88, 169, 1975.
26. **Knapp, W. and Thal, E.**, Die biochemische charakterisierung von *Yersinia enterocolitica* (syn. *"Pasteurella X")* als grundlage eines vereinfachten, *Zentralbl. Bakteriol. Parasitenkd. Infektionskr. Hyg. Abt. 1 Orig.,* 223, 88, 1972.
27. **Nicolle, P., Mollaret, H., Hamon, Y., and Vieu, J. F.**, Etude lysogénique, bacteriocinogénique et lysotypique de l'espèce *Yersinia enterocolitica, Ann. Inst. Pasteur Paris,* 112, 86, 1967.
28. **Niléhn, B. and Ericson, C.**, Studies on *Yersinia enterocolitica, Acta Pathol. Microbiol. Scand.,* 75, 177, 1969.
29. **Wauters, G., Le Minor, L., and Chalon, A. M.**, Antigènes somatiques et flagellaires des *Yersinia enterocolitica, Ann. Inst. Pasteur,* 120, 631, 1971.
30. **Winblad, S.**, Studies on serological typing of *Yersinia enterocolitica, Acta Pathol. Microbiol. Scand.,* S187, 115, 1967.
31. **Winblad, S.**, Studies on O antigen factors of *Yersinia enterocolitica,* in *Int. Symp. Pseudotuberculosis,* Regamey, R. H., Ed., S. Karger, Basel, 1968, 337.
32. **Ginoza, H. S. and Matney, T. S.**, Transmission of a resistance transfer from *Escherichia coli* to two species of *Pasteurella. J. Bacteriol.,* 85, 1177, 1963.
33. **Knapp, W. and Lebek, G.**, Übertragung des infektiösen resistenz auf Pasteurellen, *Pathol. Microbiol.,* 30, 103, 1967.
34. **Rusu, V., Baron, O., and Lazaroae, D.**, Transfert du facteur de résistance (R) des *Enterobacteriaceae* a *Yersinia enterocolitica, Arch. Roum. Pathol. Exp. Microbiol.,* 29, 571, 1970.
35. **Cornelis, G., Wauters, G., and Bruynoghe, G.**, Résistances transférables chez des souches sauvages de *Yersinia enterocolitica, Ann. Microbiol. (Paris),* 124A, 299, 1973.
36. **Fredericq, P.**, personnal communication, 1975.
37. **Kimura, S., Eda, T., Mitsui, Y., and Nakata, K.**, R plasmids from *Yersinia, J. Gen. Microbiol.,* 97, 141, 1976.
38. **Cornelis, G. and Colson, C.**, Restriction of DNA in *Yersinia enterocolitica* detected by recipient ability for a derepressed R factor from *Escherichia coli, J. Gen. Microbiol.,* 87, 285, 1975.
39. **Cornelis, G.** unpublished experiments, 1976.
40. **Wauters, G., Le Minor, L., Chalon, A. M., and Lassen, J.**, Supplément au schéma antigénique de *"Yersinia enterocolitica," Ann. Inst. Pasteur Paris,* 122, 951, 1972.

Chapter 7

HUMAN YERSINIA ENTEROCOLITICA INFECTION: LABORATORY MODELS

Philip B. Carter

TABLE OF CONTENTS

I.	Introduction	74
II.	Pathogenicity of *Y. enterocolitica* for Laboratory Animals	74
	A. Studies in Guinea Pigs	74
	B. Studies in Mice	74
	C. Studies in Rabbits	76
	D. Studies in Monkeys	76
	E. Studies in Rats	77
III.	In Vitro Correlates	77
IV.	Pathogenic Mechanisms	77
	A. Temperature and Virulence	77
	B. Invasiveness	77
	C. Enterotoxin	78
	D. Iron Dependence	78
	E. VW Antigens	78
V.	Immune Responses	79
VI.	Perspectives	79
References		79

I. INTRODUCTION

Increasing attention has been directed toward *Yersinia enterocolitica* over the last 15 years as an etiologic agent in human ileitis, mesenteric lymphadenitis, and septicemia as well as possible sequelae of the enteric infection: arthritis, Reiter's syndrome, and erythema nodosum. However, the connection between these syndromes and infection with *Y. enterocolitica* has been circumstantial to date since infection of human volunteers has not been attempted. As recently as 1970, the causative role of *Y. enterocolitica* in acute ileitis was still being questioned.[1] A serious weakness in establishing a case for a causal relationship between *Y. enterocolitica* and acute ileitis and mesenteric lymphadenitis had been the inability to confirm an early report[2] of the pathogenicity of human isolates for laboratory animals.[3,4] In recent years, however, human isolates of *Y. enterocolitica* have been shown to be virulent for laboratory animals, reproducing in them the basic characteristics of the human disease.[5-7] It is the function of this chapter to review the work which has been done on animal models of human *Yersinia* enteritis, to evaluate models for assessing the virulence of *Y. enterocolitica* for humans, and to discuss present knowledge of virulence mechanisms.

II. PATHOGENICITY OF *Y. ENTEROCOLITICA* FOR LABORATORY ANIMALS

A. Studies in Guinea Pigs

The first attempts to infect laboratory animals with *Y. enterocolitica* were made at the New York State Department of Health Laboratories in 1933. A single paragraph in their 1933 Annual Report[8] describes pathology in guinea pigs following injection with two different human isolates of the same serotype, submitted in 1923 and 1932. The route of inoculation was not indicated in the report, but it is likely to be intraperitoneal. The isolates were from superficial lesions of the face and may not have been *Y. enterocolitica*. However, later studies by Schleifstein and Coleman[2] demonstrated that these two early isolates were serologically and physiologically identical to three additional isolates from patients with enteritis. All five isolates studied by Schleifstein and Coleman produced disease in intraperitoneally infected guinea pigs but were nonpathogenic when given either subcutaneously or in food. More recent work[5] has shown that isolates of *Y. enterocolitica*, highly pathogenic in mice, are virtually nonpathogenic for guinea pigs. The susceptibility of guinea pigs to *Y. enterocolitica* is the reverse of that found for *Y. pseudotuberculosis* and thus the guinea pig is rarely used now in the study of the pathogenesis of *Y. enterocolitica* infections.

B. Studies in Mice

Primarily for economic reasons, most laboratory studies done in vivo with *Y. enterocolitica* have employed mice. Fortunately, studies in the last decade have demonstrated that the major pathological features of *Y. enterocolitica* infections in humans are also reproduced in naturally infected mice. Schleifstein and Coleman[2] were the first to employ the mouse in virulence studies on *Y. enterocolitica*, at that time only an unidentified microorganism resembling *Y. pseudotuberculosis* and *Actinobacillus lignierisii*. Their studies showed that mice were particularly susceptible to infection by the human isolates used, more so than guinea pigs, rabbits, and rats, and most interestingly could be infected via contaminated food. These workers[2] mentioned that the isolates attenuated quickly and attempted to maintain the virulence of the microorganisms by mouse-to-mouse passage. It was this attenuation[5] and inability to confirm[3,4] the early New York State work on virulence that led to the skepticism that the early work was perhaps

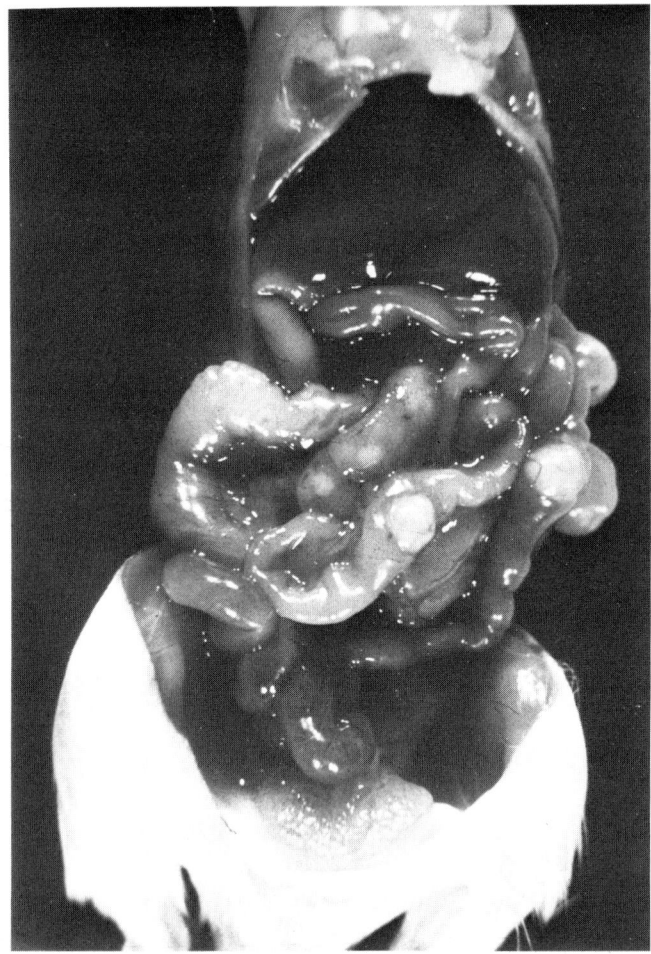

FIGURE 1. Gross pathology in an adult ICR pathogen-free mouse 7 days after intragastric infection with *Y. enterocolitica* serotype O:8. Abscesses in the Peyers' patches and mesenteric lymph nodes are readily apparent.

performed using an organism different from what is presently recognized as *Y. enterocolitica*. Recent studies in many laboratories have since confirmed the early work using both the New York serotype and biotype O:8[5,9] and the European O:3 serotype.[10-12]

The pathogen-free mouse has proven to be an excellent model for the type of human infection in the U.S. associated with the O:8 serotype, i.e., acute ileitis, mesenteric lymphadenitis, and, rarely, enteric fever.[13] Oral infection of mice with virulent human isolates of *Y. enterocolitica* causes a suppurative infection of the Peyer's patches and mesenteric lymph nodes.[14] With a high challenge dose, the microorganisms can proceed past the draining lymph nodes to cause a systemic infection of the lungs, liver, and spleen. Figure 1 shows the gross pathological changes in a mouse 7 days after intragastric infection with a human isolate of *Y. enterocolitica* recovered during an enteritis epidemic in northern New York State.[15] In orally infected pathogen-free mice, a neutrophil infiltrate appears early after infection in the Peyer's patches of the distal ileum. The Peyer's patches develop abscesses and the surrounding mucosa becomes inflamed. The infection extends to the mesenteric lymph nodes where abscesses develop, and from there to the systemic reticuloendothelial organs. A leukocytosis ensues and specific antibodies appear in the serum after 1 week. Diarrhea or loose stools are not common features of yersiniosis in mice infected with the O:8 serotype. Again, this reflects the human infection in which fever and abdominal pain without diarrhea are the common clinical manifestations of the disease.[15]

In Europe and Canada, human infection with serotype O:8 is rare; most human isolates are serotype O:3. Diarrhea, abdominal pain, but rare systemic involvement are associated with infections by this serotype. Although mice can be infected with serotypes O:9 and O:3,[10] and excrete the microorganisms in their feces for long periods,[16,17] diarrhea has not been reported in such mice. It may be, therefore, that the laboratory mouse does not represent a suitable model for gastroenteritis caused by *Y. enterocolitica* serotype O:3.

C. Studies in Rabbits

Schleifstein and Coleman[2] report no ill effects in rabbits following i.p. injection of *Y. enterocolitica* isolated from humans. However, Une[18] has recently been able to induce disease in rabbits by injecting human isolates of the O:3 and O:9 serotype intraduodenally. Such infection gives rise to an acute diarrhea 2 to 3 days after inoculation, thus resembling human gastroenteritis caused by the O:3 serotype of *Y. enterocolitica*.

Une[18] and Une and Zen-Yoji[19] have demonstrated that *Y. enterocolitica* isolates from human patients are pathogenic for rabbits when injected intraduodenally. They observed penetration of the mucosal epithelium with subsequent invasion of the draining lymphoid tissues. Unlike the mouse model in which abscess formation is commonly observed, a granulomatous response was often seen. Granulomas, characterized by an infiltrate of mononuclear phagocytes, develop in the appendix, mesenteric lymph nodes, liver, and spleen of rabbits as early as 7 days postinfection. Une and Zen-Yoji[19] showed that virulent *Y. enterocolitica* would actually multiply in rabbit macrophages in vitro whereas avirulent strains would be phagocytized and killed.

Rabbits which survived an intraduodenal infection with O:3 and O:9 serotypes of *Y. enterocolitica* were found to excrete the infecting organisms in their feces for as long as 40 days postinfection.[18] It was not determined whether the infecting agents actually colonized the intestinal tract for this period of time, or whether the organisms present in the feces issued from mucosal ulcerations of the gut-associated lymphoid tissue. The latter would be consistent with the pathology described.

D. Studies in Monkeys

In 1973, Maruyama[20] reported the establishment of an experimental *Y. enterocolitica* infection in *Macaca irus* monkeys. Oral infection with an O:3 serotype of human origin results in an acute infection of less than 2 weeks duration with minimal pathological changes.[21]

Naturally occurring *Y. enterocolitica* infections have been reported in two monkey colonies in the U.S. In 1971, McClure et al.[22] reported an outbreak in which two vervet monkeys (*Cerophithecus aethiops*) and one mangabey (*Cercocebus torquatus*) died of a pseudotuberculosis-like disease in which *Y. enterocolitica* was isolated. Two of the monkeys exhibited a granulocytosis and had lesions in the intestine (Peyer's patches), liver, and spleen. Attempts to parenterally infect six pigtailed macaque monkeys (*Macaca memestrina*) were unsuccessful. In 1976, a large outbreak occurred in a colony of owl monkeys (*Aotus trivirgatus*) in which 20 of the animals died.[23] Typical pseudotuberculosis-like lesions in the intestine, spleen, and liver were observed in the 29 animals dying of the spontaneous disease. Two out of four monkeys experimentally infected orally with the *Y. enterocolitica* strain isolated during the outbreak developed yersiniosis and one died. The serotype responsible for either of these outbreaks was not reported.

E. Studies in Rats

Rats have not commonly been used in experimental studies involving *Y. enterocolitica*. McGregor and Carter[48] showed that it was possible to infect inbred rats parenterally with *Y. enterocolitica* strain WA. However, the rats were able to withstand a much higher challenge dose than mice and there was great variation in the severity of the disease within groups, an undesirable feature for a model system. Attempts at oral infection failed to produce disease.

III. IN VITRO CORRELATES

The need to develop a means for testing the virulence of environmental isolates of *Y. enterocolitica* has led to the examination of possible in vitro correlates of the animal infection. The most popular has been the ability of an isolate to invade HeLa cells, a method used successfully by LaBrec et al.[24] to study invasiveness of *Shigella* and Gianella et al.[25] to assess invasiveness of salmonellae. In both systems, the ability of the strains to infect HeLa cells in culture correlated directly with ability to penetrate the intestinal mucosal epithelium of laboratory animals.

Lee et al.[26] and Une[27] were the first to successfully correlate HeLa cell invasiveness by *Y. enterocolitica* with animal pathogenicity. A number of laboratories have since confirmed that human isolates of *Y. enterocolitica* serotypes O:8, O:3, O:9, and others are HeLa cell invasive;[28-30] atypical forms of *Y. enterocolitica*, obtained from the environment or foodstuffs, are exclusively noninvasive.[29] The ability of *Y. enterocolitica* isolates to invade HeLa cells is temperature dependent.[26,29] The microorganisms are most invasive when grown at 22°C, less invasive if cultured at 30°C, and some strains are almost completely noninvasive when grown at 36°C.[26] Such differential temperature-related virulence has also been noted in vivo[9] (see below).

IV. PATHOGENIC MECHANISMS

A. Temperature and Virulence

The incidence of human disease caused by *Y. enterocolitica* peaks during the cold months of the year,[31] just the reverse of what is observed for *Salmonella* infections. It was, therefore, quite interesting when a temperature-dependent differential in the mouse virulence of a single *Y. enterocolitica* strain was observed.[9] Whether administered orally or parenterally, a marked lowering of the virulence of *Y. enterocolitica* WA for mice was consistently observed if the challenge inoculum was cultured at temperatures exceeding 28°C. When cultured at temperatures below 25°C (to as low as 5°C), *Y. enterocolitica* WA was optimally virulent. The less virulent organisms were cleared more quickly from the blood of infected animals and destroyed more rapidly by phagocytes.[9] A similar phenomenon has been noted for HeLa cell invasiveness.[26]

Indole production and motility are absent in *Y. enterocolitica* grown at 37°C, but it is unlikely that these characteristics have any relation to virulence. Rather, the temperature-dependent differential in virulence may be related to the presence of a higher proportion of VW antigen-negative *Y. enterocolitica* in challenge inocula grown at 37°C.[32]

B. Invasiveness

As mentioned above, several different serotypes of *Y. enterocolitica* have been shown to possess invasive properties similar to those described for *Salmonella* and *Shigella*. In addition to HeLa cell invasiveness, pathogenic *Y. enterocolitica* are capable of producing a positive Sereny reaction (conjunctivitis in the guinea pig)[33] and penetration of the rabbit[18,19,34] and mouse[14] ileal mucosa. Such a property can account for the *Salmonella*-like gastroenteritis commonly caused by the O:8 serotype of *Y. enterocolitica*, but for the acute diarrhea often associated with O:3 serotypes, a different virulence mechanism was sought.

C. Enterotoxin

At a meeting in 1977, Feeley et al.[33] and Robins-Browne et al.[34] reported the independent discovery of a heat-stable enterotoxin in human isolates of *Y. enterocolitica*. Enterotoxin was demonstrable only in the infant mouse assay (heat-stable toxin[35]) and not in the Y-1 adrenal cell assay (heat-labile toxin[35]). Heat-stable enterotoxin has been found in a large number of *Y. enterocolitica* serotypes and biotypes, but is most prevalent among those which are associated with human gastrointestinal disease.[36] The heat-stable enterotoxin produced by isolates of *Y. enterocolitica* is identical in its physical, chemical, and physiological characteristics to that produced by *Escherichia coli*.[37-39] Production of enterotoxin in vitro is temperature-dependent; it is not produced when culture temperatures exceed 30°C.[33,34,37] Since the enterotoxin is only produced at lower temperatures, its role in the pathogenesis of diarrheal disease remains a mystery and must be documented by in vivo studies.

D. Iron Dependence

Epidemiological studies in a South African black population led Rabson et al.[40] to suggest that iron may enhance virulence of *Y. enterocolitica* serotype O:3. Robins-Browne et al.[41] have shown that iron, in the form of ferric ammonium citrate, interferes with the natural bactericidal activity of fresh human serum. The ability of fresh human serum to kill as many as 10^6 *Y. enterocolitica* O:3 per milliliter in 4 hr is completely ablated by either heat inactivation (56°C/30′) or by saturation with iron. Intraperitoneal inoculation of mice with graded doses of ferric ammonium citrate increased their susceptibility to subsequent intraperitoneal challenge with *Y. enterocolitica* O:3.[41] This effect may, however, be particular to only some serotypes of *Y. enterocolitica* since Carter[49] was unable to demonstrate any enhancement of virulence or host susceptibility with human isolates of the O:8 serotype.

E. VW Antigens

Further similarity between *Y. enterocolitica* and other species in the genus *Yersinia* has recently been demonstrated by the finding of identical V and W antigens in virulent forms of *Y. enterocolitica* O:8.[32] The presence of V and W antigens was demonstrated by immunochemical and biochemical means and directly correlates with virulence for mice. The possession of common virulence antigens in all species of the genus *Yersinia* may account for the resistance against plague observed in mice experimentally infected with *Y. enterocolitica*.[42] Although the capability of isolates to produce the V and W antigens is directly related to their pathogenicity in mice, direct proof that virulence depends upon possession of the V antigen is lacking.

The high rate of attenuation of human isolates of *Y. enterocolitica* has led to the suspicion that a plasmid is involved in the virulence of this organism.[34] The existence of a plasmid in virulent strains of *Y. enterocolitica* O:8 has recently been formally proven by Zink et al.[43] and Gemski et al.[44] The plasmid is 42.2 megadaltons in size and by inference codes for the V antigen.[44] The V antigen is apparently related to invasiveness since parental *Y. enterocolitica* WA is Vwa$^+$ and both invasive and enterotoxigenic; Vwa$^-$ derivatives of this strain are noninvasive but still produce a heat-stable enterotoxin.[32,44]

The presence of such plasmid-mediated virulence factors in pathogenic *Y. enterocolitica* O:3 serotypes remain to be demonstrated.

V. IMMUNE RESPONSES

The immune response to *Y. enterocolitica* has only been analyzed in the mouse model using the O:8 serotype. The response in the mouse is similar in many respects to that observed in man, but unrecognized basic differences may exist.

The response to *Y. enterocolitica* infection in man is often characterized by fever and a granulocytosis,[15] a pyroninophilic response in the draining lymph nodes[45] which correlates with the early rise in antibody titer, and abscess formation in severe cases.[46] Such characteristics are consistent with a humoral rather than a cellular immune response. Whether such a response is protective in humans is not known since reinfection studies have not been done.

The question of which type of immune response is protective has, however, been analyzed in mice.[47] In such studies, passively transferred antibody completely protects normal mice against parenteral challenge with the homologous strain. Passive transfer of spleen cells taken from actively immunized mice was ineffectual and delayed hypersensitivity, often a correlate of cell-mediated immunity, was not demonstrable at any time postinfection.[47]

Unlike the human system with *Y. enterocolitica* O:3,[41] mouse antibody and complement is incapable of killing *Y. enterocolitica* O:8.[47] The observed protective effect of humoral factors is in enhancement of phagocytosis and killing;[47] the exact mechanism, however, remains to be defined.

VI. PERSPECTIVES

An acceptable animal model for human intestinal infections by the O:8 serotype of *Y. enterocolitica* is available. This model, however, does not reproduce the diarrheal illness which so often occurs in human disease caused by *Y. enterocolitica* O:3; an animal model which can be used to analyze the pathogenesis and immune responses in this disease remains to be developed.

The in vivo role of the heat-stable enterotoxin and V antigen in the pathogenesis of human yersiniosis awaits determination. In this connection, the exact mechanism by which *Y. enterocolitica* is able to infect and survive in cells remains to be defined.

No mention has been made in this chapter of possible sequelae to intestinal infections caused by *Y. enterocolitica*: arthritis, erythema nodosum, and Reiter's syndrome. The definition of a possible role of the immune response in the pathogenesis of these sequelae will require further epidemiological studies and development of an animal model.

REFERENCES

1. **Knapp, W., Fahrlander, H., and Hartweg, H.,** Zur Atiologie der akuten regionaren Enteritis (Ileitis), *Schweiz. Med. Wochenschr.*, 100, 364, 1970.
2. **Schleifstein, J. I. and Coleman, M. B.,** An unidentified microorganism resembling *B. lignieri* and *Past. pseudotuberculosis,* and pathogenic for man, *N. Y. State J. Med.,* 39, 1749, 1939.
3. **Knapp, W. and Thal, E.,** Untersuchungen uber die kulturellbiochemischen, serologischen, tierexperimentellen and immunologishen Eigenschaften einer vorlaufig "Pasteurella X" benannter Bakterienart, *Zentralbl. Bakteriol. Parasitenkd. Infektionskr. Hyg. Abt. 1 Orig.,* 190, 472, 1963.
4. **Mollaret, H. H. and Guillon, J. C.,** Contribution l' étude d'un nouveau group de germes proches du bacille de Malassez et Vignal. I. Caracteres culturaux et biochimique, *Ann. Inst. Pasteur (Paris)*, 109, 608, 1965.

5. **Carter, P. B., Varga, C. F., and Keet, E. E.,** A new strain of *Yersinia enterocolitica* pathogenic for rodents, *Appl. Microbiol.*, 26, 1016, 1973.
6. **Maruyama, T.,** Studies on biological characteristics and pathogenicity of *Yersinia enterocolitica*. II. Experimental infections in monkeys (in Japanese), *Jpn. J. Bacteriol.*, 28, 413, 1973.
7. **Quan, T. J., Meek, J. L., Tsuchiya, K. R., Hudson, B. W., and Barnes, A. M.,** Experimental pathogenicity of recent North American *Yersinia enterocolitica* isolates, *J. Infect. Dis.*, 129, 341, 1974.
8. **Gilbert, R.,** Interesting Cases and Unusual Specimens, Annual Report of the Division of Laboratories and Research, New York State Department of Health, 1933, 57.
9. **Carter, P. B. and Collins, F. M.,** Experimental *Yersinia enterocolitica* infection in mice: kinetics of growth, *Infect. Immun.*, 9, 851, 1974.
10. **Alonso, J. M., Bercovier, H., Destombes, P., and Mollaret, H. H.,** Pouvoir pathogéne experimental de *Yersinia enterocolitica* chez la souris athymique (nude), *Ann. Microbiol. (Paris),* 126B, 187, 1975.
11. **Bercovier, H., Alonso, J. M., Destombes, P., and Mollaret, H. H.,** Infection expérimentale de souris axéniques par *Yersinia enterocolitica, Ann. Microbiol. (Paris),* 127A, 493, 1976.
12. **Alonso, J. M., Mazigh, D., Bercovier, H., and Mollaret, H. H.,** Infection expérimentale de la souris par *Yersinia enterocolitica* (souche du chimiotype 4, du sérogroupe O:3, du lysotype VIII): devenir de l'inoculum chez des souries athymiques ou traitées par le cyclophosphamide, *Ann. Microbiol. (Paris),* 129B, 27, 1978.
13. **Carter, P. B.,** Animal model: Oral *Yersinia enterocolitica* infection of mice, *Am. J. Pathol.*, 81, 703, 1975.
14. **Carter, P. B.,** Pathogenicity of *Yersinia enterocolitica* for mice, *Infect. Immun.*, 11, 164, 1975.
15. **Black, R. E., Jackson, R. J., Tsai, T., Medvesky, M., Shayegani, M., Feeley, J. C., MacLeod, K. I. E., and Wakelee, A. M.,** Epidemic *Yersinia enterocolitica* infection due to contaminated chocolate milk, *N. Engl. J. Med.*, 298, 76, 1978.
16. **Ricciardi, I. D., Pearson, A. D., Suckling, W. G., and Klein, C.,** Long-term fecal excretion and resistance induced in mice infected with *Yersinia enterocolitica, Infect. Immun.*, 21, 342, 1978.
17. **Pearson, A. D., Ricciardi, I. D., Wright, D. H., and Suckling, W. G.,** An experimental study of the pathology and ecology of *Yersinia enterocolitica* infection in mice, in *Contributions to Microbiology and Immunology,* Vol. 5, Carter, P. B., Lafleur, L., and Toma, S., Eds., S. Karger, Basel, 1979, 336.
18. **Une, T.,** Studies on the pathogenicity of *Yersinia enterocolitica*. I. Experimental infection in rabbits, *Microbiol. Immunol.*, 21, 349, 1977.
19. **Une, T. and Zen-Yoji, H.,** Investigations on the pathogenicity of *Yersinia enterocolitica* by experimental infections in rabbits and cultured cells, in *Contributions to Microbiology and Immunology,* Vol. 5, Carter, P. B., Lafleur, L., and Toma, S., Eds., S. Karger, Basel, 1979, 304.
20. **Maruyama, T.,** Studies on biological characteristics and pathogenicity of *Yersinia enterocolitica*. II. Experimental infection in monkeys, *Jpn. J. Bacteriol.*, 28, 413, 1973.
21. **Fukai, K. and Maruyama, T.,** Histopathological studies on experimental *Yersinia enterocolitica* infection in animals, in *Contributions to Microbiology and Immunology,* Vol. 5, Carter, P. B., Lafleur, L., and Toma, S., Eds., S. Karger, Basel, 1979, 310.
22. **McClure, H. M., Weaver, R. E., and Kaufmann, A. F.,** Pseudotuberculosis in nonhuman primates: infection with organisms of the *Yersinia enterocolitica* group, *Lab. Anim. Sci.*, 21, 376, 1971.
23. **Baggs, R. B., Hunt, R. D., Garcia, F. G., Hajema, E. M., Blake, B. J., and Fraser, C. E. O.,** Pseudotuberculosis *(Yersinia enterocolitica)* in the owl monkey *(Aotus trivirgatus), Lab. Anim. Sci.*, 26, 1079, 1976.
24. **LaBrec, E. H., Schneider, H., Magnani, T. J., and Formal, S. B.,** Epithelial cell penetration as an essential step in the pathogenesis of bacillary dysentery, *J. Bacteriol.*, 88, 1503, 1964.
25. **Giannella, R. A., Washington, O., Gemski, P., and Formal, S. B.,** Invasion of HeLa cells by *Salmonella typhimurium:* a model for study of invasiveness of *Salmonella, J. Infect. Dis.*, 128, 69, 1973.
26. **Lee, W. H., McGrath, P. P., Carter, P. H., and Eide, E. L.,** The ability of some *Yersinia enterocolitica* strains to invade HeLa cells, *Can. J. Microbiol.*, 23, 1714, 1977.
27. **Une, T.,** Studies on the pathogenicity of *Yersinia enterocolitica*. II. Interaction with cultured cells *in vitro, Microbiol. Immunol.*, 21, 365, 1977.
28. **Une, T., Zen-Yoji, H., Maruyama, T., and Yanagawa, Y.,** Correlation between epithelial cell infectivity *in vitro* and O-antigen groups of *Yersinia enterocolitica, Microbiol. Immunol.*, 21, 727, 1977.
29. **Lee, W. H.,** Testing for the recovery of *Yersinia enterocolitica* in foods and their ability to invade HeLa cells, in *Contributions to Microbiology and Immunology,* Vol. 5, Carter, P. B., Lafleur, L., and Toma, S., Eds., S. Karger, Basel, 1979, 229.
30. **Maruyama, T., Une, T., and Zen-Yoji, H.,** Observations on the correlation between pathogenicity and serovars of *Yersinia enterocolitica* by the assay applying cell culture system and experimental mouse infection, in *Contributions to Microbiology and Immunology,* Vol. 5, Carter, P. B., Lafleur, L., and Toma, S., Eds., S. Karger, Basel, 1979, 318.

31. **Mollaret, H. H.**, L'infection humaine a "Yersinia enterocolitica" en 1970, a la lumiere de 642 cas recents: Aspects clinique et perspectives épidémiologiques, *Pathol. Biol. (Paris)*, 19, 189, 1971.
32. **Carter, P. B., Zahorchak, R. J., Brubaker, R. R.**, Plaque virulence antigens from *Yersinia enterocolitica*, *Infect. Immun.*, 28, 638, 1980.
33. **Feeley, J. C., Wells, J. G., Tsai, T. F., and Puhr, N. D.**, Detection of enterotoxigenic and invasive strains of *Yersinia enterocolitica*, in *Contributions to Microbiology and Immunology*, Vol. 5, Carter, P. B., Lafleur, L., and Toma, S., Eds., S. Karger, Basel, 1979, 329.
34. **Robins-Browne, R. M., Jansen van Vuuren, C. J., Still, C. S., Miliotis, M. D., and Koornhof, H. J.**, The pathogenesis of *Yersinia enterocolitica* gastroenteritis, in *Contributions to Microbiology and Immunology*, Vol. 5, Carter, P. B., Lafleur, L., and Toma, S., Eds., S. Karger, Basel, 1979, 324.
35. **Sack, R. B.**, Human diarrheal disease caused by enterotoxigenic *Escherichia coli*, *Annu. Rev. Microbiol.*, 29, 333, 1975.
36. **Pai, C. H., Mors, V., and Toma, S.**, Prevalence of enterotoxigenicity in human and nonhuman isolates of *Yersinia enterocolitica*, *Infect. Immun.*, 22, 334, 1978.
37. **Pai, C. H. and Mors, V.**, Production of enterotoxin by *Yersinia enterocolitica*, *Infect. Immun.*, 19, 908, 1978.
38. **Boyce, J. M., Evans, D. J., Jr., Evans, D. G., and DuPont, H. L.**, Production of heat-stable, methanol-soluble enterotoxin by *Yersinia enterocolitica*, *Infect. Immun.*, 25, 532, 1979.
39. **Robins-Browne, R. M., Still, C. S., Miliotis, M. D., and Koornhof, H. J.**, Mechanism of action of *Yersinia enterocolitica* enterotoxin, *Infect. Immun.*, 25, 680, 1979.
40. **Rabson, A. R., Hallet, A. F., and Koornhof, H. J.**, Generalized *Yersinia enterocolitica* infection, *J. Infect. Dis.*, 131, 447, 1975.
41. **Robins-Browne, R. M., Rabson, A. R., and Koornhof, H. J.**, Generalized infection with *Yersinia enterocolitica* and the role of iron, in *Contributions to Microbiology and Immunology*, Vol. 5, Carter, P. B., Lafleur, L., and Toma, S., Eds., S. Karger, Basel, 1979, 277.
42. **Alonso, J. M., Joseph-Francois, A., Mazigh, D., Bercovier, H., and Mollaret, H. H.**, Resistance á la peste de souris expérimentalement infectées par *Yersinia enterocolitica*, *Ann. Microbiol. (Paris)*, 129B, 203, 1978.
43. **Zink, D. L., Feeley, J. C., Wells, J. G., Vanderzant, C., Vickery, J. C., and O'Donovan, G. A.**, Possible plasmid-mediated virulence in *Yersinia enterocolitica*, *Trans. Gulf Coast Mol. Biol. Conf.*, 3, 155, 1978.
44. **Gemski, P., Lazere, J. R., and Casey, T.**, A plasmid associated with pathogenicity and calcium dependency of *Yersinia enterocolitica*, *Infect. Immun.*, 27, 682, 1980.
45. **Ahlqvist, J., Ahvonen, P., Rasanen, J. A., and Wallgren, G. R.**, Enteric infection with *Yersinia enterocolitica*. Large pyroninophilic cell reaction in mesenteric lymph nodes associated with early production of specific antibodies, *Acta Pathol. Microbiol. Scand. Sect. A*, 79, 109, 1971.
46. **Keet, E. E.**, *Yersinia enterocolitica* septicemia: source of infection and incubation period identified, *N.Y. State J. Med.*, 74, 2226, 1974.
47. **Carter, P. B., MacDonald, T. T., and Collins, F. M.**, Host responses to infection with *Yersinia enterocolitica*, in *Contributions to Microbiology and Immunology*, Vol. 5, Carter, P. B., Lafleur, L. and Toma, S., Eds. S. Karger, Basel, 1979, 347.
48. **McGregor, D. D. and Carter, P. B.**, unpublished data, 1973.
49. **Carter, P. B.**, unpublished data, 1977.

Chapter 8

YERSINIA ENTEROCOLITICA: CLINICAL OBSERVATIONS

Gary P. Wormser and Gerald T. Keusch

TABLE OF CONTENTS

I.	Description and Epidemiology	84
II.	Disease Transmission	84
III.	Clinical Presentation	85
IV.	Pathophysiology	88
V.	Methods for Diagnosis	88
VI.	Treatment	89
	References	90

I. DESCRIPTION AND EPIDEMIOLOGY

Yersinia enterocolitica is a small, Gram-negative coccobacillus in the family *Enterobacteriaceae*. Actually, *Y. enterocolitica* does not refer to a single organism, but rather to a heterogeneous species which can be separated on the basis of 33 O antigens and 19 H antigens, as well as five biotypes and five phage types. These distinctions are important for antimicrobial susceptibility, geographical distribution, natural reservoirs, laboratory growth requirements, and for an understanding of the disease producing capability of a particular strain.[1-6]

Clinical isolation of *Y. enterocolitica* (serotype O:8) was first reported by Schleifstein and Coleman[7] from New York State in 1939. By 1966 only 23 human infections had been documented.[8] Since then, however, a spectacular rise in incidence has been noted. The organism has now been isolated in many countries spanning five continents, and by the end of 1978 several thousand infections had been reported, mostly from Europe and Scandinavia.[9] Ironically, less than 5% of the cases have arisen in the U.S. Increased isolation of *Y. enterocolitica* is in large measure due to its increased recognition by the clinical microbiologist; however, data also indicate an actual increase in frequency of this disease. In a study from Finland,[10] sera which had been collected in 1969 to 1971 from 197 healthy Finnish mothers in a maternity hospital and from 50 blood donors were compared for serologic responses to *Y. enterocolitica* (serotypes O:3 and O:9) with fresh sera from 101 mothers, 542 blood donors, and 258 hospital staff members in 1973. Samples from the latter period showed a tenfold higher frequency of titers ≥ 160 ($P < .001$).

Most human infections are caused by only three O antigen types, O:3, O:9, and O:8. Inexplicably, incidence rates and distribution of these serotypes vary widely and often abruptly between neighboring regions, even within the same country.[8] Serotype O:8 is prominent in the U.S., while it is rarely seen in Canada where type O:3, biotype 4, phage type 9B predominates. In Sweden over 90% of *Y. enterocolitica* infections are caused by *Y. enterocolitica* O:3 biotype 4, phage type 8, whereas in nearby Finland this O antigen type is responsible for 60% of infections, type O:9 accounting for most of the remaining.[11,12] In South Africa type O:3, phage type 9A prevails. In Japan, most infections are related to type O:3 phage type 8.[13]

II. DISEASE TRANSMISSION

How man acquires *Y. enterocolitica* infections is unknown, although there is no reason to doubt that, like other enteric pathogens, transmission is via the oral route. *Y. enterocolitica* has been isolated from numerous nonhuman sources including monkeys, dogs, cats, pigs, chinchillas, hares, antelopes, horses, deer, guinea pigs, cows, mussels, oysters, and fleas.[8,13-19] Animals harboring this organism may appear entirely well or they may be ill, often with a diarrheal illness. *Y. enterocolitica* type O:1 is found among chinchillas and has been responsible for numerous epizootics decimating chinchilla colonies in Europe and the U.S.[15,17] Nevertheless, type O:1, like most other *Y. enterocolitica* types found in animals, is rarely incriminated in human disease. However, important exceptions exist. Of special interest are the isolates from feces, cecal contents, and mesenteric lymph nodes of swine.[20,21] These organisms, either type O:3 or O:9, are of similar phage and biotype to the predominating local *Y. enterocolitica* human pathogen. In some areas peak incidence of human infection has coincided with the seasonal pig slaughter[22] and there exist anecdotal accounts of simultaneous infection in pigs and pig farmers.[23,24] Occasionally dogs, cats, and deer harbor *Y. enterocolitica* strains with disease-producing potential for humans.[14,18,23,25] In some instances diarrheal disease in a dog or cat, possibly related to *Y. enterocolitica*, has antedated by 1 to 2 weeks yersinial infection in the pet owner.[23,26] However, the majority of human yersiniosis appears to be unrelated to direct animal contact.

Y. enterocolitica and *Y. enterocolitica*-like strains have also been isolated from nonchlorinated well and stream water, raw meat, vacuum-packed beef and lamb, unpasteurized milk, pasteurized chocolate milk, and ice cream, but these are predominantly strains that are nonhuman pathogens.[2,3,8,15,27-32] However, chocolate milk and well water may have been responsible for two large outbreaks of yersinial disease in the U.S.[29,30] and mountain stream water was strongly implicated in a single case.[33] *Y. enterocolitica* is capable of surviving in water for more than 150 days, and, unlike most other *Enterobacteriaceae*, grows well at refrigeration temperature (4°C).[33]

Human-to-human transmission has been suggested as the route of inter- and intrafamilial and intrahospital spread noted in some outbreaks of *Y. enterocolitica*.[22,26,34-36] Asymptomatic fecal carriers are well-documented and they may also be important in transmission.[35,37,38] Therefore, it is prudent to institute enteric isolation precautions in patients hospitalized with yersinial disease.[39]

III. CLINICAL PRESENTATION

Convincing documentation of pathogenicity is isolation of an organism in pure culture from otherwise sterile sites such as blood, lymph node, or tissue. Thus, the less common presentations of clinical illness due to *Y. enterocolitica* including sepsis, mesenteric adenitis, and abscess formation are the best documented infections due to this organism. Less rigorous but considerable circumstantial evidence, such as a positive stool culture with a fourfold serologic response to the isolate, have implicated *Y. enterocolitica* as a cause of acute gastroenteritis and terminal ileitis. Most large series of patients considered to have yersinial enteritis fail to satisfy both criteria for diagnosis. This is particularly unfortunate as the clinical features of yersinial disease have largely been based on these less than optimally proven cases. Even less well substantiated are the presumed late sequelae of yersinial disease such as arthritis, carditis, and erythema nodosum. In most instances these complications have been ascribed to *Y. enterocolitica* solely on the basis of a single elevation in yersinial antibody titer.

Many reports have found yersinial enteritis to peak in late autumn and early winter.[11,22,36,40,41] This seasonal incidence could possibly be related to the organism's enhanced growth and virulence at cooler temperatures (see below). There is no sex predilection for the various clinical manifestations of *Y. enterocolitica* with the exception of erythema nodosum. This lesion is found 7 times more often in women than in men.[23] There is little information on the incubation period of *Y. enterocolitica* infection. In a small hospital outbreak of *Y. enterocolitica* serotype O:9 in Finland in which the index case was clearly identified, the incubation time was approximately 10 days.[34] On the other hand, the incubation period in one patient in upstate New York, who developed *Y. enterocolitica* serotype O:8 bacteremia from drinking contaminated water, was less than 5 days,[33] and in a volunteer who ingested 3.5×10^9 organisms, diarrhea began the same day.[42]

The majority of patients with *Y. enterocolitica* disease are young and the most common presentation is self-limited diarrheal disease associated with fever and abdominal pain.[1,23,36,39,41,43] However, the diarrhea can vary from a few loose stools per day to a fulminant enterocolitis with ulcerative lesions involving the entire gastrointestinal tract.[26,44] Fever is generally low grade, but may be spiking or entirely absent.[45] Nausea and vomiting occur but these symptoms are usually not prominent.[23] Occasionally blood may be present in the stool, detected either grossly or by laboratory examination, although massive intestinal bleeding has been reported.[23,26,45-47] Fecal leukocytes have been described in a few patients but the prevalence of this finding is unknown.[26,48] Approximately one half of patients will have leukocytosis (WBC > 10,000/mm^3) and/or elevation of the erythrocyte sedimentation rate above 50 mm/hr.[23] On sigmoidoscopic examination, the colonic mucosa in adult patients with marked diarrhea can vary in appearance from normal to grossly abnormal. The tissue can be diffusely swollen, erythematous, and friable, and small ulcerations are occasionally observed.[45] Microscopic examination of biopsies in the latter instances shows signs of chronic, nonspecific inflammation with infiltration of mononuclear cells and lymphocytes predominating. Dilated crypts near sites of ulceration, however, contain numerous polymorphonuclears. Giant cells and granuloma formation are not found.[45]

Gastrointestinal disease due to *Y. enterocolitica* is usually short-lived. Whether or not antimicrobials are administered, most patients improve within 2 to 3 days, although some will continue to experience mild diarrhea and abdominal discomfort for additional 1 to 2 weeks.[42,49] In a few patients fever and/or diarrhea with or without abdominal pain persists with a remittent course for months.[1,23,45,50,51] Although mortality in this illness is negligible, occasional deaths may occur due to *Y. enterocolitica* septicemia or secondary to intestinal perforation.[35,52]

The one feature of *Y. enterocolitica* disease most helpful in distinguishing this organism from other causes of acute gastroenteritis is the sharp localization of abdominal pain to the right lower quadrant (the right iliac fossa syndrome) which occurs in 3 to 15% of cases.[22,41] Unfortunately, this predilection is also responsible for considerable iatrogenic morbidity because it leads to unnecessary removal of normal appendixes. This is not necessarily poor medicine; in certain patients with the right iliac fossa syndrome, particularly those with fever and right lower quadrant pain alone, it is impossible short of exploration to differentiate this entity from acute appendicitis. In others, the possibility of yersinial disease may be suggested by the persistence of diarrhea after localization of pain, or by a concomitant exudative pharyngitis negative for group A streptococci, or by the presence of similar gastrointestinal illness in family members or friends.[26,49] The right iliac fossa syndrome rarely occurs in children less than 5 years of age. This does not imply a different pathogenesis of yersinial disease for this age group or necessarily a different incidence of terminal ileitis or mesenteric adenitis. Most children 5 years and under have difficulty in localizing pain and as a result there may be no right lower quadrant pain, even with documented acute appendicitis. As many as 70% of these young patients may have already perforated the appendix when the diagnosis is first suspected.[53] Although the right iliac fossa syndrome is encountered most often in the adolescent and young adult, up to 40% of older patients will localize pain to this area as well.[45] Exploratory laparotomy performed because of suspicion of acute appendicitis will in fact reveal in a small percentage a type of acute or subacute appendicitis culture positive for *Y. enterocolitica* and presumably caused by this organism.[23,29,47] In the majority, however, there is either mesenteric lymphadenitis (40%), terminal ileitis (40%), or no abnormality at all (20%).[49] Whether or not there is gross evidence for appendicitis, the removed appendix may show histological evidence of microabscess formation in the germinal centers of the lymphatic tissue, sometimes in combination with mucosal abscesses.[54] These microscopic changes are similar to those seen in other sections of the gastrointestinal tract in yersinial disease.

Mesenteric adenitis may be either over or under diagnosed, as there are no definitive criteria for what is normal, and if laparotomy is not done the diagnosis can be missed. At times the mesenteric nodes form such a large conglomerate mass that it is readily palpated preoperatively, usually in the right lower quadrant.[49,55] Histologically, the nodes are said to show proliferation of large pyroninophilic cells or focal necrosis, but rarely demonstrate granulomas or giant cells.[23,44]

In patients with terminal ileitis the surgeon finds the most distal 20 cm of ileum to be inflamed and edematous with a hardened mesentery, sharply demarcated from the adjacent healthy intestine.[49] Consistent with these findings, barium contrast X-rays of the small bowel performed in adults with bacteriologically confirmed *Y. enterocolitica* disease demonstrate diffuse thickening of mucosal folds, nodular filling defects, and ulcerations extending over the terminal 10 to 20 cm of ileum.[45] Serial radiographic studies show gradual resolution over several months. In contrast to Crohn's disease, there are no skip lesions, fibrotic stenosis, or fistula formation. The antemortem pathological findings in terminal ileitis due to *Y. enterocolitica* have been reported in a single case in which 30 cm of terminal ileum and 30 cm of ascending colon were surgically removed because of massive gastrointestinal bleeding, 3 days prior to the patient's demise. Superficial ulcerations with necrosis were present along the entire mucosal surface of the resected ileum, and the wall was diffusely inflamed with ulceration extending to the muscularis. No areas of mural fibrosis or granulomatous inflammation were present and no crypt abscesses were found within the mucosal glands.[44] In a second case examined postmortem by the same authors there was ulceration in the ileum overlying the anatomic sites of intestinal lymphoid tissue.[44]

Long-term follow-up studies suggest that patients with acute ileitis who have either serologic or bacteriologic evidence of *Y. enterocolitica* do not go on to develop Crohn's disease.[49] Conversely, patients with Crohn's disease have no serologic or bacteriologic evidence for *Yersinia*.[56]

Bacteremia may occur with *Y. enterocolitica* usually (but not invariably) in patients with serious pre-existing disease.[25,33,52,57-72] *Y. enterocolitica* sepsis carries a high mortality outcome, approximately 40%, in general correlating with severity of the underlying condition. A surprising number of these patients have had cirrhosis and/or hemachromatosis.[52,66,69] Bacteremia without diarrhea has been associated with liver or splenic abscesses.[66,69] Other sites of focal *Y. enterocolitica* infection presumably seeded by bacteremic spread have included the meninges,[65] synovium,[52,61] skin,[63,65] lung,[61,62] bone,[57] eye,[60,65] and kidney.[65]

Y. enterocolitica intestinal infection is occasionally accompanied by a nonspecific macular-papular cutaneous eruption.[23] In some cases, particularly in Scandinavia, erythema nodosum or erythema multiforme develops, usually 1 to 2 weeks after the intestinal symptoms have resolved. This complication occurs predominantly in adult females and subsides in 3 to 4 weeks.[73-75] Erythema nodosum is rarely observed following *Y. enterocolitica* infection in Canada or the U.S., which suggests that this event may be related to specific strains or to certain genetic traits of the host, or both.[76]

Another intriguing late complication of *Yersinia* infection is oligoarthritis with or without Reiter's syndrome or carditis.[23,77-80] This occurs most frequently in patients with HLA-B27 histocompatibility antigen.[1,14,81,82] Recently, acute proliferative glomerulonephritis has been associated with *Yersinia* infection as well.[83,84]

Y. enterocolitica has been isolated from feces, blood, abscess cavities, synovial fluid, bile, sputum, conjunctival secretions, urine, cerebrospinal fluid, anterior chamber fluid, skin lesions, and operative wounds.[23,29,33,60,62,63,85,89] This organism has been responsible for pneumonia,[61,69] conjunctivitis,[85,88] Parinaud's oculo-glandular syndrome,[60] facial abscess,[86] lung abscess,[62] furuncles,[87] cellulitis,[63] osteomyelitis,[57,62] and fevers of unknown origin.[40,90] Of interest is a recent report from New York City implicating atypical strains of *Y. enterocolitica* as a cause of conjunctivitis in six patients.[88] In some of these infections the organism could be acquired through direct inoculation or by aerosol. Spread by air has also been suggested by Pahl and cited by Rabson.[38] In studying *Y.*

enterocolitica infected pigs, he isolated *Y. enterocolitica* from neighboring animals who were housed in such a way as to exclude all other routes of transmission.

IV. PATHOPHYSIOLOGY

The pathophysiology of *Y. enterocolitica* disease in humans is poorly understood. *Y. enterocolitica* strains grown in vitro at 25°C differ remarkably from the parent strains cultivated at 37°C.[76] At the lower temperature, flagella and a heat-stable enterotoxin are produced.[91,92] This toxin resembles *E. coli* stable toxin (ST) in that both cause fluid secretion in the suckling mouse and the rabbit ileum, but its role, if any, in human disease needs further clarification.[91] Certain strains grown at low temperatures are more resistant to antimicrobials and to the bactericidal effect of human serum and they are more virulent for mice.[92,93] These temperature-dependent phenotypic variations have led to the suggestion that enhanced virulence may occur when the organism passes through an intermediate cold cycle.[76]

Carter, using a human isolate, established an animal model of *Y. enterocolitica* disease by intragastric inoculation into mice.[94,95] In less than 24 hr there is neutrophil infiltration in Peyers' patches of the distal ileum. By 5 days the infection spreads to mesenteric lymph nodes, causing medullary abscess formation. Thereafter infection becomes systemic. Abscesses develop in the liver, spleen, and lung and in some instances abscesses originating in the Peyers' patches may ulcerate into the lumen of the intestine.[94] Agglutinating antibody to the infecting strain of *Y. enterocolitica* appears 2 weeks after inoculation. Une[96] inoculated various strains of *Y. enterocolitica* intraduodenally into rabbits. Some strains were found to cause enterocolitis, but in contrast to the mouse experiment, there was a predominantly mononuclear inflammatory response with granuloma formation. Une was able to demonstrate that disease-producing strains of *Y. enterocolitica* rapidly penetrate intestinal epithelial cells, become engulfed in cytoplasmic vesicles, and then promptly enter the lamina propria and lymph follicles. Strains nonpathogenic in this system are rapidly excreted without penetration of epithelial linings. Further in vitro experiments by this investigator have shown that pathogenicity correlates with ability to penetrate HeLa cells and to survive or multiply in macrophages.[97] Of 178 strains of *Y. enterocolitica* of 16 different O antigen types, as well as 25 nongroupable strains, only those belonging to O antigen types frequently pathogenic for man, (O:3, O:8, O:9, and O:5B strains) were found to be invasive for HeLa cells.[98] Similar findings were reported by Lee et al.[99] who also noted that some strains were invasive only when cultivated at temperatures less than 36°C. This system appears to provide a valuable tool for assessing the potential virulence of a given isolate.[48]

V. METHODS FOR DIAGNOSIS

Bacteriologic diagnosis requires considerable expertise in primary isolation and enrichment techniques (particularly when working with fecal specimens) and in identification procedures. Standard processing of stool specimens designed for isolation of *Salmonella* and *Shigella* will in many instances overlook *Y. enterocolitica*.[76,100]

Marked geographical variation in isolation rates without doubt reflects in part interest in and awareness of the organism by clinical microbiology personnel. It is certainly helpful when the clinician is suspicious of a *Y. enterocolitica* infection to alert the laboratory personally. Stool cultures are generally positive during the first 2 weeks of illness regardless of the nature of the gastrointestinal manifestations and even when the patient has received antimicrobials to which the isolate is susceptible.[101,102] In some patients stool cultures remain positive for 2 to 4 months, but whether or not truly chronic carriers (> 1 year) exist is unknown.[22,101] The probability of isolating *Y. enterocolitica* is increased by repeated culturing, that is at least three samples of stool.[51] It is important to remember that other potential pathogens may be found simultaneously in approximately 10% of patients, and that the organism can be recovered occasionally from asymptomatic individuals.[41] Isolation therefore does not prove involvement in disease. However, in conjunction with a high serum antibody level, preferably with a fourfold rise in titer, isolation can be considered to be strong evidence.

Serology alone has been extensively used in diagnosis outside the U.S. This has been a practical and reliable modality because (1) only one or two serotypes (O:3, O:9) are responsible for more than 90% of *Y. enterocolitica* disease in these areas, (2) clinical disease due to these serotypes is nearly always accompanied by a brisk antibody response within 1 to 4 weeks of onset,[37,47,101] and (3) prevalence and height of background titer is already known from screening large numbers of controls.[47,103,104] Unfortunately, in the U.S. serologic evaluation is much less informative where there is greater heterogeneity in the number of serotypes isolated.[8,15,105,106] Because there is no single common antigen, serologic screen requires a battery of different antigens, and this may be impractical. Furthermore, serotypes other than O:3 and O:9 apparently do not reliably elicit an antibody response.[88]

In serologic diagnosis it is important to remember that a single high-titer specimen cannot be considered proof of recent infection. Serial samples are needed for this purpose, although titers ≥ 160 suggest relatively recent infection and elevated IgM titers indicate infection within 3 months.[101,107] Certain serologic cross-reactions are known, most importantly between *Y. enterocolitica* serotype O:9 and *Brucella* species.[12,101,108-110] Cross-reactivity with other members of *Enterobacteriaceae*, especially *Morlanella (Proteus) morganii* and *Salmonella* are reported.[76,111] In addition, patients with thyroid disease may have persistently elevated *Y. enterocolitica* O:3 antibody titers, as high as 1:256, thought to be due to antigenic similarity with the membrane of the thyroid epithelial cell.[112-114]

VI. TREATMENT

Y. enterocolitica strains are susceptible in vitro to tetracycline, aminoglycosides, trimethoprim-sulfamethoxazole, and chloramphenicol.[6,115-117] Most strains show considerable resistance to penicillin, and many to cephalothin, although cefamandole may still be active.[115] Two distinct β-lactamases have been discovered in *Y. enterocolitica* which may account for the observed variable sensitivity pattern to the β-lactam antibiotics.[118] Rare strains of *Y. enterocolitica* contain R plasmids coding for resistance to chloramphenicol, sulfonamides, streptomycin, kanamycin, and tetracycline.[76,119]

No data exist as to the efficacy of antimicrobials in either acute yersinial enteritis, which is generally self-limited and unnecessary to treat, or the right iliac fossa syndrome, which in fact might be desirable to treat. Antimicrobials have no effect on the late complications of disease such as erythema nodosum and arthritis.[23] Patients with bacteremia or sites of infection other than the gastrointestinal tract should be treated, but data are too limited to suggest the agent of choice or the optimum duration of therapy. An occasional patient with *Y. enterocolitica* bacteremia will clear the infection without specific treatment.[64] Gentamicin, streptomycin, tetracycline, colistin, and chloramphenicol have been used successfully.[58,60,65,69,71] One patient with a TMP-SMX-sensitive organism failed to respond to 60 hr of therapy with this medication intravenously, later responding to tetracycline.[67] In two other patients, gentamicin eradicated *Y. enterocolitica* from the blood, but was unable to sterilize a joint[52] or a cutaneous lesion, respectively.[63]

REFERENCES

1. **Leirisalo, M.,** Human *Yersinia enterocolitica* infections, *Ann. Clin. Res.,* 10, 63, 1978.
2. **Schiemann, D. A. and Toma, S.,** Isolation of *Yersinia enterocolitica* from raw milk, *Appl. Environ. Microbiol.,* 35, 54, 1978.
3. **Lassen, J.,** *Yersinia enterocolitica* in drinking water, *Scand. J. Infect. Dis.,* 4, 125, 1972.
4. **Knapp, W. and Thal, E.,** Differentiation of *Yersinia enterocolitica* by biochemical reactions in *Contributions to Microbiology and Immunology,* Vol. 2, Winblad, S., Ed., S. Karger, Basel, 1973, 10.
5. **Winblad, S.,** Studies on the O-serotypes of *Yersinia enterocolitica* in *Contributions to Microbiology and Immunology,* Vol. 2, Winblad, S., Ed., S. Karger, Basel, 1973, 27.
6. **Hausnerová, S., Hausner, O., and Paučkova, V.,** Antibiotic sensitivity of *Yersinia enterocolitica* strains isolated in two regions of Czechoslovakia in *Contributions to Microbiology and Immunology,* Vol. 2, Winblad, S., Ed., S. Karger, Basel, 1973, 76.
7. **Schleifstein, J. I. and Coleman, M. B.,** An unidentified microorganism resembling *B. lignieri* and *Past. pseudotuberculosis,* and pathogenic for man, *N.Y. State J. Med.,* 39, 1749, 1939.
8. **Morris, G. K. and Feeley, J. C.,** *Yersinia enterocolitica:* a review of its role in food hygiene, *Bull. W.H.O.,* 54, 79, 1976.
9. **Anon.,** The spectacular rise of *Yersinia enterocolitica, Can. Med. Assoc. J.,* 108, 1097, 1973.
10. **Toivanen, P., Toivanen, A., Olkkonen, L., and Aantaa, S.,** Is the incidence of *Yersinia enterocolitica* infection increasing?, *Acta Pathol. Microbiol. Scand.,* 82B, 303, 1974.
11. **Ahvonen, P.,** *Yersinia enterocolitica* infections in Finland in *Contributions to Microbiology and Immunology,* Vol. 2, Winblad, S., Ed., S. Karger, Basel, 1973, 133.
12. **Esseveld, H. and Goudzwaard, C.,** Serological diagnosis in proven human infections with *Yersinia enterocolitica* Type 9 in *Contributions to Microbiology and Immunology,* Vol. 2, Winblad, S., Ed., S. Karger, Basel, 1973, 146.
13. **Toma, S. and Lafleur, L.,** Survey on the incidence of *Yersinia enterocolitica* infection in Canada, *Appl. Microbiol.,* 28, 469, 1974.
14. **Wilson, H. D., McCormick, J. B., and Feeley, J. C.,** *Yersinia enterocolitica* infection in a four month old infant associated with infection in household dogs, *J. Pediatr.,* 89, 767, 1976.
15. **Anon.,** Worldwide spread of infections with *Yersinia enterocolitica, W.H.O. Chron.,* 30, 494, 1976.
16. **Quan, T. J., Meek, J. L., Tsuchiya, K. R., Hudson, B. W. and Barnes, A. M.,** Experimental pathogenicity of recent North American isolates of *Yersinia enterocolitica, J. Infect. Dis.,* 129, 341, 1974.
17. **Hubbert, W. T.,** Yersiniosis in mammals and birds in the United States. Case reports and review, *Am. J. Trop. Med. Hyg.,* 21, 458, 1972.
18. **Ahvonen, P., Thal, E., and Vasenius, H.,** Occurrence of *Yersinia enterocolitica* in animals in Finland and Sweden in *Contributions to Microbiology and Immunology,* Vol. 2, Winblad, S., Ed., S. Karger, Basel, 1973, 135.
19. **Kaneko, K., Hamada, S., Kasai, Y., and Kato, E.,** Occurrence of *Yersinia enterocolitica* in house rats, *Appl. Environ. Microbiol.,* 36, 314, 1978.
20. **Esseveld, H. and Goudzwaard, C.,** On the epidemiology of *Yersinia enterocolitica* infections: Pigs as the source of infections in man in *Contributions to Microbiology and Immunology,* Vol. 2, Winblad, S., Ed., S. Karger, Basel, 1973, 99.
21. **Zen-Yoji, H., Sakai, S., Maruyama, T., and Yanagawa, Y.,** Isolation of *Yersinia enterocolitica* and *Yersinia pseudotuberculosis* from swine, cattle and rats at an abattoir, *Jpn. J. Microbiol.,* 18, 103, 1974.
22. **Rakovsky, J., Paučkova, V., and Aldová, E.,** Human *Yersinia enterocolitica* infections in Czechoslovakia in *Contributions to Microbiology and Immunology,* Vol. 2, Winblad, S., Ed., S. Karger, Basel, 1973, 93.
23. **Ahvonen, P.,** Human yersiniosis in Finland. II. Clinical features, *Ann. Clin. Res.,* 4, 39, 1972.
24. **Rabson, A. R. and Koornhof, H. J.,** *Yersinia enterocolitica* infections in South Africa in *Contributions to Microbiology and Immunology,* Vol. 2, Winblad, S., Ed., S. Karger, Basel, 1973, 102.
25. **Baier, R., Puppel, H., and Hein, J.,** Septicaemia caused by *Yersinia enterocolitica* Dtsch Med. Wochenschr., 102, 55, 1977.
26. **Gutman, L. T., Ottesen, E. A., Quan, T. J., and Katz, S. L.,** An inter-familial outbreak of *Yersinia enterocolitica* enteritis, *N. Engl. J. Med.,* 288, 1372, 1973.
27. **Kapperud, G.,** *Yersinia enterocolitica* and Yersinia like microbes isolated from mammals and water in Norway and Denmark, *Acta Pathol. Microbiol. Scand. B,* 85, 129, 1977.
28. **Harvey, S., Greenwood, J. R., Pickett, M. J., and Mah, R. A.,** Recovery of *Yersinia enterocolitica* from streams and lakes of California, *Appl. Environ. Microbiol.,* 32, 352, 1976.
29. **Black, R. E., Jackson, R. J., Tsai, T., Medvesky, M., Shayegani, M., Feeley, J. C., Macleod, K. I. E., and Wakelee, A. M.,** Epidemic *Yersinia enterocolitica* infection due to contaminated chocolate milk, *N. Engl. J. Med.,* 298, 76, 1978.

30. **Eden, K. V., Rosenberg, M. L., Stoopler, M., Wood, B. T., Highsmith, A. K., Skaliy, P., Wells, J. G., and Feeley, J. C.**, Waterborne gastrointestinal illness at a ski resort — isolation of *Yersinia enterocolitica* from drinking water, *Public Health Rep.*, 92, 245, 1977.
31. **Schiemann, D. A.**, Association of *Yersinia enterocolitica* with the manufacture of cheese and occurrence in pasteurized milk, *Appl. Environ. Microbiol.*, 36, 274, 1978.
32. **Hanna, M. O., Zink, D. L., Carpenter, Z. L., and Vanderzant, C.**, *Yersinia enterocolitica*-like organisms from vacuum packaged beef and lamb, *J. Food Sci.*, 41, 1254, 1976.
33. **Keet, E. E.**, *Yersinia enterocolitica* septicemia. Source of infection and incubation period identified, *N.Y. State J. Med.*, 74, 2226, 1974.
34. **Toivanen, P., Toivanen, A., Olkkanen, L., and Aantaa., S.**, Hospital outbreak of *Yersinia enterocolitica* infection, *Lancet*, 1, 801, 1973.
35. **Kohl, S., Jacobsen, J. A., and Nahmias, A.**, *Yersinia enterocolitica* infections in children, *J. Pediatr.*, 89, 77, 1976.
36. **Mollaret, H. H.**, Un domaine pathologigue nouveau; l' infection à *Yersinia enterocolitica*, *Ann. Biol. Clin. (Paris)*, 30, 1, 1972.
37. **Asakawa, Y., Akahane, S., Kagata, N., Noguchi, M., Sakazaki, R., and Tamura, K.**, Two community outbreaks of human infection with *Yersinia enterocolitica*, *J. Hyg.*, 71, 715, 1973.
38. **Rabson, A. R. and Koornhof, H. J.**, *Yersinia enterocolitica* infections in South Africa, *S. Afr. Med. J.*, 46, 798, 1972.
39. **Randall, C. and Bannatyne, R. M.**, Experience with *Yersinia enterocolitica* at the Hospital for Sick Children 1972—1974, *Can. Med. Assoc. J.*, 113, 542, 1975.
40. **Winblad, S., Arvastson, B., Damgaard, K., and Winblad, S.**, Clinical symptoms of infection with *Yersinia enterocolitica*, *Scand. J. Infect. Dis.*, 3, 37, 1971.
41. **Vandepitte, J., Wauters, G., and Isebaert, A.**, Epidemiology of *Yersinia enterobolicita* infections in Belgium in *Contributions to Microbiology and Immunology*, Vol. 2, Winblad, S., Ed., S. Karger, Basel, 1973, 111.
42. **Szita, J., Káli, M., and Rédey, B.**, Incidence of *Yersinia enterocolitica* infections in Hungary in *Contributions to Microbiology and Immunology*, Vol. 2, Winblad, S., Ed., S. Karger, Basel, 1973, 106.
43. **Toma, S. and Diedrick, V.R.**, Incidence of *Yersinia enterocolitica* and *pseudotuberculosis* infections in Canada; 1975 semi-annual report, *Can. Med. Assoc. J.*, 114, 16, 1976.
44. **Bradford, W. D., Noce, P. S., and Gutman, L. T.**, Pathologic features of enteric infection with *Yersinia enterocolitica*, *Arch. Pathol.*, 98, 17, 1974.
45. **Vantrappen, G., Agg, H. O., Ponette, E., Geboes, K., and Bertrand, P. H.**, *Yersinia* enteritis and enterocolitis: gastroenterological aspects, *Gastroenterology*, 72, 220, 1977.
46. **Bergstrand, C. G. and Winblad, S.**, Clinical manifestations of infection with *Yersinia enterocolitica* in children, *Acta Paediatr. Scand.*, 63, 875, 1974.
47. **Niléhn, B.**, Studies on *Yersinia enterocolitica* with special reference to bacterial diagnosis and occurrence in human acute enteric disease, *Acta Pathol. Microbiol. Scand.*, Suppl. 206, 5, 1969.
48. **Mäki, M., Grönroos, P., and Vesikari, T.**, In vitro invasiveness of *Yersinia enterocolitica* isolated from children with diarrhea, *J. Infect. Dis.*, 138, 677, 1978.
49. **Sjöström, B.**, Surgical aspects of infection with *Yersinia enterocolitica* in *Contributions to Microbiology and Immunology*, Vol. 2, Winblad, S., Ed., S. Karger, Basel, 1973, 137.
50. **Saebø, A.**, Liver affection associated with *Yersinia enterocolitica* infection, *Acta Chir. Scand.*, 143, 445, 1977.
51. **Delorme, J., Laverdiere, M., Martineau, B., and Lafleur, L.**, Yersiniosis in children, *Can. Med. Assoc. J.*, 110, 281, 1974.
52. **Spira, T. J. and Kabins, S. A.**, *Yersinia enterocolitica* septicemia with septic arthritis, *Arch. Int. Med.*, 136, 1305, 1976.
53. **Touloukian, R. J., Seashore, J. H., and Pickett, L. K.**, Appendicitis in *Textbook of Pediatrics*, Vaughan, V. C., III., McKay, R. J., Nelson, W. E., Eds., W. B. Saunders, Philadelphia, 1975, 864.
54. **Sternby, N. H.**, Morphologic findings in appendix in human *Yersinia enterocolitica* infections in *Contributions to Microbiology and Immunology*, Vol. 2, Winblad, S., Ed., S. Karger, Basel, 1973, 141.
55. **Saebø, A.**, Some surgical manifestations of mesenterial lymphadenitis associated with infections of *Yersinia enterocolitica*, *Acta Chir. Scand.*, 140, 655, 1974.
56. **Persson, S., Danielsson, D., Kjellander, J., and Wallensten, S.**, Studies on Crohn's disease. I. The relationship between *Yersinia enterocolitica* infection and terminal ileitis, *Acta Chir. Scand.*, 142, 84, 1976.
57. **Thirumoorthi, M. C. and Dajani, A. S.**, *Yersinia enterocolitica* osteomyelitis in a child, *Am. J. Dis. Child.*, 132, 578, 1978.

58. **Narasimhan, S. L., Schleven, B. C., and Campsall, E. W. R.,** Septicemia caused by *Yersinia enterocolitica, Can. Med. Assoc. J.,* 118, 682, 1978.
59. **Caplan, L. M., Dobson, M. L., and Dorkin, H.,** *Yersinia enterocolitica* septicemia, *Am. J. Clin. Pathol.,* 69, 189, 1978.
60. **Chin, G. N., and Noble, R. C.,** Ocular involvement in *Yersinia enterocolitica* infection presenting as Parinaud's oculoglandular syndrome, *Am. J. Ophthalmol.,* 83, 19, 1977.
61. **Taylor, B. G., Zafarzai, M. Z., Humphreys, D. W., and Manfredi, F.,** Nodular pulmonary infiltrates and septic arthritis associated with *Yersinia enterocolitica* bacteremia, *Am. Rev. Respir. Dis.,* 116, 525, 1977.
62. **Sebes, J. I., Mabry, E. H., Jr., and Rabinowitz, J. G.,** Lung abscess and osteomyelitis of ribs due to *Yersinia enterocolitica, Chest,* 69, 546, 1976.
63. **Abramovitch, H. and Butas, C. A.,** Septicemia due to *Yersinia enterocolitica, Can. Med. Assoc. J.,* 109, 1112, 1973.
64. **Chessum, B., Frengley, J. D., Fleck, D. G., and Mair, N. S.,** Case of septicemia due to *Yersinia enterocolitica, Br. Med. J.,* 3, 466, 1971.
65. **Sonnenwirth, A. C.,** Bacteremia with and without meningitis due to *Yersinia enterocolitica, Edwardsiella tarda, Comamonas terrigena,* and *Pseudomonas maltophilia, Ann. N.Y. Acad. Sci.,* 174, 488, 1970.
66. **Reinicke, V. and Korner, B.,** Fulminant septicemia caused by *Yersinia enterocolitica, Scand. J. Infect. Dis.,* 9, 249, 1977.
67. **Eriksson, M. and Olcén, P.,** Septicaemia due to *Yersinia enterocolitica* in a non-compromised host, *Scand. J. Infect. Dis.,* 7, 78, 1975.
68. **Josefsson, K. and Lindber, A.,** Fatal *Yersinia enterocolitica* septicaemia, *Scand. J. Infect. Dis.,* 7, 76, 1975.
69. **Rabson, A. R., Hallett, A. F., and Koornhof, H. J.,** Generalized *Yersinia enterocolitica* infection, *J. Infect. Dis.,* 131, 447, 1975.
70. **Hassig, A., Karrer, J., and Pusterla, F.,** Über Pseudotuberkulose beim menschem, *Schweiz Med. Wochenschr.,* 971, 1949.
71. **Butzler, J. P., Alexander, M., Segers, A., Cremer, N., and Blum, D.,** Enteritis, abscess, and septicemia due to *Yersinia enterocolitica* in a child with thalassemia, *J. Pediatr.,* 93, 619, 1978.
72. **Janosek, J., Kleibl, K., and Valkova, M.,** *Yersinia enterocolitica* as a causative agent of septicaemia, *J. Hyg Epidemiol. Microbiol. Immunol.,* 19, 254, 1975.
73. **Debois, J., Vandepitte, J., and Degreef, H.,** *Yersinia enterocolitica* as a cause of erythema nodosum, *Dermatologica,* 156, 65, 1978.
74. **Hannuksela, M. and Ahvonen, P.,** Skin manifestations in human yersiniosis, *Ann. Clin. Res.,* 7, 368, 1975.
75. **Hannuksela, M.,** Human yersiniosis: a common cause of erythematous skin eruptions, *Int. J. Dermatol.,* 16, 665, 1977.
76. **Bottone, E. J.,** *Yersinia enterocolitica:* a panoramic view of a charismatic microorganism, *CRC Crit. Rev. Microbiol.,* 5, 211, 1977.
77. **Agner, E., Larsen, J. H., Leth, A.,** *Yersinial enterocolitica* carditis as a differential diagnosis — and the prognosis of this disease. *Scand. J. Rheumatol.,* 7, 26, 1978.
78. **Ford, D. K., Henderson, E., Price, G. E., Stein, H. B.,** Yersinia related arthritis in the Pacific Northwest, *Arthritis Rheum.,* 20, 1226, 1977.
79. **Anon.,** Polyarthritis and *Yersinia enterocolitica* infection, *Br. Med. J.,* 2, 404, 1975.
80. **Ahvonen, P. and Dickhoff, K.,** Uveitis, episcleritis and conjunctivitis associated with Yersinia infection, *Acta Ophthalmol.,* Suppl. 123, 209, 1974.
81. **Laitinen, O., Leirisalo, M., and Skylv, G.,** Relation between HLA-B27 and clinical features in patients with Yersinia arthritis, *Arthritis Rheum.,* 20, 1121, 1977.
82. **Aho, K., Ahvonen, P., Lassus, A., Sievers, K., and Tiilikainen, A.,** HLA 27 in reactive arthritis. A study of Yersinia arthritis and Reiter's disease. *Arthritis Rheum.,* 17, 521, 1974.
83. **Friedberg, M., Larsen, S., and Denneberg, T.,** *Yersinia enterocolitica* and glomerulonephritis, *Lancet,* 1, 498, 1978.
84. **Forsström, J., Viander, M., Lehtonen, A., and Ekfors, T.,** *Yersinia enterocolitica* infection complicated by glomerulonephritis, *Scand. J. Infect. Dis.,* 9, 253, 1977.
85. **Crichton, E. P.,** Suppurative conjunctivitis caused by *Yersinia enterocolitica, Can. Med. Assoc. J.,* 118, 22, 1978.
86. **Lewis, J. F. and Alexander, J.,** Facial abscess due to *Yersinia enterocolitica, Am. J. Clin. Pathol.,* 66, 1016, 1976.
87. **Lawrence, M. R., Ting, S. K., and Neilly, S.,** Furuncle caused by *Yersinia enterocolitica, Can. Med. Assoc. J.,* 112, 1289, 1975.

88. **Bottone, E. J.**, Atypical *Yersinia enterocolitica:* clinical and epidemiological parameters, *J. Clin. Microbiol.,* 7, 562, 1978.
89. **Greenstein, A. J. and Dreiling, D. A.**, Postoperative combined undermining infection of abdominal wound due to *Yersinia enterocolitica, Mt. Sinai J. Med. N.Y.,* 41, 665, 1974.
90. **Bliddel, J. and Kaliszan, S.**, Prolonged monosymptomatic fever due to *Yersinia enterocolitica, Acta Med. Scand.,* 201, 387, 1977.
91. **Pai, C. H. and Mors, V.**, Production of enterotoxin by *Yersinia enterocolitica, Infect. Immun.,* 19, 908, 1978.
92. **Niléhn, B.**, The relationship of incubation temperature to serum bactericidal effect, pathogenicity, and in vivo survival of *Yersinia enterocolitica* in *Contributions to Microbiology and Immunology,* Vol. 2, Winblad, S., Ed., S. Karger, Basel, 1973, 85.
93. **Carter, P. B. and Collins, F. M.**, Experimental *Yersinia enterocolitica* infection in mice: kinetics of growth, *Infect. Immun.,* 9, 851, 1974.
94. **Carter, P. B.**, Animal model: oral *Yersinia enterocolitica* infection of mice, *Am. J. Pathol.,* 81, 703, 1975.
95. **Carter, P.**, Pathogenicity of *Yersinia enterocolitica* for mice, *Infect. Immun.,* 11, 164, 1975.
96. **Une, T.**, Studies on the pathogenicity of *Yersinia enterocolitica.* I. Experimental infection in rabbits, *Microbiol. Immunol.,* 21, 349, 1977.
97. **Une, T.**, Studies on the pathogenicity of *Yersinia enterocolitica.* II. Interaction with cultured cells in vitro, *Microbiol. Immunol.,* 21, 365, 1977.
98. **Une, T., Zen-Yoji, H., Maruyama, T., and Yanagawa, Y.**, Correlation between epithelial cell infectivity in vitro and O-antigen groups of *Yersinia enterocolitica, Microbiol. Immunol.,* 21, 727, 1977.
99. **Lee, W. H., McGrath, P. P., Carter, P. H., and Eide, E. L.**, The ability of some *Yersinia enterocolitica* strains to invade HeLa cells, *Can. J. Microbiol.,* 23, 1714, 1977.
100. **Wauters, G.**, Improved methods for the isolation and the recognition of *Yersinia enterocolitica* in *Contributions to Microbiology and Immunology,* Vol. 2, Winblad, S., Ed., S. Karger, Basel, 1973, 68.
101. **Ahvonen, P.**, Human yersiniosis in Finland. I. Bacteriology and serology, *Ann. Clin. Res.,* 4, 30, 1972.
102. **Doraiswamy, N. V., Currie, A. B. M., Gray, J., Lynton-Moll, C., and Mair, N. S.**, Terminal ileitis: *Yersinia enterocolitica* isolated from faeces, *Br. Med. J.,* 2, 23, 1977.
103. **Leino, R. and Kalliomäki, J. L.**, Yersiniosis as an internal disease, *Ann. Int. Med.,* 81, 458, 1974.
104. **Winblad, S., Niléhn, B., and Steenby, N. H.**, *Yersinia enterocolitica (Pasteurella X)* in human enteric infections, *Br. Med. J.,* 2, 1363, 1966.
105. **Bissett, M. L.**, *Yersinia enterocolitica* isolates from humans in California, 1968–1975, *J. Clin. Microbiol.,* 4, 137, 1976.
106. **Tsai, T., Feeley, J. C., and Shayegani, M.**, Yersinia infection, *N. Engl. J. Med.,* 298, 977, 1978.
107. **Granfors, K., Viljanen, M. K., Ahvonen, P., and Toivanen, P.**, Measurement of IgM and IgG antibodies to Yersinia by solid phase radioimmunoassay, *J. Infect. Dis.,* 138, 232, 1978.
108. **Ahvonen, P., Jansson, E., and Aho, K.**, Marked cross-agglutination between *Brucellae* and a subtype of *Yersinia enterocolitca, Acta Pathol. Microbiol. Scand.,* 75, 291, 1969.
109. **Fribourg-Blanc, A.**, Immunofluorescence study of antigenic relationships between Brucella and some strains of *Yersinia enterocolitica, Ann. N.Y. Acad. Sci.,* 177, 37, 1971.
110. **Hurvell, B. and Lindberg, A. A.**, Immunochemical studies on the cross reactions between Brucella species and *Yersinia enterocolitica* type 9 in *Contributions to Microbiology and Immunology,* Vol. 2, Winblad, S., Ed., S. Karger, Basel, 1973, 159.
111. **Lysy, J., Knapp, W.**, Serological studies with *Yersinia enterocolitica* in *Contributions to Microbiology and Immunology,* Vol. 2, Winblad, S., Ed., S. Karger, Basel, 1973, 42.
112. **Shenkman, L. and Bottone, E. J.**, Antibodies to *Yersinia enterocolitica* in thyroid disease, *Ann. Int. Med.,* 85, 735, 1976.
113. **Lidman, K., Eriksson, V., Fagraeus, A., and Norberg, R.**, Antibodies against thyroid cells in *Yersinia enterocolitica* infection, *Lancet,* 2, 1449, 1974.
114. **Lidman, K., Eriksson, V., Norberg, R., and Fagraeus, A.**, Indirect immunofluorescence staining of human thyroid by antibodies occurring in *Yersinia enterocolitica* infection, *Clin. Exp. Immunol.,* 23, 429, 1976.
115. **Raevuori, M., Harvey, S. M., Pickett, M. J., and Martin, W. J.**, *Yersinia enterocolitica:* in vitro antimicrobial susceptibility, *Antimicrob. Agents Chemother.,* 13, 888, 1978.
116. **Hammerberg, S., Sorger, S., and Marks, M. I.**, Antimicrobial susceptibilities of *Yersinia enterocolitica* biotype 4, serotype 3, *Antimicrob. Agents Chemother.,* 11, 566, 1977.
117. **Gutman, L. T., Wilfert, C. M., and Quan, T.**, Susceptibility of *Yersinia enterocolitica* to trimethoprim-sulfamethoxazole, *J. Infect. Dis.,* 128(S), 538, 1973.
118. **Cornelis, G.**, Distribution of beta-lactamases A and B in some groups of *Yersinia enterocolitica* and their role in resistance, *J. Gen. Microbiol.,* 91, 391, 1975.
119. **Kimura, S., Ikeda, T., Eda, T., Mitsui, Y., and Nakata, K.**, R plasmids from Yersinia, *J. Gen. Microbiol.,* 97, 141, 1976.

Chapter 9

YERSINIA ENTEROCOLITICA GASTROENTERITIS IN CHILDREN AND THEIR FAMILIES

Melvin I. Marks, Chik H. Pai, and Lucette Lafleur

TABLE OF CONTENTS

I. Introduction 96

II. Incidence 96
   A. General 96
   B. Bacteriology 96
   C. Seasonal Distribution 97

III. Clinical Manifestations 97
   A. Gastroenteritis 97
   B. Abdominal Pain 99
   C. Bacteremia 99
   D. Rash 99

IV. Serologic Response 99

V. Communicability 100

VI. Summary 102

References 102

I. INTRODUCTION

Yersinia enterocolitica infections are being recognized more frequently as we become increasingly familiar with their clinical symptomatology and the bacteriologic characteristics of the microorganism. The original descriptions of infection in the 1930s include children with diarrhea and abdominal pain, the most common manifestations of this illness.[1] As described in other chapters of this book, the literature also contains many reports of localized infections, bacteremia, and arthritis. However, these uncommon complications of *Y. enterocolitica* infection are more readily diagnosed than gastroenteritis because of the isolation of *Y. enterocolitica* from normally sterile body fluids or tissues.

As the alimentary tract is the usual portal of entry in human colonization and infection with these bacteria, careful bacteriologic techniques are needed to isolate and identify *Y. enterocolitica* from gastrointestinal tissues or contents. This is mostly due to the slow growth of these bacteria at the usual 37°C incubation temperature as well as their morphologic and biochemical features which may be easily confused with common fecal flora.

Much remains to be learned about the distribution of *Yersinia* in man and his environment and its relationship to disease. Some information has already been derived from sero-epidemiologic studies. Tests employed for these studies usually rely upon bacterial agglutination by serum antibody produced against somatic antigens of *Y. enterocolitica*. The method is rather insensitive and cross-reactions have been reported with *Salmonella*[2] and *Brucella*.[3]

II. INCIDENCE

A. General

Protean clinical manifestations as well as complicated bacteriologic and serologic methods have doubtless underestimated the contribution of *Y. enterocolitica* to human infection. Occasional outbreaks have hinted that this is true by virtue of their size and/or severity.[4-7] Similarly, several reports of endemic disease, albeit retrospective in design, have included surprisingly large numbers of patients.[8-11] A recent serological survey of 15,968 blood samples in Copenhagen revealed a high ($\geq$ 1:40) *Y. enterocolitica* antibody titer in 1152 (7.2%) subjects.[12]

A prospective evaluation of 6364 children with diarrhea in Montreal yielded *Y. enterocolitica* in 181 (2.8%).[13] *Y. enterocolitica* was not isolated from 545 children free of gastrointestinal symptoms. Much of the following description of the clinical and epidemiologic characteristics of childhood gastroenteritis is derived from this study. Careful bacteriologic, serologic, and epidemiologic techniques were used to study the course and spread of *Y. enterocolitica* gastroenteritis in these children over a 15-month period. The study population represented mostly children seeking medical care at the two pediatric hospitals in Montreal, a city with a population of approximately 2½ million. The frequency of *Y. enterocolitica* gastroenteritis is only slightly less than that determined in the same study for *Salmonella* (4.4%) and *Campylobacter* (4.3%) and considerably greater than for *Shigella* (1.1%).[14] This confirms the impression of frequency previously reported from this city.[9]

B. Bacteriology

Most of *Y. enterocolitica* gastroenteritis in Canada,[15] Europe,[16,17] and Japan[18] is due to serotype O:3, biotype 4. In contrast, serotype O:9, biotype 2 is recovered almost exclusively in Scandinavian countries[5] and serotype O:8, biotype 1 in the U.S.[19] The 181

Y. enterocolitica isolates in our study consisted of 163 serotype O:3, biotype 4; the remaining 18 isolates belonged to other serotypes. We did not encounter any serotypes O:8 or O:9. Furthermore, all serotype O:3 isolates were of Canadian phage type, 9B, whereas those isolated in Europe are type 8 and in South Africa type 9A.[20]

Y. enterocolitica serotypes other than O:3, O:8, and O:9 are atypical in many of their laboratory characteristics,[21] and are widely distributed, without geographical limitation, in natural waters,[22] common domestic and wild animals,[23] and processed foods.[24] The ability of these bacteria to grow at refrigerator and room temperatures suggests that opportunities for infection abound. However, these atypical strains are only infrequently isolated from human infections[25] and, in contrast to typical clinical isolates (serotypes O:3, O:8, and O:9), fail to produce diarrhea in experimental infections.[26] Most of these isolates are also devoid of invasiveness in tissue culture.[27] Perhaps the slow growth of these bacteria at body temperature, lack of motility at 37°C, and other characteristics reduce their virulence. Although 18 of 181 isolates in our studies were serotypes other than O:3 (and O:8 and O:9), their role in gastroenteritis appeared to be minimal since we could not demonstrate serologic responses in infected hosts nor could we demonstrate communicability in household contacts. The majority of these isolates were recovered only after a prolonged period of cold enrichment in buffered saline and were excreted from stool for only a very short period of time, suggesting a transient colonization of the intestinal tract with a small number of bacteria. Pai et al.[28] have reported that heat-stable enterotoxin is produced by *Y. enterocolitica* from human and environmental isolates with atypical serotypes and biochemical characteristics. However, the in vivo role of these enterotoxin-producing atypical *Y. enterocolitica* is not clear at this time. Perhaps, specific surface antigens or colonization factors similar to those found in enterotoxigenic *Escherichia coli* may also be a prerequisite for strains causing gastroenteritis.

C. Seasonal Distribution

Although several retrospective reports had suggested a seasonal variation in the incidence of *Y. enterocolitica* gastroenteritis (more common in the cold months), our prospective study failed to confirm this impression (Figure 1). In fact, *Y. enterocolitica* gastroenteritis tended to be more common in the warm summer months than in winter in Montreal. Spread of illness is often attributed to food or water ingestion and/or contact with pigs or other animals. This, however, has rarely been documented. A recent outbreak of *Y. enterocolitica* infections in school children in New York was caused by *Y. enterocolitica* contamination of chocolate milk,[7] but most cases in other communities have been sporadic and common source exposures were not discovered.

III. CLINICAL MANIFESTATIONS

A. Gastroenteritis

The clinical and laboratory features of *Y. enterocolitica* infection of children may be extremely variable (Table 1). These manifestations bear some relationship to host factors including age and predisposing illness. For example, gastroenteritis is more common in children under 5 years of age, whereas abdominal pain syndromes predominate in school-age children. Children with thalassemia, malnutrition, and very young infants may be unduly susceptible to septicemia and abscess formation. The roles of iron excess and splenic dysfunction may be important in predisposing these children to infection as may be the adherent and invasive characteristics of the infecting strain of *Yersinia*.

Diarrhea, fever, vomiting, and abdominal pain characterize the clinical presentation of gastroenteritis. In our experience, most children with *Y. enterocolitica* gastroenteritis are under 5 years of age with the median age 24 months.[13] This is in contrast to the median

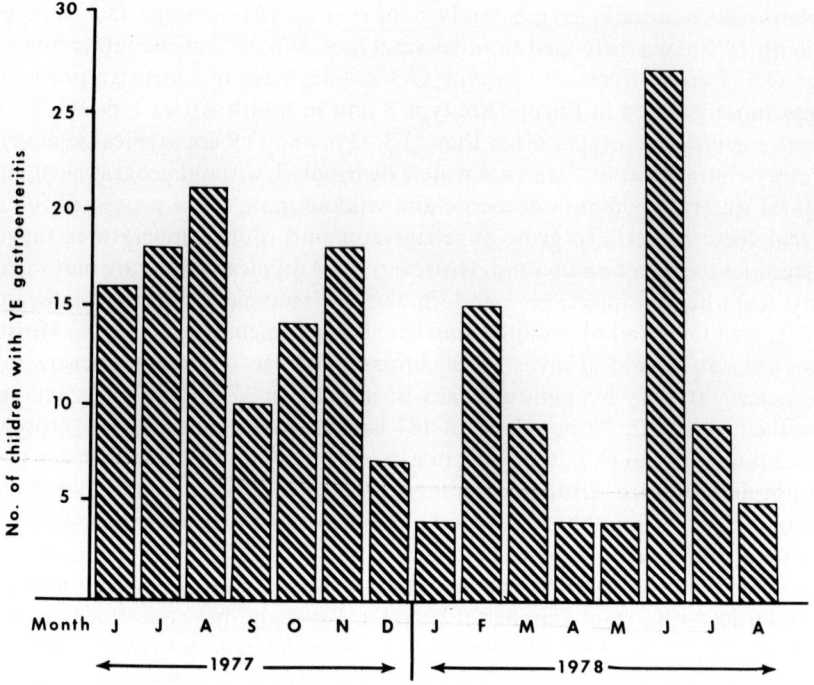

FIGURE 1. Seasonal distribution of *Y. enterocolitica* in Montreal children from July 1977 to August 1978.

Table 1
MANIFESTATIONS OF
Y. ENTEROCOLITICA INFECTIONS
IN CHILDREN

Gastroenteritis
Lymphadenitis
Appendicitis
Terminal ileitis
Septicemia
Abscess
Arthritis
Osteomyelitis
Exanthem
Convulsion

ages for *Salmonella* (30 months), *Shigella* (41 months), and *Campylobacter* (41 months).[14] No sex predilection is obvious in this condition. The clinical manifestations of illness are often pronounced in the first few days of illness, with fever, abdominal pain, and diarrhea most frequent. Vomiting is seen in approximately one third of the cases. Hospitalization of the child at this stage may be prompted by toxicity, abdominal pain, and, in approximately 25% of children, bloody diarrhea. Fever and vomiting usually disappear within 2 to 4 days and the child's appetite improves. Diarrhea is prolonged, however, with the average duration being 14 days and the range 1 to 46 days. Abdominal pain may last as long as 4 weeks, but usually dissipates by 10 days. Chronic diarrhea may be the presenting syndrome in infancy.[29] There are generally no complications of *Y. enterocolitica* gastroenteritis in otherwise healthy children.

B. Abdominal Pain

Abdominal pain is often the most striking feature of *Y. enterocolitica* infection.[30] In preschoolers this usually accompanies diarrhea; however in older children the pain may suggest the diagnosis of appendicitis and diarrhea is absent. Fever, diffuse or right lower quadrant abdominal tenderness, and occasionally vomiting may be present. Thus appendectomies were performed in 16 of 36 children hospitalized in a recent outbreak of *Y. enterocolitica* infection in New York (see Chapter 19). The source of the abdominal pain may be mesenteric lymphadenitis, suppurative appendicitis, terminal ileitis, or, rarely, periappendiceal abscess.[31,32] Soft tissue abscesses may also be associated with localized lymphadenitis. *Y. enterocolitica* was cultured from the inguinal lymph node in one such case, a 4-month-old infant with a labial abscess.[33]

C. Bacteremia

Bacteremia is apparently rare in normal children with *Y. enterocolitica* infection. We cultured the blood of 11 children with acute *Y. enterocolitica* gastroenteritis with negative results. As mentioned above, however, septicemia has been reported both in normal children and in those with predisposing illnesses or malnutrition.[32,34] It is possible that transient bacteremia may accompany gastroenteritis, as found in children with *Salmonella* gastroenteritis for example, but is rarely detected because of the timing of blood sampling. An outbreak of *Y. enterocolitica* infection was recently described with two fatalities.[4] A particularly virulent strain of *Y. enterocolitica* may be responsible in such situations. Localized abscesses of the skin, spleen, colon, peritoneum, and genitourinary and biliary tracts have also been reported. The pathogenesis of these lesions and the rare cases of osteomyelitis, lung abscess, etc. is unconfirmed but probably involves gastrointestinal colonization followed by hematogenous spread. Other complications such as erythema nodosum, Reiter's syndrome, migratory polyarthritis, myocarditis, etc. probably represent postinfectious immunologic tissue injury as may occur, for example, with circulating immune complexes. An exception to this might be arthritis, which may accompany the acute manifestations of *Y. enterocolitica* infection (with a marked polymorphonuclear reaction in the synovial fluid)[35] or present in the convalescent stage of illness.[10,36] In such circumstances, arthritis may represent bacteremia or an acute inflammatory response of the host. Manifestations, such as suppurative conjunctivitis, may be due to direct inoculation of the infected site with contaminated material derived from the patient, his contacts, or environmental sources.

D. Rash

Several types of rashes have been noted in children with *Y. enterocolitica* infection.[9,32,37] These include an evanescent macular erythematous form, morbilliform, target-like lesions, and erythema nodosum. These are uncommon in children with gastroenteritis, in our experience. Convulsions have been reported in young children with *Yersinia* gastroenteritis suggesting an initial clinical diagnosis of *Shigella* enteritis.[10,37] Although the pathogenesis of these seizures in yersiniosis in unknown and fever is the probable cause, meningitis should be ruled out in these cases.

IV. SEROLOGIC RESPONSE

Children with gastroenteritis due to *Y. enterocolitica* serotype O:3 demonstrate a serologic response to infection in most cases. Bacterial agglutination titers are the most

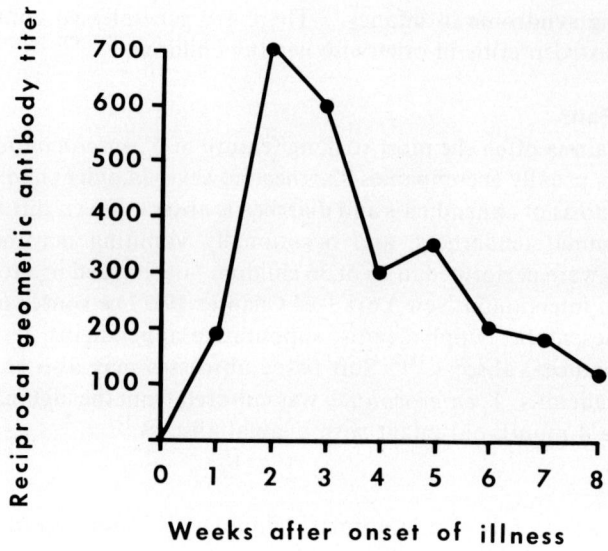

FIGURE 2. Serologic (bacterial agglutination) response of 67 children with *Y. enterocolitica* (serotype O:3) gastroenteritis.

commonly measured and are serotype specific. None of 252 sera from 177 children without gastrointestinal illness had serum antibody titers above 1:50, whereas all but two young patients (2 and 3 months of age) had titers of 1:200 or greater in association with *Y. enterocolitica* gastroenteritis.[13]

Antibody response is rapid, often occurring by the 5th to 10th day of illness, and frequently begins to decline by the 4th week of infection (Figure 2). Demonstration of a significant rise in antibody titer is often difficult in these children unless the first serum sample is obtained within the first 5 days of illness. Lacking this response, a single serum bacterial agglutination antibody titer of 1:200 or more should suggest the diagnosis of *Yersinia* infection in children.[13] The decline in antibody is surprising as most children continue to excrete *Y. enterocolitica* in their stools for an average of 27 days after their diarrhea subsides. Although some children may excrete bacteria for as long as 97 days from the onset of infection, carriage and excretion of *Y. enterocolitica* in the stools for more prolonged periods have not been documented. We have found a bacteriologic technique called cold enrichment to be of considerable use in documenting bacterial excretion of *Y. enterocolitica* in the stools of convalescent patients and in the diagnosis of asymptomatic carriage.[38]

V. COMMUNICABILITY

Familial outbreaks of *Y. enterocolitica* gastroenteritis and/or abdominal pain syndromes are not uncommon (Table 2). In one report, arthritis was the predominant syndrome.[35] Children seem to be at higher risk of infection and this may relate to poor hygienic practices, lack of acquired immunity, or other factors. Strains pathogenic for man have been isolated from pigs and epidemiologic features have suggested dogs may be occasional sources of human infection.[4,35] Both adults and children probably have close contact with environmental substances, such as water and food products, which may be contaminated with *Y. enterocolitica*. As discussed above, the characteristics of these strains are quite different from human isolates and rarely infect man. In our recently

Table 2
THE PREVALENCE OF *Y. ENTEROCOLITICA* INFECTIONS IN FAMILIES

Ref.	No. of families studied	No. of adult contacts infected[a]/No. tested	No. of children contacts infected[a]/No. tested
Ahvonen[35]	2	3/4	6/7
Szita[17]	4	3/13	6/11
Gutman[4]	4	3/4	6/17
Marks[13]	66	22/134	15/45
Total	76	31/155 (20%)	33/80 (41%)

[a] Bacteriologic and/or serologic evidence of infection with the same *Y. enterocolitica* serotype.

completed prospective study of *Y. enterocolitica* gastroenteritis of children, bacteriologic spread within the household of infected children was common and occurred in 30% of 66 families with 29% of 45 exposed children and 8.4% of 134 adult contacts becoming infected.[13] Spread was also suspected by serological evidence in an additional 18% of the families. Approximately one half of infected contacts had diarrhea. Person-to-person spread was suggested by the pattern, although not proven. The interval between infection of one family member and spread to other household members varied from a few days to 3 weeks. The spread of *Y. enterocolitica* within families of children ill with *Y. enterocolitica* gastroenteritis may be caused, in part, by the difficulty of observing strict hygienic procedures in young infants and children with copious diarrhea. It seems worthwhile to stress handwashing and to attempt some isolation procedures in view of the demonstrated frequency of spread.

Intrafamilial spread of *Y. enterocolitica* also offers insight into the relationship between the age of the host and the response to infection. Young school children and adolescents usually manifest abdominal pain and fever in contrast to the gastroenteritis described above in children and infants under 5 years of age. The abdominal pain may suggest the diagnosis of appendicitis.[30,39] The appendix is often normal in appearance, although it, the terminal ileum, and/or the mesenteric lymph nodes may be inflamed and infected with *Y. enterocolitica*. Elderly family contacts and those with debilitating conditions or underlying disease seem at increased risk for invasive infection due to *Y. enterocolitica* with resultant bacteremia, abscess formation, a spectrum of focal infections, and, occasionally, death.

Although *Y. enterocolitica* are highly susceptible in vitro to cotrimoxazole and aminoglycoside antibiotics and moderately sensitive to tetracycline, the role of antibiotic therapy in the treatment of the clinical illness and in the prevention of bacterial spread in childhood *Y. enterocolitica* gastroenteritis is unclear.[40] As in the case of *Salmonella* gastroenteritis, it is possible that such therapy will be of little benefit to normal children with uncomplicated gastroenteritis and may increase the duration of bacteriologic excretion. Studies of the pathogenesis of *Y. enterocolitica* infection indicate two possible mechanisms for inducing gastroenteritis, i.e., invasion of mucosal epithelial cells and the production of enterotoxin. The relative roles of these mechanisms of pathogenesis and the incidence of bacteremia in acute *Y. enterocolitica* gastroenteritis need to be determined.

We have begun to define a common cause of childhood gastroenteritis, enterocolitis, and mesenteric lymphadenitis with many characteristics in common with disease due to other members of the family *Enterobacteriaceae*. Increased awareness of the clinical syndromes and the bacteriologic diagnosis should assist in the recognition of these infections. Technological improvements in the serodiagnosis should also be helpful in this regard as well as for the elucidation of postinfectious complications of *Y. enterocolitica* infection.

VI. SUMMARY

Y. enterocolitica is a common cause of bacterial gastroenteritis in Montreal children. It is probably more widespread in North American communities than has been previously recognized. Diarrhea with abdominal pain and fever predominates in preschool children and febrile abdominal pain syndromes in those over 5 years of age. The diarrhea often lasts 2 weeks or longer and *Yersinia* is excreted in the stool of these patients well into convalescence. Abdominal pain may also be prolonged. Communicability in family units appears moderately high with approximately 30% of exposed household contacts becoming infected and one half of these developing symptoms. Biotype 4, serotype O:3 is the predominant pathogenic strain in Canada and stimulates a consistent host immune response in the form of bacterial agglutinating antibody usually in excess of 1/200 by the first week of illness. Although domestic and farm animals have been thought to be the source of infection in certain cases, the relationship of infection of these animals to human disease is unclear.

Other strains of *Yersinia* are occasionally isolated from children and their families in association with diarrheal syndromes. These bacteria are only transiently present in the stool, rarely evoke a serological response, and have not been demonstrated to spread within family units.

Treatment of *Yersinia* gastroenteritis consists of hygienic methods and elucidation and elimination of common source exposures, if they exist. Antibiotic therapy with cotrimoxazole or aminoglycoside antibiotics is indicated for disease outside the gastrointestinal tract. However, the role of drug therapy in *Y. enterocolitica* gastroenteritis is unknown. It may be appropriate to identify the infecting bacterial strain and patients at high risk for extragastrointestinal yersiniosis. These include very young infants, those with malnutrition, thalassemia, and other predisposing illnesses. It is predicted that *Y. enterocolitica* will be recognized with increased frequency in the forthcoming years as a cause of gastroenteritis in children and adults in North America.

REFERENCES

1. **Schleifstein, J. I. and Coleman, M. B.,** An unidentified microorganism resembling *B. lignieri* and *Past. pseudotuberculosis*, and pathogenic for man, *N. Y. State J. of Med.*, 39, 1749, 1939.
2. **Moeland, J. A. and Digranes, A.,** Common enterobacterial antigens in *Yersinia enterocolitica*, *Acta Pathol. Microbiol. Scand. Sect. B*, 83, 382, 1975.
3. **Ahvonen, P., Jansson, E., and Aho, K.,** Marked cross-agglutination between *Brucella* and a sub-type of *Yersinia enterocolitica*, *Acta Pathol. Microbiol. Scand.*, 75, 291, 1969.
4. **Gutman, L. T., Ottesen, E. A., Quan, T. J., Noce, P. S., and Katz, S. L.,** An interfamilial outbreak of *Yersinia enterocolitica* enteritis, *N. Engl. J. Med.*, 288, 1372, 1973.
5. **Toivanen, P., Olkkonen, L., Toivanen, A., and Aantaa, S.,** Hospital outbreak of *Yersinia enterocolitica* infection, *Lancet*, 1, 801, 1973.
6. **Zen-Yoji, H., Maruyama, T., Sakei, S., Mizuno, T., and Momose, T.,** An outbreak of enteritis due to *Yersinia enterocolitica* occurring at a junior high school, *Jpn. J. Microbiol.*, 17, 220, 1973.
7. **Black, R. E., Jackson, R. J., Tsai, T., Medvesky, M., Shayegani, M., Feeley, J. C., MacLeod, K. I. E., and Wakelee, A. M.,** Epidemic *Yersinia enterocolitica* infection due to contaminated chocolate milk, *N. Engl. J. Med.*, 298, 74, 1978.

8. **Winblad, S.**, The clinical panorama of human *Yersiniosis enterocolitica*, in *Contributions to Microbiology and Immunology*, Vol. 2, Winblad, S., Ed., S. Karger, Basel, 1973, 129.
9. **Delorme, J., Laverdiere, M., Martineau, B., and Lafleur, L.**, Yersiniosis in children, *Can. Med. Assoc. J.*, 110, 281, 1974.
10. **Randall, C. and Bannatyne, R. M.**, Experience with *Yersinia enterocolitica* at The Hospital for Sick Children, 1972-74, *Can. Med. Assoc. J.*, 113, 542, 1975.
11. **Lafleur, L., Hammerberg, O., Delage, G., and Pai, C. H.**, *Yersinia enterocolitica* infections in children: 4 years experience in the Montreal urban community, in *3rd Int. Symp. Yersinia*, S. Karger, Basel, in press.
12. **Larsen, J. H.**, Yersiniosens kliniske betydning, *Videnskab Praksis*, 3, 565, 1975.
13. **Marks, M. I., Pai, C. H., Lafleur, L., Lackman, L., and Hammerberg, O.**, *Yersinia enterocolitica* gastroenteritis. A prospective study of clinical, bacteriologic, and epidemiologic features, *J. Pediatr.*, 96, 26, 1980.
14. **Pai, C. H., Sorger, S., Lackman, L., Sinai, R., and Marks, M. I.**, *Campylobacter gastroenteritis* in children, *J. Pediatr.*, 94, 589, 1979.
15. **Toma, S. and Lafleur, L.**, Survey on the incidence of *Yersinia enterocolitica* infection in Canada, *Appl. Microbiol.*, 28, 469, 1974.
16. **Vandepitte, J., Wauters, G., and Isebart, A.**, Epidemilogy of *Yersinia enterocolitica* infections in Belgium, in *Contributions to Microbiology and Immunology*, Vol. 2, Winblad, S., Ed., S. Karger, Basel, 1973, 111.
17. **Szita, J., Kali, M., and Redey, B.**, Incidence of *Yersinia enterocolitica* infection in Hungary, in *Contributions to Microbiology and Immunology*, Vol. 2, Winblad, S., Ed., S. Karger, Basel, 1973, 106.
18. **Zen-Yoji, H. and Maruyama, T.**, The first successful isolations and identification of *Yersinia enterocolitica* from human cases in Japan, *Jpn. J. Microbiol.*, 16, 493, 1972.
19. **Bissett, M. L.**, *Yersinia enterocolitica* isolates from humans in California, 1968—1975, *J. Clin. Microbiol.*, 4, 137, 1976.
20. **Nicolle, P., Mollaret, H. H., and Brault, J.**, Recherches sur la lysogenie, la lysocytie et la serologie de *Yersinia enterocolitica*, in *Contributions to Microbiology and Immunology*, Vol. 2, Winblad, S., Ed., S. Karger, Basel, 1973, 54.
21. **Chester, B., Stotzky, G., Bottone, E. J., Malowany, M. S., and Allerhand, J.**, *Yersinia enterocolitica*: biochemical, serological and gas-liquid chromatographic characterization of rhamnose-, raffinose-, melibiose-, and citrate-utilizing strains, *J. Clin. Microbiol.*, 6, 461, 1977.
22. **Highsmith, A. K., Feeley, J. C., Shaliy, P., Wells, J. G., and Wood, B. T.**, Isolation of *Yersinia enterocolitica* from well water and growth in distilled water, *Appl. Environ. Microbiol.*, 34, 745, 1977.
23. **Zen-Yoji, H., Sakai, S., Maruyama, T., and Yanagawa, Y.**, Isolation of *Yersinia enterocolitica* and *Yersinia pseudotuberculosis* from swine, cattle and rats at an abattoir, *Jpn. J. Microbiol.*, 18(1), 103, 1974.
24. **Schiemann, D. A.**, Association of *Yersinia enterocolitica* with the manufacture of cheese and occurrence in pasteurized milk, *Appl. Environ. Microbiol.*, 36, 274, 1978.
25. **Bottone, E. J.**, Atypical *Yersinia enterocolitica*: clinical and epidemiological parameters, *J. Clin. Microbiol.*, 7, 562, 1978.
26. **Une, T.**, Studies on the pathogenicity of *Yersinia enterocolitica*. I. Experimental infection in rabbits, *Microbiol. Immun. (Japan)*, 21, 349, 1977.
27. **Une, T., Zen-Yoji, H., Maruyama, T., and Yanagawa, Y.**, Correlation between epithelial cell infectivity *in vivo* and O-antigen groups of *Yersinia enterocolitica*, *Microbiol. Immun. (Japan)*, 21, 727, 1977.
28. **Pai, C. H., Mors, V., and Toma, S.**, Prevalence of enterotoxigenicity in human and non-human isolates of *Yersinia enterocolitica*, *Infect. Immun.*, 22, 334, 1978.
29. **Toma, S., Lior, H., Quinn-Hill, M., Sher, N., and Walker, W. A.**, *Yersinia enterocolitica* infection: report of two cases, *Can. J. Public Health*, 63, 433, 1972.
30. **Jepsen, O. B., Korner, B., Lauritsen, K. B., Hancke, A. B., Andersen, L., Henrichsen, S., Brenoe, E., Christiansen, P. M., and Johansen, A.**, *Yersinia enterocolitica* infection in patients with acute surgical abdominal disease, *Scand. J. Infect. Dis.*, 8, 189, 1976.
31. **Doraiswamy, N. V., Currie, A. B. M., Gray, J., and Lynton-Moll, C.**, Terminal ileitis: *Yersinia enterocolitica* isolated from faeces, *Br. Med. J.*, 2, 23, 1977.
32. **Butzler, J. P., Alexander, M., Segers, A., Cremer, N., and Blum, D.**, Enteritis, abscess, and septicemia due to *Yersinia enterocolitica* in a child with thalassemia, *J. Pediatr.*, 93, 619, 1978.
33. **Wilson, H. D., McCormick, J. B., and Feeley, J. C.**, *Yersinia enterocolitica* infection in a 4-month-old infant associated with infection in household dogs, *J. Pediatr.*, 89, 767, 1976.
34. **Caplan, L. M., Dobson, M. L., and Dorkin, H.**, *Yersinia enterocolitica* septicemia, *Am. J. Clin. Pathol.*, 69, 189, 1978.

35. **Ahvonen, P. and Rossi, T.**, Familial occurrence of *Yersinia enterocolitica* infection and acute arthritis, *Acta Paediatr. Scand. Suppl.*, 206, 121, 1970.
36. **Jacobs, J. C.**, *Yersinia enterocolitica* arthritis, *Pediatrics*, 55, 236, 1975.
37. **Kohl, S., Jacobson, J. A., and Nahmias, A.**, *Yersinia enterocolitica* infections in children, *J. Pediatr.*, 89, 77, 1976.
38. **Sorger, S., Pai, C. H., Lafleur, L., Lackman, L., Hammerberg, O., and Marks, M. I.**, Is cold enrichment necessary for isolation of *Yersinia enterocolitica* (YE) from stool in a clinical laboratory?, in Proc. 18th Interscience Conf. Antimicrobial Agents and Chemotherapy, Atlanta, October 2, 1978.
39. **Mayer, L. and Greenstein, A. J.**, Acute yersinial ileitis: a distinct entity, *Am. J. Gastroenterol.*, 65, 548, 1976.
40. **Hammerberg, S., Sorger, S., and Marks, M. I.**, Antimicrobial susceptibilities of *Yersinia enterocolitica* biotype 4, serotype O:3, *Antimicrob. Agents Chemother.*, 11, 566, 1977.

Chapter 10

YERSINIA ENTERITIS AND CROHN'S DISEASE

Jacob S. Walfish and David B. Sachar

TABLE OF CONTENTS

I.	Introduction	106
II.	Acute Terminal Ileitis — A Historical Perspective	106
III.	Differential Diagnosis	108
	A. Clinical Features	108
	B. Radiology and Endoscopy	108
	C. Pathology	109
IV.	Pathophysiology	110
V.	Conclusion	110
References		111

I. INTRODUCTION

Although there is no evidence for an etiologic link between Yersinia infection and the idiopathic chronic inflammatory bowel disease known as Crohn's disease, a comparison of the two disorders is worthwhile from at least two standpoints. First, both entities must be considered in the differential diagnosis of the syndrome of "acute terminal ileitis". Second, clarification of the pathophysiologic mechanisms involved in Yersinia enteritis, for which there exist good experimental models, may provide helpful insights into the pathogenesis of Crohn's disease.

II. ACUTE TERMINAL ILEITIS — A HISTORICAL PERSPECTIVE

In 1932 Crohn, Ginzburg, and Oppenheimer separated out from a heterogeneous group of nonspecific granulomatous diseases of the bowel, a distinct entity which they called regional ileitis.[1] This disorder was characterized by mucosal and transmural inflammation of the terminal ileum, often culminating in chronic cicatrization, fistulization, and obstruction. Their landmark paper included a subgroup of patients with the acute onset of symptoms mimicking appendicitis, who at surgery were found to have an inflamed and edematous terminal ileum. Although the pathologic specimens reviewed were from patients who had been ill at least a year, it was the authors' opinion that the subgroup with acute illness, in whom resection was not performed, represented patients in the very early phase of the disease. They speculated that some cases of acute ileitis spontaneously recovered while other developed the chronic form of the disease.

Over the years, Crohn advanced the view that regional ileitis could arise acutely without antecedent symptoms, and in such cases carried a better prognosis than the chronic form of the disease. In Crohn's follow-up study of 33 such patients with "acute regional ileitis", approximately half showed spontaneous clinical and radiologic resolution while the other half went on to develop chronic regional ileitis.[2,3] Many other authors accepted this concept that Crohn's disease could appear in either an acute or chronic form, with a high incidence of spontaneous resolution in the acute variety. Consequently, there was general agreement that resection should not be performed in acute cases.[4-6]

While the several varieties of *chronic* regional enteritis are now recognized as forms of the clinical entity described by Crohn, Ginzburg, and Oppenheimer, the classification of "*acute* terminal ileitis" has been a debated and often confusing subject.* In 1967 Gump et al.[7] noted that reports in the literature concerning the frequency of complete resolution of acute terminal ileitis varied from 25 to 100%. They concluded that terminal ileitis was a nonspecific condition that could result from a variety of causes, including tuberculosis, allergy, and mesenteric lymphadenitis, as well as Crohn's disease. Furthermore, they criticized the lack of rigid criteria for defining acute illness, noting that some cases with a more prolonged preoperative course has been included inappropriately among the "acute" cases, thus beclouding the prognosis of truly acute disease. On the basis of their own series of acute terminal ileitis in which 23 out of 24 cases recovered completely, and

---

* In this discussion, the designation *terminal ileitis* will refer to the purely descriptive and nonspecific diagnosis made at exploration, usually for suspected appendicitis, when a hyperemic and edematous terminal ileum is found. *Regional enteritis* denotes the idiopathic clinical and pathologic entity described by Crohn, Ginzburg, and Oppenheimer and known as Crohn's disease. (Crohn had initially proposed the term terminal ileitis for the entity he described but he changed the name to regional ileitis on the advice of Dr. Bargen, who felt that terminal conveyed the meaning of agonal. A subsequent modification to regional enteritis was made because the disease was not invariably confined to the ileum.)

review of the literature of genuinely acute cases with a brief appendicitis-like history, Gump and his associates concluded that acute terminal ileitis arising in previously healthy patients rarely developed into a chronic process and was therefore unrelated to Crohn's disease.

Other authors also challenged the concept of "acute Crohn's disease"[8-10] but, like Gump et al., they could offer no alternative etiologies for their cases of acute terminal ileitis. In the absence of any specific etiologic markers, it remained impossible to prove that acute ileitis was a separate entity, and not simply a case of acute Crohn's disease undergoing spontaneous resolution. In the mid-1960s, however, a number of Scandinavian investigators made the connection between acute terminal ileitis and *Y. enterocolitica* and *Y. pseudotuberculosis* infection. It was not until the recognition of Yersinia infection as a major specific cause of acute terminal ileitis that one could finally make clear-cut distinctions between this entity and Crohn's disease.

It should be noted that even before the recent interest in Yersinia ileitis, some authors did implicate specific etiologies, other than Crohn's disease, in the pathogenesis of acute terminal ileitis. In 1935 Felsen[11] called attention to the appendicular form of bacillary dysentery, in which a number of patients who underwent surgery for suspected appendicitis were found to have a normal appendix in association with mesenteric adenitis and terminal ileitis. These patients manifested bacteriologic or serologic evidence of *Shigella* infection, which Felsen believed to be the cause of their syndrome. His further contention, however, that dysentery-producing organisms had a central role in the pathogenesis of chronic ileitis and colitis,[12] was not borne out by later investigations.[2]

Austin[13] speculated that certain cases of terminal ileitis were secondary to mesenteric lymphadenitis, a self-limited process whose cause is in itself unclear but which has been associated with various pyogenic and viral etiologies.[14] Isolated cases of acute terminal ileitis have also been attributed to such diverse causes as allergy, vasculitis, anisikiasis, and tuberculosis,[7,15] but none of these specific etiologies has been found responsible for large series of patients with acute terminal ileitis.

Prompted by earlier reports implicating Yersinia organisms in acute terminal ileitis and mesenteric adenitis,[16,17] Winblad et al.[18] and later Niléhn and Sjöström[19] retrospectively studied the incidence of Yersinia infection in a group of surgical patients. Of 581 patients who underwent surgery for suspected appendicitis, 22 had positive cultures for *Y. enterocolitica*. Among these 22 patients, the highest correlation with Yersinia infection was in terminal ileitis (8 out of 10), followed by a lesser incidence in mesenteric adenitis (9 out of 69). In contrast, only 2 of 337 patients with appendicitis and only 1 of 974 controls had positive Yersinia cultures. The prevalence of Yersinia infection in patients with terminal ileitis clearly pointed to an etiologic relationship which has since been amply corroborated.[20-22]

By demonstrating the absence of significant Yersinia titers in a group of 22 patients with typical chronic regional enteritis, the Swedish investigators further showed that Yersiniosis was a distinct entity unrelated to Crohn's disease. Moreover, among 29 cases of surgically documented acute terminal ileitis, 21 patients with bacteriologic or serologic evidence of *Y. enterocolitica* infection all recovered completely, with no clinical or radiologic sequelae noted on follow-up extending to 5 years.[23] Five of the remaining eight non-Yersinia or "idiopathic" cases were, on microscopic examination, later proven to be Crohn's disease with complications. Of the three remaining unknown cases, one developed fistulas and diarrhea despite normal small bowel X-rays, one was attributed to a foreign body perforation, and one did not have a serologic determination. A more recent study, also from Sweden, of 52 patients with acute, subacute, and chronic ileitis yielded similar results, with complete resolution of disease occurring predominantly among Yersinia-positive patients while disease progression occurred in Yersinia-negative patients.[24] (Four Yersinia-negative patients also seemed to recover completely, but whether these were cases of resolved Crohn's disease or yet another unidentified entity remains undetermined.)

These studies indicated that there are at least two distinct diseases presenting with the clinical picture of acute terminal ileitis: (1) Yersinia-related cases, which resolve completely and (2) non-Yersinia cases, which are likely to represent an early stage in the development of chronic regional enteritis (Crohn's disease). A case of acute terminal ileitis encountered at laparotomy therefore poses a diagnostic problem of both prognostic and therapeutic importance.

III. DIFFERENTIAL DIAGNOSIS

A. Clinical Features

In their classic presentations, there is little ground for confusion between Crohn's disease and Yersinia enteritis. Typically, the former is associated with chronic intestinal obstruction, fistulization, and sometimes in advanced cases with malabsorption, malnutrition, and debility. Yersinia enteritis, on the other hand, more characteristically appears as a brief, self-limited diarrheal illness, occasionally as part of an infectious outbreak.[25,26] Nonetheless, problems of differential diagnosis may arise, as we have seen, in cases of surgically discovered acute terminal ileitis, and even in nonsurgical cases of Yersinia enteritis when the clinical course is prolonged.[27-29] Both diseases may manifest abdominal pain, fever, diarrhea with leukocytes in the stool, peripheral arthritis, ankylosing spondylitis, uveitis, erythema nodosum, leukocytosis, and elevated sedimentation rate.[20,29-31] The peripheral arthritis is of similar distribution in Yersinia enteritis and Crohn's disease — knees, ankles, and hands — and in both diseases ankylosing spondylitis also occurs, especially in association with the HLA B27 antigen. It is noteworthy, however, that the peripheral arthritis in Crohn's disease, which is not B27-associated, tends to parallel the activity of the underlying bowel disease,[32] whereas peripheral arthritis in Yersinia enteritis (like spondylitis in both diseases) is B27-associated and, also like spondylitis, follows a chronic course in the absence of active bowel disease.[33,34]

There are a number of other differential points between Crohn's disease and Yersinia enteritis. Pharyngitis and cervical adenopathy, which may occur in Yersinia infection,[25,26] are not found with regional enteritis. Conversely, the findings of pyoderma gangrenosum and perianal fistulas or abscesses should suggest Crohn's disease and have not been reported in Yersinia infection. Likewise, the presence of a right lower quadrant abdominal mass, so highly characteristic of Crohn's disease, has been reported only rarely with Yersinia ileitis.[21]

B. Radiology and Endoscopy

Whereas X-ray findings have always been the principal means of diagnosis for Crohn's disease, the radiologic changes in Yersinia enteritis have been described only recently. Even Gump et al., who were among the first to perceive the distinction between chronic regional enteritis and acute terminal ileitis, mistakenly believed that radiologic changes were associated only with the chronic process of Crohn's disease.[7] Two European studies have recently demonstrated that the great majority of patients with Yersinia enteritis will initially manifest radiologic abnormalities in the terminal ileum which may mimic those seen in Crohn's disease. In Vantrappen's series,[25] 21 of 24 patients with medically treated Yersinia enteritis showed radiologic changes in the distal 10 to 20 cm of terminal ileum. The most prominent findings were nodularity, coarseness, or irregularity of the mucosa;

frequently, luminal dilation and mucosal ulceration were also noted. These abnormalities persisted for several months. Similar findings, with the notable exception of ulceration, were reported by Ekberg[35] in a detailed study of 25 patients with acute ileitis who underwent surgery. Comparing the radiologic appearances of Yersinia enteritis and Crohn's disease, Ekberg observed that the greatest resemblance occurs during the first or "nodular stage", which lasts up to 3 weeks, and that within 10 weeks there is usually complete resolution of radiologic abnormalities in yersinosis. In contrast, complete radiologic resolution is rarely, if ever, seen in Crohn's disease.[36]

There are a number of other X-ray features by which Crohn's disease and Yersinia enteritis can be differentiated. Fistulas and strictures, occurring frequently in Crohn's disease, are not seen in Yersinia ileitis. Also, Yersinia infection does not often produce changes in more proximal segments of the bowel.[35]

Some endoscopic features may be shared by both diseases. Among 13 patients with Yersinia enteritis studied endoscopically,[25] colitis was seen in six and aphthous ulcers, the probable "earliest lesion of Crohn's disease",[37] in two. The classic "cobblestone" pattern of Crohn's colitis, however, has not been described in Yersinia infection, either endoscopically or radiologically.

C. Pathology

The rigid, strictured "hose-pipe" terminal ileum characteristic of advanced regional enteritis would scarcely be mistaken for Yersinia enteritis, but this late finding is not often encountered in patients with an acute presentation and no prior symptomatology. In these latter cases of acute terminal ileitis, gross pathology as visualized at laparotomy is of little help in differentiating between Yersinia enteritis and Crohn's disease. In Sjöström's series, for example, the gross appearance of the inflamed terminal ileum gave no hint as to the duration or underlying cause of the process.[23]

Associated mesenteric lymphadenopathy is a nonspecific finding in both Crohn's disease and Yersinia ileitis, but its frequency and probably its pathophysiologic significance are different in the two conditions. In acute Yersinia infection, mesenteric adenitis is often seen in the absence of ileitis, whereas the converse situation does not often occur; Yersinia-induced ileitis is seldom found without concomitant inflammation in the adjacent mesenteric nodes.[18,21,38,39] Thus, in concordance with Austin, who had postulated earlier that some forms of terminal ileitis may occur as a secondary consequence of a primary mesenteric adenitis,[13] Mollaret considered Yersinia adenitis and ileitis to be part and parcel of the same process, and not two distinct entities as others had believed.[40] In Crohn's disease, on the other hand, it is the bowel which is primarily affected with or without accompanying lymphatic involvement.[41] In contrast to Yersiniosis, Crohn's disease rarely presents as pure mesenteric lymphadenitis. Furthermore, if granulomas are found in the intestinal wall in Crohn's disease, they may also occur in the draining lymph nodes in about 25% of cases, but granulomas are not found in lymph nodes unless they are also present in the bowel.[42] Thus, lymph node involvement seems to be primary in Yersinia infection, and intestinal involvement secondary, whereas the reverse appears to be true in Crohn's disease.

Other gross findings which may be helpful in making a diagnosis of Crohn's disease as opposed to Yersinia ileitis include: the creeping of serosal fat to the antimesenteric border of the intestine, fissures, fistulas, perforation of the bowel with abscess formation, and segmental involvement of other part of the bowel.

With regard to intestinal histopathology in acute Yersinia ileitis, little is known as resection or biopsy is usually not performed. In a study of a number of cases of fulminant Yersinia infection, Bradford reported that the primary lesion was a transmural neutrophilic and mononuclear process with necrosis and ulceration of the bowel.[43] No granulomas or giant cells have been seen in these limited studies of intestinal specimens in Yersinia ileitis, nor in more extensive examinations of ileocecal lymph nodes. On the other hand, microabscesses are a common finding in the mesenteric adenopathy associated with acute Yersinia infection.[44] A characteristic but not pathognomonic lesion consisting of a large pyroninophilic cell reaction has also been described in Yersiniosis by Ahlquist et al.[45]

In the last analysis, the diagnosis of Crohn's disease depends upon a constellation of clinical, radiologic, and pathologic findings, whereas a diagnosis of Yersinia enteritis can be definitively established by cultures and serology as described elsewhere in this volume.

IV. PATHOPHYSIOLOGY

Identification of Yersinia as the cause of acute terminal ileitis has provided students of Yersiniosis with a tool for pathophysiologic investigation not available to workers in the field of Crohn's disease. On the other hand, the recently developed experimental models of Yersinia infection might provide some insights into the pathogenesis of Crohn's disease. The rabbit model of Une, for example, has shown that granulomas can develop acutely in lymphoid follicles after penetration of reticuloendothelial cells by the Yersinia organism.[46] Although granulomas are not a characteristic feature of human Yersiniosis, this model, nevertheless, does demonstrate that granuloma formation can be induced by acute bacterial infection of the intestine. Perhaps analogous is the demonstration by Orr that a granulomatous enteritis can develop in the rabbit upon experimental infection with L-form bacteria.[47]

The mouse model of Carter,[48] like Bradford's studies of human Yersiniosis,[43] demonstrated that ulceration occurs principally at the site of intestinal lymphoid follicles. Pathologic studies of the "aphthoid ulcer" of Crohn's disease suggest a similar pattern of ulceration. Light and scanning electron microscopic examinations have demonstrated that the typical aphthoid ulcer of Crohn's disease is usually closely associated with an underlying hyperplastic lymphoid nodule.[37,49] It is thus tempting to hypothesize, in analogy to the Yersinia model, that the "aphthoid ulcer" represents the site of entry of an offending agent which elicits a primary lymphoid reaction and additional secondary epithelial damage. Regardless of the exact mechanism, however, the close relationship of the mucosal and lymphoid lesions may explain in part the predilection of both Yersinia and Crohn's enteritis for the terminal ileum, which is especially well-endowed with the lymphoid tissue that appears to play an important role in both processes.

With regard to the extra-intestinal manifestations of Yersinia infection, such as arthritis, erythema nodosum, and uveitis, local infection by the organism does not appear to play a role.[20] Since similar extra-intestinal phenomena occur in Crohn's disease, one may speculate whether there is mediation through a common final pathway such as antigen-antibody complexes related to intestinal injury.[50] Different immunologic and genetic mechanisms may play a role in the more chronic, HLA-B27-related arthritides which, as noted earlier, may be independent of active intestinal disease. For the present, however, mechanisms of extraintestinal manifestations of both diseases remain obscure.

V. CONCLUSION

The recent delineation of Yersinia enteritis as a distinct clinical entity reminds us that the bowel may have a limited repertoire of responses to various pathogenic stimuli. From the recognition that acute terminal ileitis may result from various diverse causes, we may wonder whether all cases of chronic granulomatous ileocolitis, collectively classified as

Crohn's disease, may not also represent a collection of various pathologic entities, or perhaps, as in the case of Yersinia infection, Crohn's disease is caused by one agent with strains of differing pathogenicity giving rise to various clinical pictures. Moreover, the role of host defenses in determining different clinical responses has yet to be fully explored. In any event, the identification of Yersinia ileitis should strengthen our resolve to separate out other specific etiologic entities from the constellation of disorders that we now lump together under the broad rubric of inflammatory bowel disease.

REFERENCES

1. **Crohn, B. B., Ginzburg, L. and Oppenheimer, G. D.**, Regional ileitis: a pathologic and clinical entity, *JAMA*, 99, 1323, 1932.
2. **Crohn, B. B. and Yarnis, H.**, *Regional Ileitis*, 2nd ed., Grune & Stratton, New York, 1958, 148.
3. **Crohn, B. B.**, The pathology of acute regional ileitis, *Am. J. Dig. Dis.*, 10, 565, 1965.
4. **Sneirson, H. and Ryan, J.**, Regional ileitis, *Am. J. Surg.*, 52, 424, 1941.
5. **Eckel, J. H. and Ogilivie, J. B.**, Regional enteritis, *Am. J. Surg.*, 53, 345, 1941.
6. **Smith, H. G.**, Conservatism in the surgical management of acute regional enteritis, *Surgery*, 13, 122, 1943.
7. **Gump, M. D., Lepore, M., and Barker, H. G.**, A revised concept of acute regional enteritis, *Ann. Surg.*, 166, 942, 1967.
8. **Armitage, G. and Wilson, M.**, Crohn's disease, *Br. J. Surg.*, 38, 182, 1950.
9. **Storrs, R. C. and Hoekelman, R. A.**, Acute regional enteritis in children, *N. Engl. J. Med.*, 248, 320, 1953.
10. **Thomasson, B. and Havia, T.**, Is acute terminal ileitis a precursor of Crohn's disease?, *Acta Chir. Scand.*, 139, 192, 1973.
11. **Felsen, J.**, Appendicular form of bacillary dysentery, with notes on mesenteric adenitis and inflammation of distal portion of ileum, *Am. J. Dis. Child.*, 50, 661, 1935.
12. **Felsen, J.**, The relationship of bacillary dysentery to distal ileitis, chronic ulcerative colitis, and nonspecific intestinal granuloma, *Ann. Int. Med.*, 10, 645, 1936.
13. **Austin, W. E.**, Acute regional ileitis, *Can. Med. J.*, 74, 289, 1956.
14. **Blattner, R. J.**, Acute mesenteric lymphadenitis, *J. Pediatr.*, 74, 479, 1969.
15. **Baerloher, C.**, *Regional Enteritis*, Skandia Int. Sym., Nordiska Bokhandelns Forlag, Stockholm, 1971, 77.
16. **Knapp, W.**, *Pasteurella pseudotuberculosis* als erreger einer mesenterialen lymphadenitis beimmenschen, *Zentralbl. Bakteriol.*, 161, 422, 1954.
17. **Carlsson, M. G., Ryd, H., and Sternby, N. H.**, A case of human infection with *Pasteurella pseudotuberculosis* X, *Acta Pathol. Microbiol. Scand.*, 62, 128, 1964.
18. **Winblad, S., Niléhn, B., and Sternby, H. G.**, *Yersinia enterocolitica (Pasteurella X)* in human enteric infections, *Br. Med. J.*, 2, 1363, 1966.
19. **Niléhn, B. and Sjöström, B.**, Studies on *Yersinia enterocolitica*: occurrence in various groups of acute abdominal disease, *Acta Pathol. Microbiol. Scand.*, 71, 612, 1967.
20. **Ahvonen, P.**, Human Yersiniosis in Finland. II. Clinical features, *Ann. Clin. Res.*, 4, 39, 1972.
21. **Gurry, J. F.**, Acute terminal ileitis and Yersinia infection, *Br. Med. J.*, 4, 264, 1974.
22. **Mayer, M. and Greenstein, A. J.**, Acute Yersinia ileitis: a distinct entity, *Am. J. Gastroenterol.*, 65, 548, 1976.
23. **Sjöström, B.**, Acute terminal ileitis and its relation to Crohn's disease in *Regional Enteritis*, Skandia Int. Sym., Nordiska Bokhandelns Forlag, Stockholm, 1971, 73.
24. **Persson, S., Danielsson, D., Kjellander, J., and Wallensten, S.**, Studies on Crohn's disease. I. The relationship between *Yersinia enterocolitica* infection and terminal ileitis, *Acta Chir. Scand.*, 142, 84, 1976.
25. **Vantrappen, G., Agg, H. O., Ponette, E., Geboes, K., and Bertrand, P. H.**, Yersinia enteritis and enterocolitis: gastroenterological aspects, *Gastroenterology*, 72, 220, 1977.
26. **Gutman, L. T., Ottesen, E. A., Quan, T. J., Nore, P. S., and Katz, S. L.**, An inter-familial outbreak of *Yersinia enterocolitica* enteritis, *N. Engl. J. Med.*, 288, 1372, 1973.
27. **Hinderaker, S., Liavaag, I., and Lassen, J.**, *Yersinia enterocolitica* infection, *Lancet*, 2, 322, 1973.
28. **Delorme, J., Laverdiére, M., Martineau, B., and Lafleur, L.**, Yersiniosis in children, *CMA J.*, 110, 281, 1974.
29. **Leino, R. and Kalliomaki, J. L.**, Yersiniosis as an internal disease, *Ann. Int. Med.*, 81, 458, 1974.

30. **Ahvonen, P., Sievers, K., and Aho, K.,** Arthritis associated with *Yersinia enterocolitica* infection, *Acta Rheum. Scand.,* 15, 232, 1969.
31. **Greenstein, A. J., Janowitz, H. D., and Sachar, D. B.,** The extraintestinal complications of Crohn's disease and ulcerative colitis: a study of 700 patients, *Medicine,* 55, 401, 1976.
32. **Haslock, I. and Wright, V.,** The musculo-skeletal complications of Crohn's disease, *Medicine,* 52, 217, 1973.
33. **Laitenen, O., Leirisalo, M., and Skylv, G.,** Relation between HLA-B27 and clinical features in patients with Yersinia arthritis, *Arthritis Rheum.,* 20, 1121, 1977.
34. **Winblad, S.,** Arthritis associated with *Yersinia enterocolitica* infections, *Scand. J. Infect. Dis.,* 7, 191, 1975.
35. **Ekberg, O., Sjöström, B., and Brahme, F.,** Radiological findings in Yersinia ileitis, *Radiology,* 123, 15, 1977.
36. **Stiegman, F.,** Can regional enteritis resolve completely?, *Am. J. Gastroenterol.,* 63, 464, 1975.
37. **Morson, B. C.,** The early histological lesion of Crohn's disease, *Proc. R. Soc. Med.,* 65, 71, 1972.
38. **Black, R. E., Jackson, R. J., Tsai, T., Medvesky, M., Shayegani, M., Feeley, J. C., MacLeod, K. I. E., and Wakelee, A. M.,** Epidemic *Yersinia enterocolitica* infection due to contaminated chocolate milk, *N. Engl. J. Med.,* 298, 76, 1978.
39. **Rodgers, B. and Gordon, K.,** Yersinia enterocolitis, *J. Pediatr. Surg.,* 10, 497, 1975.
40. **Mollaret, H. H.,** *Yersinia enterocolitica* infection: a new problem in pathology, *Ann. Biol. Clin.,* 30, 1, 1972.
41. **Morson, B. C. and Dawson, I. M. P.,** *Gastrointestinal Pathology,* Blackwell Scientific, Oxford, 1972, 265.
42. **Morson, B. C. and Dawson, I. M. P.,** *Gastrointestinal Pathology,* Blackwell Scientific, Oxford, 1975, 270.
43. **Bradford, W. D., Noce, P. S., and Gutman, L. T.,** Pathologic features of enteric infection with *Yersinia enterocolitica, Arch. Pathol.,* 98, 17, 1974.
44. **Morson, B. C. and Dawson, I. M. P.,** *Gastrointestinal Pathology*, Blackwell Scientific, Oxford, 1975, 265.
45. **Ahlquist, J., Ahvonen, P., Rosanen, J., and Wallgren, G. R.,** Enteric infection with Yersinia infection: large pyroniophilic cell reaction in mesenteric lymph nodes associated with early production of specific antibodies, *Acta Pathol. Microbiol. Scand. Sect. A,* 79, 109, 1971.
46. **Une, T.,** Studies on the pathogenicity of *Yersinia enterocolitica, Microbiol. Immunol.,* 21, 349, 1977.
47. **Orr, M. M.,** Experimental intestinal granulomas, *Proc. R. Soc. Med.,* 66, 34, 1975.
48. **Carter, B.,** Animal model: oral *Yersinia enterocolitica* infection of mice, *Am. J. Pathol.,* 81, 703, 1975.
49. **Rickert, R. R. and Carter, H. W.,** The gross, light microscopic, and scanning electron microscopic appearance of the early lesions of Crohn's disease, in *Scanning electron Microscopy,* Vol. 2, Proc. Workshop Biomedical Applications, MT Research Institute, Chicago, 1977, 174.
50. **Hodgson, H. J. F., Potter, B. J., and Jewell, D. P.,** Immune complexes in ulcerative colitis and Crohn's disease, *Clin. Exp. Immunol.,* 29, 187, 1977.

Chapter 11

ARTHRITIS ASSOCIATED WITH *YERSINIA ENTEROCOLITICA* INFECTION

Kimmo Aho, Paavo Ahvonen, Ossi Laitinen, and Marjatta Leirisalo

TABLE OF CONTENTS

I.	Introduction	114
II.	Genetics	114
III.	Pathogenesis	115
IV.	Diagnosis	118
V.	Laboratory Findings	118
VI.	Clinical Features	119
	A. Preceding Symptoms	119
	B. Joint Symptoms	119
	C. Carditis	119
	D. Other Manifestations	120
	E. Sequelae	120
VII.	Treatment	121
VIII.	Summary	121
References		122

I. INTRODUCTION

During the past 10 years nonpurulent arthritis associated with *Yersinia enterocolitica* infections has been described in several countries. In fact, it seems to be one of the most common forms of reactive arthritis.[1-19] So far the disease has been reported especially in Scandinavian countries, but it is probably present unrecognized in wide areas of the world. There seem to be differences in the occurrence of the condition between different countries. For example, it is most likely rare in the U.S.[20] and Great Britain.

II. GENETICS

There is a large group of rheumatic diseases strongly associated with the histocompatibility antigen HLA-B27.[21] They have the following characteristics in common:

1. Inflammatory involvement of the spine, the sacroiliac joints and peripheral joints
2. Mucocutaneous abnormalities, inflammatory disease of the eye, and occasionally carditis
3. Negative tests for rheumatoid factors and absence of subcutaneous nodules
4. Overlap of the clinical features in individual patients and tendency of familial aggregation

This group, frequently referred to as seronegative spondyloarthropathies, includes ankylosing spondylitis, spondylitis associated with psoriasis and chronic inflammatory bowel diseases, Reiter's disease, and reactive arthritis following *Yersinia*, *Salmonella*, and *Shigella* infections. A subset of patients with juvenile chronic polyarthritis and adult seronegative peripheral polyarthritis is also in this category.

It can be expected that the frequency of the B27 antigen in reactive arthritis will depend on the means of selection of the patient series and, to some degree, on the frequency of the antigen in the basic population. In Finland the B27 antigen occurs in about 80% of patients with *Yersinia* and *Salmonella* arthritis and with postdysenteric Reiter's disease.[10,19,22] This means that persons with the B27 antigen incur a 25 times greater risk of developing arthritis as a result of the above infections than those individuals who lack this histocompatibility antigen.

There is some evidence that the disease is milder in B27-negative individuals.[10,19] Thus, on the average, these patients have arthritic symptoms which have a shorter duration and affect fewer joints than those of their B27-positive counterparts. Extra-articular manifestations such as ocular inflammation, carditis, and signs of urologic inflammation are rare in the B27-negative individuals (Table 1). In contrast, patients with *Yersinia* arthritis associated with erythema nodosum are largely confined to the B27-negative group. There seems to be no other HLA-antigen characteristic of the erythema nodosum-group of arthritis patients.

In discussions about the associations between B27 and rheumatic diseases, the question why only a small proportion of individuals with this antigen develop the diseases has been raised frequently. Several hypotheses have been put forward to explain the findings. Thus, there may be a rare disease-susceptibility gene which, when present, is closely linked to B27. Other explanations could involve epistasis or the requirement for further genes, possibly located on other chromosomes.[23]

Data obtained from patients with *Yersinia* are not especially informative in this respect, since most cases of *Yersinia* infection, although capable of triggering the arthritis, are mild and the patients do not seek medical care for the infection. Figures available concerning *Salmonella*[24] and *Shigella*[25] epidemics suggest that 20 to 30% of the

Table 1
SYMPTOMS IN B27+ AND B27− PATIENTS
WITH *YERSINIA* ARTHRITIS[a]

	HLA-B27+ (N = 80)	HLA-B27− (N = 19)	Level of significance (χ^2-test with Yates' correction)[b]
Acute disease			
Acute arthritis			
Monoarthritis	7	3	
Oligoarthritis	39	9	
Polyarthritis	34	7	N.S.[c]
Carditis	6	1	N.S.
Ocular complications	15	0	N.S.
Nephrourological abn.	26	0	$p < 0.01$
Erythema nodosum	3	7	$p < 0.001$
Sequelae			
Chronic peripheral joint symptoms	41/71	12/19	N.S.
Thoracic or lumbar back pain	32/71	1/19	$p < 0.01$

[a] Leirisalo, M., Laitinen, O., Suoranta, H., Nissilä, M., and Voipio, L.-M., unpublished results.[50]
[b] Comparison between B27+ and B27− groups
[c] Mono- and oligoarthritis vs. polyarthritis in HLA-B27+ and HLA-B27− groups.

B27-positive individuals will develop arthritis, although in some epidemics the figure may be lower.

A significant association of any disease with a good genetic marker such as an HLA antigen is per se evidence of the existence of a genetic component in the etiology of the disease. Family studies are necessary to clarify the inheritance of the various elements of the disease spectrum and to elucidate their relationship with the genetic markers.

A family study of Reiter's disease[26] serves to illustrate this. In most of the 12 families selected for the study there, in addition to the proband, were two or three affected members. The manifestations included acute polyarthritis, which freuqently followed urethritis or occurred as a complication of *Yersinia* or *Shigella* infection, and chronic arthritis, either ankylosing spondylitis or peripheral arthritis. The latter characteristically had a remitting course, affecting mainly the large joints. Not a single subject had seropositive rheumatoid arthritis. The HLA-B27 antigen was detected in all the 12 families, and it served as the main indicator of the familial trait for developing arthritis. In individual patients, however, the association was not especially close, since there were members with this antigen who did not have arthritis despite an apparently adequate triggering stimulus and others who had arthritis but not the antigen. Reasons for these findings are not entirely clear, but they can hardly be explained on the basis of crossovers of susceptibility genes.

III. PATHOGENESIS

Undoubtedly the B27 antigen (or a closely linked disease-susceptibility gene) is an important determinant in the pathogenesis of the seronegative spondyloarthropathies. In contrast to rheumatoid arthritis, these diseases are characterized by a paucity of immunological abnormalities. Therefore, immune mechanisms probably do not play an essential role in their pathogenesis, or the defect may be a rather specific one.

In addition to the B27 antigen there are several other factors, both genetic and environmental, which contribute to the diversity of the clinical manifestations of the disease. Individual symptoms in the disease pattern, such as the position of the acute anterior uveitis, have already been explained to some extent. Brewerton et al.[27] recently demonstrated an association between acute uveitis and the α_1-antitrypsin phenotype MZ. This association was similar in patients with uveitis as an isolated phenomenon and in patients with associated rheumatic manifestations, whereas patients with ankylosing spondylitis without any uveitis episodes and healthy individuals had the same low frequency of the MZ phenotype. These findings strongly imply that uveitis is not a complication of ankylosing spondylitis or ulcerative colitis; the diseases are associated because their genetic factors are related.

Follow-up studies indicate that many patients with Reiter's disease following *Shigella* infection[28] and possibly with *Yersinia* arthritis (see later) exhibit symptoms and signs of spondylitis. However, this does not necessarily mean that the above bacterial infections initiate the lengthy process leading to ankylosing spondylitis in susceptible subjects. It is more likely that the acute peripheral arthritis associated with the infection and the spondylitis are independent manifestations of the same underlying factors.[1,19]

On the other hand, the pathogenesis of the disease is apparently the same in reactive arthritis following *Yersinia*, *Salmonella*, and *Shigella* infections. Reiter's disease following nonspecific urethritis probably belongs to the same disease category. Further knowledge of the range of microbial infections that can lead to reactive arthritis similar to *Yersinia* arthritis may give clues to its pathogenesis.

An additional point of interest is that *Salmonella*, *Shigella* and *Yersinia* bacteria all have the enterobacterial common antigen.[29] There is evidence that chlamydiae, frequently implicated as an etiological agent for nonspecific urethritis, are degenerated bacteria, possibly enterobacteria, adapted to intracellular parasitism. It is not known whether they also possess this antigen.

On the basis of the above, connecting any laboratory abnormality observed in patients with *Yersinia* arthritis with the pathogenesis of the disease would require that (a) the abnormality does not occur in uncomplicated *Yersinia* infections and (b) it is present in *Salmonella* or *Shigella* arthritis.

Several mechanisms have been proposed to explain the associations between histocompatibility antigens and diseases. It has been suggested that:

1. Histocompatibility antigens cross-react with antigenic organisms, such as a virus, resulting in partial tolerance, i.e., the development of an inadequate or inappropriate immune response (molecular mimicry).
2. The histocompatibility antigens serve as a receptor for a specific microbe.
3. A pathogenic agent may incorporate into its coat a portion of target cell membrane, which may influence the response of another host attacked by the pathogen.
4. There is an association with immune response genes which determine excessive or poorly regulated response to the pathogen.
5. There is a pathological alteration of a metabolic gene linked to HLA.

These are still only attractive hypotheses, since none has become firmly established as a definite explanation of the underlying pathogenetic process. In a related field, however, the lack of a receptor (Duffy blood group antigen) has proved to be the reason for genetic resistance to infection by *Plasmodium vivax*.[30]

As pointed out previously, *Yersinia* arthritis does not result from a direct microbial invasion of the joint, but is more likely to be related to some host response. The association with the B27 antigen is manifested in the arthritis and not in the triggering infection.[31] Thus, the question is not of genetically determined susceptibility to infection, but rather of a genetically based reactivity pattern to infection. These findings are not compatible with the receptor hypothesis, according to which the majority of infections would be restricted to B27-positive hosts. Nonetheless, the possibility of some specific interaction between the infectious organism and the molecular structure of the B27 antigen cannot be completely ruled out.[32]

There is some fragmentary data pertinent to the molecular mimicry hypothesis. Dausset and Hors[33] briefly mention some experiments in which they could not observe any specific absorption of anti-B27 antibodies by *Y. enterocolitica*. This indicates that the B27 determinant in its complete form is not present in *Y. enterocolitica*, but does not rule out a minor cross-reactivity. Ebringer et al.[34] have presented some preliminary evidence suggesting cross-reactivity between *Klebsiella* and lymphocytes in B27-positive individuals. However, the antisera were made in rabbits, whereas specific anti-HLA antisera are of human origin, and it is not known whether the reactivity was related to the B27 determinant.

Patients with *Yersinia* infection exhibit a number of tissue antibodies in higher frequencies than healthy controls.[35] Antibodies against thyroid epithelium are especially frequent.[36] However, they occur both in patients with arthritis and in patients with uncomplicated abdominal disease, and they have not been found in patients with *Salmonella* infections. Thus, it is improbable that these antibodies are intimately related to the pathogenesis of *Yersinia* arthritis.

It could be assumed that a mutant immune response gene allows the propagation and perpetuation of an infectious agent, resulting in an immune complex process. Recently a hemagglutination technique based on the inhibition of anti-antibody was described for the measurement of immune complexes.[37] Sera from 46 patients with *Yersinia* infection were tested by this technique;[38] 14 patients had uncomplicated abdominal disease, 9 had erythema nodosum, 21 had arthritis, and 2 had prolonged fever. Circulating immune complexes in low titers were detected in two patients with abdominal disease and in one with arthritis. Thus, these findings, although still very preliminary, do not support the concept that *Yersinia* arthritis is an immune complex disease. It is of interest that patients with ankylosing spondylitis usually do not have increased amounts of immune complexes in their sera,[39] although opposing views have also been presented.[40]

Humoral antibodies and the in vitro parameters of cell-mediated immunity do not always run parallel in infectious diseases. Thus, for example, there is some evidence in syphilis that changes in the lymphocyte responsiveness to treponemal antigens reflect the spectrum of clinical manifestations of the disease, whereas a corresponding relationship does not exist for humoral antibodies.[41] With regard to rheumatic diseases there seems to be a selective depression of cell-mediated immunity to rubella virus in subjects with arthritis associated with rubella vaccination.[42]

There is some evidence of T-cell abnormalities in patients with B27-associated rheumatic diseases such as reduced responses to a low dose of phytohemagglutinin in ankylosing spondylitis and Reiter's disease;[43] diminished mixed lymphocyte reaction in patients with ankylosing spondylitis, their relatives and normal individuals with B27;[44] and decreased percentage of T cells in patients with ankylosing spondylitis irrespective of their HLA type.[45] Such nonspecific abnormalities are also observed in patients with rheumatoid arthritis. Therefore, it is difficult to assess the value of these findings.

In order to investigate more specific immunological abnormalities, studies were performed on the transformation of peripheral blood lymphocytes by *Yersinia* antigen and certain other bacterial antigens. The reactivities observed in *Yersinia* arthritis were compared with those in patients with *Yersinia* infection without arthritis.[46] The occurrence of a cell-mediated immune response to the yersinia antigen could be demonstrated in a considerable proportion of the patients, whereas significant responses to the other antigens were rarely observed. However, the response in the arthritis patients was very similar to that in the *Yersinia*-infected patients without arthritis.

Patients with seropositive rheumatoid arthritis show an increased frequency of *Yersinia* agglutinins, compared with healthy controls,[16] but the same also holds true for many other bacterial antibodies.[47] Reasons for this include the hypergammaglobulinemia in rheumatoid arthritis and the augmenting effect of rheumatoid factors on the agglutination. Thus, there is no evidence that *Y. enterocolitica* plays a role in the etiology of chronic collagen disease.

IV. DIAGNOSIS

A definite diagnosis of a current *Y. enterocolitica* infection can be made only by the isolation of *Y. enterocolitica* from the patient together with the demonstration of a rising antibody titer. *Y. enterocolitica* can usually be cultured from stools within the first 2 weeks from the onset of symptoms. Since the arthritis patients seldom consult a physician during the prodromal phase, bacteriological diagnosis is not always possible. At the time of the first blood sample obtained, the antibody titer has usually reached its highest value. The maximum reciprocal agglutinin titers are 160 or higher in the vast majority of verified cases. In arthritis cases the titers are usually 320 to 20,000.[1,14,15] In a recent infection, a significant drop in the titer is to be expected within a few months. However, in some cases the titer remains high for years. Thus, a high titer alone is not diagnostic without a positive culture or the demonstration of a rapid significant decrease in the titer. The serological cross-reactions between *Y. enterocolitica* serotype O:9 on one hand and *Brucella* and certain salmonellae on the other must also be kept in mind.

Reactive arthritis following gastrointestinal symptoms can also occur in *Y. pseudotuberculosis*, *Salmonella*, and *Shigella* infections. These are recognized by appropriate bacterial cultures and serological tests. Rheumatic fever is probably the most important disease to be differentiated from *Yersinia* arthritis because of different therapy, prognosis, and prophylactic measures. Therefore, evidence for Group A streptococcal infection should be studied by throat cultures and serological tests.

In some instances, evidence of two or some infections can be shown, and it may be difficult to draw conclusions concerning the triggering infections of arthritis. In these cases it is most important to diagnose or exclude rheumatic fever.[11,15] Diagnostic difficulties may also arise between *Yersinia* arthritis and some other arthritides, for instance, early rheumatoid arthritis, pyogenic arthritis, and gouty arthritis.

V. LABORATORY FINDINGS

The synovial fluid in *Yersinia* arthritis is yellow or greenish yellow and can amount to 100 mℓ in one joint.[1,5,7,17] The white cell count of the fluid usually varies between 10,000 and 60,000/mm^3, of which 60 to 90% are polymorphonuclears. The bacterial cultures of the synovial fluid and blood are negative. Conclusive data on the existence or nonexistence of *Yersinia* antigen in the inflamed joints have not been presented.

The erythrocyte sedimentation rate is usually very high, often exceeding 100 mm/hr. Blood white cell counts are slightly elevated in about half of the cases. Often a slight to moderate anemia develops. Serum complement (C3) levels are normal. Tests for rheumatoid factors are usually negative. In a series of 104 cases of *Yersinia* arthritis the latex test was positive in six and the sensitized sheep cell agglutination test in two cases.[5]

VI. CLINICAL FEATURES

A. Preceding Symptoms

Most patients with *Yersinia* arthritis have preceding symptoms. About 20% have none or have only negligible complaints. Fever, diarrhea, abdominal pain, headache, tiredness, and feeling of weakness are the most common symptoms occurring prior to the joint symptoms.[1,4,5,7,11,13-15,17] The severity of the diarrhea or other symptoms bears no correlation to the development of arthritis. Many patients also have muscular pains, usually simultaneously with the arthritis, but sometimes before the joint symptoms are manifest. About one third of the patients have signs of upper respiratory infection before arthritis. The preceding symptoms are usually of short duration, and arthritis usually develops about 1 week after the onset of the disease.

B. Joint Symptoms

Yersinia arthritis is somewhat more common in females than in males. It rarely affects children and the child patients are mostly boys. Arthritis associated with *Yersinia* infection is usually oligo- or polyarticular, although monoarticular disease is also possible.[1-7,10,11,14,16-20] In a typical case, the joints are involved in rapid succession within a period of time varying from a few days to about 2 weeks. In general, no further joints are affected thereafter. The number of joints affected is usually between three and eight. The joint symptoms are more asymmetrical than in several other inflammatory rheumatic diseases, such as rheumatoid arthritis. The most commonly involved joints are the weight-bearing joints of the lower extremities. In this respect, *Yersinia* arthritis differs somewhat from rheumatic fever.[48] In order of frequency, the most commonly affected joints are knees, ankles, and toes, but the involvement of fingers, wrists, elbows, and shoulders is also relatively common. In exceptional cases almost any joint may be affected.[1,19]

In addition to joints with arthritis, defined as definite swelling and tenderness, in a usual case there are arthralgias in many joints (pain on motion or tenderness without apparent swelling). In several cases the onset of joint symptoms is "migratory" in the sense that at the onset of arthritis the joints are affected in succession within a few days. The pain and swelling do not subside in one joint when another becomes involved, but later the inflammatory reaction subsides more or less simultaneously in all the inflamed joints. In some cases a more prolonged history of joint disease has been observed.[7,14-16,18,49] One third of patients have back pain, sometimes severe, in lumbar or thoracic area.[17,50]

C. Carditis

Acute nonbacterial carditis has previously been regarded as a typical manifestation of rheumatic fever. In fact, there have been no reports on carditis in association with other reactive forms of arthritis, with the exception of Reiter's disease. Several groups of investigators have described ECG changes in patients with *Yersinia* infection.[7,11,14,51]

During a close observation of patients with *Yersinia* arthritis it has become obvious that some of the patients also have signs of definite carditis.[7,19,50,52] Furthermore, carditis may occur also without simultaneous arthritis.[50,52]

Symptoms indicative of cardiac involvement include radiological evidence of variation in heart volume, significant murmurs on auscultation, such as apical systolic murmurs and persistent parasternal systolic murmurs, and pericarditis and pericardial friction rubs. In ECG, in addition to ST and T changes, some patients develop QRS complex deformations. Furthermore, several patients have precordial pain, tachycardia unrelated to fever, and frequent extrasystoles. The cardiac involvement, although present in about 10% of patients with *Yersinia* arthritis, seems to be benign in all cases, and so far there have been no reports on patients who have developed cardiac decompensation or some other form of progressive cardiac disease. The practical significance of cardiac symptoms in *Yersinia* arthritis is, however, considerable. These findings contribute to the difficulty in the differential diagnosis between rheumatic fever and *Yersinia* arthritis.

D. Other Manifestations

In addition to carditis, many patients with *Yersinia* arthritis develop some other extra-articular manifestations. Iritis or conjunctivitis occur in 10 to 15% of patients.[10,14] Erythema nodosum is present in 4 to 10% of patients with *Yersinia* arthritis.[5,48]

Dysuria, urethral discharge, and/or abnormal urinalysis — mostly white cells in excess — indicate the presence of inflammation in the urethra or in the kidney. Nephrourological abnormalities are observed in 5 to 25% of patients.[14,19] Urological symptoms with ocular ones in a patient with *Yersinia* arthritis may fulfill the criteria of Reiter's syndrome. *Y. enterocolitica* thus seems to be an organism capable of triggering Reiter's syndrome.[1,4,19] The frequency of complete Reiter's triad in *Yersinia* arthritis is about 5 to 10%.

In a group of 99 patients with *Yersinia* arthritis (Table 1), one patient had acute glomerulonephritis with gross hematuria and transient deterioration of renal function.[50] Glomerulonephritis during *Yersinia* infection has also been observed by other authors.[11,53] Friedberg et al.[54] showed that in some patients with acute glomerulonephritis (without arthritis) there was serological evidence of *Y. enterocolitica* infection, and in some of the cases there was a positive immunofluorescence for *Yersinia* antigen in the glomerulus.

E. Sequelae

Acute arthritis usually takes 1 to 3 months to heal, although in one third of the patients the duration is more than 3 months.[10,14,19] In some patients, however, the disease runs a course of prolonged joint disease with relapsing periarticular swellings and even effusions in several joints.[7,14,18]

In a 3- to 5-year follow-up study of 64 patients,[49] two thirds still had some joint symptoms. Five patients used some antirheumatic medication daily and nine considered their working capacity to be at least moderately impaired. A total of 17 patients had experienced one or more episodes of acute arthritis, either before or after *Yersinia* arthritis. In two instances the attack could be attributed to *Salmonella* infection, once to *Shigella* infection, and once to an infection by another serotype of *Y. enterocolitica*. One female patient had developed symmetrical erosive polyarthritis of the peripheral joints and positive tests for rheumatoid factors within 1 year after the acute arthritic attack associated with *Yersinia* infection.

In another follow-up study in Helsinki, 90 patients with *Yersinia* arthritis were diagnosed and then followed at the outpatient departments of two Helsinki hospitals during the period 1968 to 1978[50] (Table 1). The follow-up time ranged from 3 months to 9 years (mean follow-up time 4 years). It appeared that after that time one third of the patients had pain in the lumbar or thoracic area, the pain increasing at rest. One patient showed clinical signs suggesting the development of ankylosing spondylitis (stiffness in the lumbar and thoracic area, persistent low back pain with elevated erythrocyte sedimentation rate, and recurrent episodes of iritis). He also had aortic regurgitation before the *Yersinia* arthritis. X-rays of the lumbar area and sacroiliac joints were, however, normal both at the acute phase and at the follow-up examination.

The X-ray abnormalities of the lumbosacral area of 29 patients were thoroughly analyzed by comparing the X-ray findings during the acute phase and after the follow-up period ranging from 17 to 113 months. Twenty-two of the patients initially had normal X-rays. One of them developed a syndesmophyte between the 12th thoracic and the first lumbar vertebra. Seven patients had abnormal X-rays at the acute phase: six of them had sacroiliitis and one classical ankylosing spondylitis with ankylosis of the sacroiliac joints and syndesmophytes of the lumbar vertebrae. At the follow-up study, five of the patients showed progression of the changes, while the findings of one patient remained unchanged, as also did the advanced changes of the ankylosing spondylitis patient. Although these results indicate beyond doubt that spondylitic processes are clustered in B27 positive diseases,[19] it is impossible to draw any firm etiological inferences regarding possible triggering action of any infectious agents in ankylosing spondylitis.

Cardiac sequels seem to be mild. In a group of six patients with acute carditis, the murmur gradually disappeared in two out of four patients with mitral regurgitation at the acute phase. Cardiac volume became normal in both of the two patients, with transient cardiac enlargement during the acute phase.[55] The possible development of chronic rheumatic heart disease, such as mitral stenosis, has not been demonstrated in these patients, and on the basis of results obtained by HLA analysis, similar cardiac sequels in rheumatic fever and *Yersinia* arthritis seem highly improbable.[56]

VII. TREATMENT

Most strains of *Y. enterocolitica* are susceptible to tetracyclines, aminoglycosides, and sulfonamides. However, the use of antiobiotic therapy in *Yersinia* arthritis is controversial. It is recommended that all patients with acute rheumatic fever should be treated with a course of penicillin to eliminate the hemolytic streptococci from the pharynx. By analogy, it may be advisable to treat *Yersinia* arthritis patients with tetracycline, for example, although there is a fundamental difference between streptococci and *Yersinia*: penicillin is the drug of choice in the treatment of streptococcal pharyngitis, whereas symptomatic treatment is usually sufficient for the gastrointestinal disease caused by *Y. enterocolitica*.

The consensus is that antibiotic therapy does not have any distinct effect on the course of *Yersinia* arthritis, although there are no controlled clinical studies. It is not known whether early treatment of gastrointestinal disease might prevent arthritis.

The joint symptoms of most patients are relieved by nonsteroidal anti-inflammatory agents, although the arthritis does not respond to acetosalicylic acid as readily as that in rheumatic fever. In joint effusion, local corticosteroid injection often gives permanent or prolonged help. Similar treatment of other severely affected joints with periarticular swelling also helps to prevent contraction deformities and chronic capsular thickening. In some patients with severe symptoms of both peripheral arthritis and lumbosacral inflammation or with otherwise prolonged course, systemic corticosteroid treatment with lowering dosage has been used with success. Physiotherapy is obligatory in severe cases, both to relieve pain and to prevent contraction deformities.

VIII. SUMMARY

It is commonly believed that rheumatic diseases might represent genetically based immunopathological responses to infection. In this respect, *Yersinia* arthritis is of special interest, since the triggering microbe is known and the genetic analysis of the disease is facilitated by the existence of a closely linked Mendelian marker HLA-B27.

Clinically, *Yersinia* arthritis has an acute onset, usually affecting three to eight joints. It is mostly self-limiting, although sometimes it lasts for several months. Prolonged arthralgias are common, and some patients later have new attacks of reactive arthritis. A few patients also have a benign form of carditis, or other extra-articular manifestations. Chronic cardiac sequelae seem to be mild, whereas chronic joint symptoms seem to be quite common. Some patients with *Yersinia* arthritis develop Reiter's syndrome during the acute phase of the arthritis. In *Yersinia* arthritis patients there is also an excess of other B27-positive rheumatic diseases, either in their history or after the acute *Yersinia* infection.

The different occurrence of *Yersinia* arthritis between different countries can be partly explained by epidemiological and genetic factors and possibly by differences in the arthritogenic properties of the *Yersinia* strains.

REFERENCES

1. **Ahvonen, P., Sievers, K., and Aho, K.**, Arthritis associated with *Yersinia enterocolitica* infection, *Acta Rheum. Scand.*, 15, 232, 1969.
2. **Arvastson, B., Damgaard, K., and Winblad, S.**, Clinical symptoms of infection with *Yersinia enterocolitica, Scand. J. Infect. Dis.*, 3, 37, 1971.
3. **Mollaret, H. H.**, L'infection humaine à *"Yersinia enterocolitica"* en 1970, à la lumière de 642 cas rècents, *Pathol. Biol.*, 19, 189, 1971.
4. **Solem, J. H. and Lassen, J.**, Reiter's disease following *Yersinia enterocolitica* infection, *Scand. J. Infect. Dis.*, 3, 83, 1971.
5. **Ahvonen, P.**, Human yersiniosis in Finland. II. Clinical features, *Ann..Clin. Res.*, 4, 39, 1972.
6. **Hällström, K., Sairanen, E., and Ohela, K.**, A pilot clinical study on yersinioses in South-Eastern Finland, *Acta Med. Scand.*, 191, 485, 1972.
7. **Laitinen, O., Tuuhea, J., and Ahvonen, P.**, Polyarthritis associated with *Yersinia enterocolitica* infection. Clinical features and labotatory findings in nine cases with severe joint symptoms, *Ann. Rheum. Dis.*, 31, 34, 1972.
8. **Sievers, K., Ahvonen, P., and Aho, K.**, Epidemiological aspects of *Yersinia* arthritis, *Int. J. Epid.*, 1, 45, 1972.
9. **Osnes, M. and Lassen, J.**, Artritter og polymyalgier ved *Yersinia enterocolitica* infeksjoner, *T. Norske Laegeforen.*, 94, 225, 1974.
10. **Aho, K., Ahvonen, P., Lassus, A., Sievers, K., and Tiilikainen, A.**, HL-A 27 in reactive arthritis. A study of *Yersinia* arthritis and Reiter's disease, *Arthritis Rheum.*, 17, 521, 1974.
11. **Leino, R. and Kalliomäki, J. L.**, Yersiniosis as an internal disease, *Ann. Intern. Med.*, 81, 458, 1974.
12. **Toma, S. and Lafleur, L.**, Survey on the incidence of Yersinia enterocolitica infection in Canada, *Appl. Microbiol.*, 28, 469, 1974.
13. **Thomas, A. F., Solomon, L., and Rabson, A.**, Polyarthritis associated with *Yersinia enterocolitica* infection, *S. Afr. Med. J.*, 49, 18, 1975.
14. **Winblad, S.**, Arthritis associated with *Yersinia enterocolitica* infections, *Scand. J. Infect. Dis.*, 7, 191, 1975.
15. **Laitinen, O., Leirisalo, M., and Allander, E.**, Rheumatic fever and *Yersinia* arthritis. Criteria and diagnostic problems in a changing disease pattern, *Scand. J. Rheumatol.*, 4, 145, 1975.
16. **Larsen, J. H.**, *Yersinia enterocolitica* infection and arthritis, in *Infection and Immunology in the Rheumatic Diseases,* Dumonde, D. C., Ed., Blackwell, Oxford, 1976, 133.
17. **Schilling, F.**, *Yersinia-Arthritis*, *Dtsch. Med. Wochenschr.*, 101, 1515, 1976.
18. **Jarner, D., Jarløv, N. V., and Larsen, J. H.**, *Yersinia*-artrit og kronisk kollagenose. II. Tilfaelde af *Yersinia*-artrit med langvarigt forløb og udvikling af reumatoid artrit, *Ugeskr. Laeg.*, 139, 1481, 1977.
19. **Laitinen, O., Leirisalo, M., and Skylv, G.**, Relation between HLA-B27 and clinical features in patients with *Yersinia* arthritis, *Arthritis Rheum.*, 20, 1121, 1977.
20. **Ford, D. K., Henderson, E., Price, G. E., and Stein, H. B.**, *Yersinia*-related arthritis in the Pacific Northwest, *Arthritis Rheum.*, 20, 1226, 1977.
21. **Brewerton, D. A.**, HLA-B27 and the inheritance of susceptibility to rheumatic disease, *Arthritis Rheum.*, 19, 656, 1976.
22. **Sairanen, E. and Tiilikainen, A.**, HL-A27 in Reiter's disease following shigellosis. *Scand. J. Rheum.,* Suppl. 8, Abstr. 30/11, 1975.

23. **Kidd, K. K., Bernoco, D., Carbonara, A. O., Daneo, V., Steiger, U., and Ceppellini, R.**, Genetic analysis of HLA-associated diseases: the "illness-susceptible" gene frequency and sex ratio in ankylosing spondylitis, in *HLA and Disease,* Dausset, J. and Svejgaard, A., Eds., Munksgaard, Copenhagen, 1977, 72.
24. **Håkansson, U., Löw, B., Eitrem, R., and Winblad, S.**, HL-A27 and reactive arthritis in an outbreak of salmonellosis, *Tissue Antigens,* 6, 366, 1975.
25. **Calin, A. and Fries, J. F.**, An "experimental" epidemic of Reiter's syndrome revisited. Follow-up evidence on genetic and environmental factors, *Ann. Intern. Med.,* 84, 564, 1976.
26. **Kousa, M., Lassus, A., Karvonen, J., Tiilikainen, A., and Aho, K.**, Family study of Reiter's disease and HLA B27 distribution, *J. Rheumatol.,* 4, 95, 1977.
27. **Brewerton, D. A., Webley, M., Murphy, A. H., and Ward, A. M.**, The α_1-antitrypsin phenotype MZ in acute anterior uveitis, *Lancet,* 1, 1103, 1978.
28. **Sairanen, E., Paronen, I., and Mähönen, H.**, Reiter's syndrome: a follow-up study, *Acta Med. Scand.,* 185, 57, 1969.
29. **Mäkelä, P. H. and Mayer, H.**, Enterobacterial common antigen, *Bacteriol. Rev.,* 40, 591, 1976.
30. **Miller, L. H., Mason, S. J., Clyde, D. F., and McGinniss, M. H.**, The resistance factor to *Plasmodium vivax* in blacks. The Duffy-blood-group genotype, FyFy, *N. Engl. J. Med.,* 295, 302, 1976.
31. **Aho, K., Ahvonen, P., Lassus, A., Sievers, K., and Tiilikainen, A.**, HL-A antigen 27 and reactive arthritis, *Lancet,* 2, 157, 1973.
32. **Kemple, K. and Bluestone, R.**, The histocompatibility complex and rheumatic diseases, *Med. Clin. North Am.,* 61, 331, 1977.
33. **Dausset, J. and Hors, J.**, Some contributors of the HL-A complex to the genetics of human diseases, *Transplant. Rev.,* 22, 44, 1975.
34. **Ebringer, A., Cowling, P., Ngwa Suh, N., James, D. C. O., and Ebringer, R. W.**, Crossreactivity between *Klebsiella aerogenes* species and B27 lymphocyte antigens as an aetiological factor in ankylosing spondylitis, in *HLA and Disease,* Dausset, J. and Svejgaard, A., Eds., INSERM, Paris, 1976, 27.
35. **Gripenberg, M., Miettinen, A., Kurki, P., and Linder, E.**, Humoral immune stimulation and anti-epithelial antibodies in *Yersinia* infection, *Arthritis Rheum.,* 21, 904, 1978.
36. **Lidman, K., Eriksson, U., Norberg, R., and Fagraeus, A.**, Indirect immunofluorescence staining of human thyroid by antibodies occurring in *Yersinia enterocolitica* infections, *Clin. Exp. Immunol.,* 23, 429, 1976.
37. **Kano, K., Nishimaki, T., Palosuo, T., Loza, U., and Milgrom, F.**, Detection of circulating immune complexes by the inhibition of anti-antibody, *Clin. Immunol. Immunopathol.,* 9, 425, 1978.
38. **Palosuo, T., Aho, K., Kano, K., and Milgrom, F.**, Circulating immune complexes in *Yersinia* infections as studied by inhibition of anti-antibody, in 6th Int. Convocation Immunology, Niagara Falls, N.Y., 1978.
39. **Gabay, R., Zubler, R. H., Nydegger, U. E., and Lambert, P. H.**, Immune complexes and complement catabolism in ankylosing spondylitis, *Arthritis Rheum.,* 20, 913, 1977.
40. **Corrigall, V., Panayi, C. S., Unger, A., Poston, R. N., and Williams, B. D.**, Detection of immune complexes in serum of patients with ankylosing spondylitis, *Ann. Rheum. Dis.,* 37, 159, 1978.
41. **Friedmann, P. S. and Turk, J. L.**, A spectrum of lymphocyte responsiveness in human syphilis, *Clin. Exp. Immunol.,* 21, 59, 1975.
42. **Chiba, Y., Sadeghi, E., and Ogra, P. L.**, Abnormalities of cellular immune response in arthritis induced by rubella vaccination, *J. Immunol.,* 117, 1684, 1976.
43. **Froebel, K., Sturrock, R. D., Dick, W. C., and MacSween, R. N. M.**, Cell-mediated immunity in the rheumatoid diseases. I. Skin testing and mitogenic responses in sero-negative arthritides, *Clin. Exp. Immunol.,* 22, 446, 1975.
44. **Nikbin, B., Brewerton, D. A., James, D. C. O., and Hobbs, J. R.**, Diminished mixed lymphocyte reaction in ankylosing spondylitis, relatives, and normal individuals all with HL-A27, *Ann. Rheum. Dis.,* 35, 37, 1976.
45. **Fan, P. T., Clements, P. J., Yu, D. T. Y., Opelz, G., and Bluestone, R.**, Lymphocyte abnormalities in ankylosing spondylitis, *Ann. Rheum. Dis.,* 36, 471, 1977.
46. **Aho, K., Ahvonen, P., Juvakoski, T., Kousa, M., Leirisalo, M., and Laitinen, O.**, Immune responses in *Yersinia*-associated polyarthritis, in Int. Symp. Reiter's Syndrome and Related B27-Associated Diseases, Lausanne, 1978.
47. **Aho, K., Ahvonen, P., von Essen, R., and Isomäki, H.**, Bacterial antibodies in arthritis, in Int. Symp. on Reiter's Syndrome and Related B27-Associated Diseases, Lausanne, 1978.
48. **Leirisalo, M.**, Rheumatic fever. Clinical picture, differential diagnosis and sequels, *Ann. Clin. Res.,* 9(Suppl. 20), 1977.

49. **Aho, K., Ahvonen, P., Lassus, A., Sievers, K., and Tiilikainen, A.**, *Yersinia* arthritis and related diseases: clinical and immunogenetic implications, in *Infection and Immunology in the Rheumatic Diseases,* Dumonde, D. C., Ed., Blackwell, Oxford, 1976, 341.
50. **Leirisalo, M., Laitinen, O., Suoranta, H., Nissilä, M., and Vuopio, L. M.**, unpublished results.
51. **Ahvonen, P., Hiisi-Brummer, L., and Aho, K.**, Electrocardiographic abnormalities and arthritis in patients with *Yersinia enterocolitica* infection, *Ann. Clin. Res.,* 3, 69, 1971.
52. **Agner, E., Larsen, H. J., and Leth, A.**, *Yersinia enterocolitica* carditis as a differential diagnosis — and the prognosis of this disease, *Scand. J. Rheum.,* 7, 26, 1978.
53. **Forsström, J., Viander, M., Lehtonen, A., and Ekfors, T.**, *Yersinia enterocolitica* infection complicated by glomerulonephritis, *Scand. J. Infect. Dis.,* 9, 253, 1977.
54. **Friedberg, M., Larsen, S., and Denneberg, T.**, *Yersinia enterocolitica* and glomerulonephritis, *Lancet,* 1, 498, 1978.
55. **Leirisalo, M., Laitinen, O., and Kentala, E.**, Carditis and yersinia arthritis, in 17th Scandinavian Rheumatology Congr., Elsinoor, 1978.
56. **Leirisalo, M., Laitinen, O., and Tiilikainen, A.**, HLA phenotypes in patients with rheumatic fever, rheumatic heart disease, and *Yersinia* arthritis, *J. Rheumatol.,* 4(Suppl. 3), 78, 1977.

Chapter 12

ERYTHEMA NODOSUM ASSOCIATED WITH INFECTION WITH *YERSINIA ENTEROCOLITICA*

Sten Winblad

TABLE OF CONTENTS

I.	Introduction	126
II.	Diagnostic Methods	126
	A. Serological	126
	B. Cultural	127
III.	Clinical Remarks of Erythema Nodosum Associated with Infection with *Yersinia enterocolitica*	127
IV.	Comparison Between Erythema Nodosum Associated with *Y. enterocolitica* and Cases without Such an Association	127
V.	Clinical Symptoms of Infections Probably Connected to Erythema Nodosum	130
	A. Joint Pain in Connection with Erythema Nodosum	131
References		132

I. INTRODUCTION

The cutaneous syndrome of erythema nodosum and often also erythema multiforme is a common complication to infection with *Yersinia enterocolitica*.

Erythema nodosum has historically been linked to many infections. Primary pulmonary tuberculosis was suggested among the earlier etiologies, but by 1940 it was realized that infection with β-hemolytic streptococci could also result in this syndrome. Sarcoidosis is also known to have a connection with erythema nodosum. Infections such as gonorrhea, mycosis, histoplasmosis, coccidiodomycosis, and tularemia have also caused erythema nodosum.

Monographs from 1945 to 1946, however, never reported intestinal infections associated with erythema nodosum, but recently such reports have been more common. Kelley[1] found 12 of 119 cases with colitis ulcerosa associated with erythema nodosum and Girard et al.[2] described 2 of 10 cases. Crohn's disease has also been observed with erythema nodosum.

Infection with *Y. pseudotuberculosis* has often been associated with erythema nodosum.[3-5] Kerzoncuf[6] has reported 52 such cases in 1967. In 1964, Winblad[7] noted that erythema nodosum was a rather common complication following infection with *Y. enterocolitica*, and by 1969 had observed that 73 of 333 patients with *Y. enterocolitica* infection went on to develop erythema nodosum. The high frequency of erythema nodosum in yersiniosis has also been noted in Finland.[8-11] Table 1 lists a summary of published cases of erythema nodosum associated with yersiniosis. In Scandinavia, the frequency of erythema nodosum linked with yersiniosis is between 16.4%[10] and 21.9%.[7]

II. DIAGNOSTIC METHODS

A. Serological

Somatic (O) and Flagella (OH) antigens for agglutination — *Y. enterocolitica* human type O:3, O:9, and, for use in the U.S. and Canada, serotype O:8 should be cultivated on blood agar for 48 hr at 22° C and controlled for growing in smooth (S) form. The harvested growth is suspended in saline and divided into two aliquots: one part is autoclaved for 1 hr at 120°C for O antigens and one part formalized in 0.4% formalin overnight, centrifuged, and resuspended in saline for OH antigen. For assaying sera against *Y. enterocolitica* O:9 the OH antigen is recommended and for O:3 the O antigen suspension. Both forms of the antigen should be used for testing for antibody to serotype O:8. The density of suspended cultures should be suitable for a Widal reaction.

Agglutination reaction — The patient's sera should be inactivated for 30 min in 56° C and then diluted, with saline or other neutral buffered solution, from 1/10 to 1/5120; 0.25 mℓ antigen may be added to 0.25 mℓ of diluted serum. Overnight incubation at 56°C should ensue when using O antigens and 37°C when using OH antigens.

B. Cultural

Bacterial culture — Cultivation of feces from patients with a suggestive agglutination titer (1/80 or higher) often results in detection of especially *Y. enterocolitica*. Cultivation may be undertaken by using *Salmonella-Shigella* (SS), MacConkey, or lysine sucrose urea (LSU) agars.[20,29] Small pinpoint colonies (0.1 mm) developing after 25-hr incubation at 22°C, which become larger after subsequent 24-hr incubation at the same temperature, should be isolated and identified. A good selection method is to have the feces in ordinary peptone broth kept in the cold (4°C) for 4 to 8 days after which *Y. enterocolitica* of human pathogenic types grows out over other bacteria.[25] Enrichment cultivation in Rappaport's broth[30] for 1 day at 22°C or in the magnesium chloride-

Table 1
REPORTED CASES OF ERYTHEMA NODOSUM ASSOCIATED WITH
Y. ENTEROCOLITICA SEROTYPE O:3 AND O:9 AND
Y. PSEUDOTUBERCULOSIS

	Y. enterocolitica O:3	*Y. enterocolitica* O:9	*Y. pseudotuberculosis*
Sweden			
Winblad[7]	73		2
Finland			
Hannekusela et al.[8]	5	5	
Helander et al.[9]	18	1	1
Hannekusela et al.[10]	19	17	3
Niemi et al.[11]	9	10	1
Belgium			
Debois et al.[12]	8		
Other countries			
Ref. 13—19	9	1	

malachite green-carbenicillin broths of Wauters[21] may also be used. Suspect isolates, which are usually lactose-negative and urease-positive, may be confirmed as *Y. enterocolitica* by fermentation and slide agglutination using anti- *Yersinia* O antisera.[22-27]

Blood samples — Sera from patients with erythema nodosum should be tested for antibodies against hemolytic streptococcal exoenzymes (streptolysin O, DNAse, NADase, etc.) as well as assayed for antibodies against *Y. enterocolitica* and *Y. pseudotuberculosis*.

III. CLINICAL REMARKS OF ERYTHEMA NODOSUM ASSOCIATED WITH INFECTION WITH *Y. ENTEROCOLITICA*

Skin manifestation — The erythematous lesions resemble other eruptions in their appearance. Sometimes, however, the erythema is not only localized to the legs or arms, but may also occur on hands, feet, or neck. Studies in Finland by Niemi et al.[11] showed many patients with clinical manifestations resembling erythema multiforme and erythema figuratum. Hannekusela and Ahvonen[8] have described cockade-like erythema multiforme, vesicular form of erythema multiforme, erythema figuratum, and drug eruption-like exanthema, all caused by *Y. enterocolitica* serotype O:3 and serotype O:9.

IV. COMPARISON BETWEEN ERYTHEMA NODOSUM ASSOCIATED WITH *Y. ENTEROCOLITICA* AND CASES WITHOUT SUCH AN ASSOCIATION

Such comparison has been undertaken by Winblad[7] who studied 73 patients with erythema nodosum who showed serum agglutinins against *Y. enterocolitica* O:3, and compared three such patients with 260 control subjects without any serologic reactivity against *Y. enterocolitica*.

It is clear from Table 2 that the titers were rather high. The cultures were positive in only 12 of 64 cases, probably because of the fairly long interval between the onset of the syndrome and the time for culture, during which the bacteria may have disappeared from the fecal flora. In some cases, antibiotic therapy may have been responsible for the lack of isolation on culture.

Sex and age — In both groups erythema nodosum was more common in females

Table 2
ANTIBODY TITERS AND RESULTS OF CULTURES IN CASES OF ERYTHEMA NODOSUM RELATED TO *Y. ENTEROCOLITICA*

Results of cultures	Maximal O-agglutinin titer							
	80	160	320	640	1280	2560	>5120	Total
Positive culture	—	1	—	—	6	1	4	12
Negative culture	—	1	12	7	13	10	9	52
No culture	3	2	—	1	1	—	2	9
Total	3	4	12	8	20	11	15	73

Table 3
SEX DISTRIBUTION OF THE CASES OF ERYTHEMA NODOSUM

Sex	Stimulated by *Y. enterocolitica*	Control group
Men	10 (13.7%)	36 (13.8%)
Women	63 (86.3%)	224 (86.2%)
Total	73 (100%)	260 (100%)

Table 4
AGE DISTRIBUTION OF THE CASES OF ERYTHEMA NODOSUM

Age	Stimulated by *Y. enterocolitica*		Control group	
0—10	12		12	
—20	4		37	
—30	13	(56.1%)	55	(73.5%)
—40	10		37	
—50	13		50	
—60	13		34	
—70	17		28	
—80	2	(43.9%)	6	(26.5%)
—90	1		1	
Total	73		260	

(Tables 3 and 4). It should, perhaps, be pointed out that the material included rather few children. The age distribution of patients in the series is given in Table 4. It appears that approximately one half of the patients with associated *Y. enterocolitica* infection were above 50 years of age as compared with only one fourth of those without this infection.

Erythrocyte sedimentation rate (ESR) — The distribution of the highest ESR values recorded during the observation period is given in Table 5. The ESR tended to be higher in the presence of *Y. enterocolitica* infection.

Antistreptolysin titer (AST) (Table 6) — Titers above 180 ASU on beta-lipoprotein-free serum were about 3 times as frequent in the control group than in the *Yersinia-*

Table 5
ESR[a] DISTRIBUTION IN THE CASES OF ERYTHEMA NODOSUM

Maximal ESR (min/hr)	Stimulated by Y. enterocolitica	Control group
0—20	1	52
21—40	3	78
41—60 (>40)	20 (93.7%)	30 (48.6%)
61—80 (>60)	10 (64.1%)	33 (36.5%)
81—100 (>80)	18 (48.4%)	31 (22.8%)
101—120 (>100)	8 (20.3%)	15 (10.6%)
121—140	5	12
Total	65	251

[a] Erythrocyte sedimentation rate.

Table 6
MAXIMAL AST[a] IN THE CASES OF ERYTHEMA NODOSUM

Maximal AST	Stimulated by Y. enterocolitica	Control group
<35	15	29
35—80	27	62
90—150	18	80
180—250	5 ⎫	37 ⎫
260—480	1 ⎬ (13.0%)	20 ⎬ (31.6%)
>500	3 ⎭	22 ⎭
Total	69	250

[a] Antistreptolysin titer.

Table 7
TUBERCULIN REACTION IN CASES OF ERYTHEMA NODOSUM

Tuberculin reaction	Stimulated by Y. enterocolitica	Control group
Positive	28 (38.4)	92 (35.4)
Negative (1 mg)	17 (23.3)	101 (38.8)
Negative but not fulfilled until 1 mg	11 (15.1)	18 (6.9)
Not done	17 (23.3)	49 (18.9)
Total	73	260

Note: Percentages in parentheses.

Table 8
CHEST X-RAY IN CASES OF ERYTHEMA NODOSUM

Chest X-ray	Stimulated by *Y. enterocolitica*	Control group
No pathological observations	62 (89.9)	173 (69.8)
Unilateral enlargement of hilar glands	3 (4.3)	11 (4.4)
Bilateral enlargement of hilar glands	0	51 (20.6)
Other pathological parenchymal observations	4 (5.8)	13 (5.2)
Total	69	248

Note: Percentages in parentheses.

stimulated group, probably because the control group included the cases associated with infections with haemolytic streptococci.

Tuberculin reaction (Table 7) — The two groups did not differn in frequency of positive or negative tuberculin reactions to 1 mg of tuberculin.

Chest X-ray (Table 8) — All the bases with bilateral enlargement of the hilar lymph glands belonged to the control group. Only three of the *Yersinia*-stimulated cases had suspected unnatural enlargement. The control group included thus almost all the cases of sarcoidosis. Co-existing sarcoidosis and *Y. enterocolitica* infection seems to be rare. Most of the cases in the *Y. enterocolitica* group showed no roentgenological abnormalities.

V. CLINICAL SYMPTOMS OF INFECTIONS PROBABLY CONNECTED TO ERYTHEMA NODOSUM

Since erythema nodosum generally is believed to be stimulated by various forms of infection, it was of interest to registrate the symptoms of such infections in the two groups compared. The following groups were distinguished: (1) infections with sore throat otitis media, and cases in which beta-hemolytic streptococci were cultured from throat swabs, (2) infections with symptoms from the upper and lower respiratory tracts, (3) infections with symptoms from the gastrointestinal tract, (4) infections with symptoms from other organs, and (5) cases without any symptoms of infection at all. The results are given in Table 9.

Practically all infections of the throat were found in the control group. Infections of the respiratory tract were more common in the control group. In only one case (belonging to the control group) pulmonary tuberculosis was suspected. In the group of erythema nodosum stimulated by *Y. enterocolitica* symptoms from the intestinal tract were predominant (58%). Such symptoms were much less common in the control group. Of the 18 cases with gastrointestinal symptoms in the control group, 4 were classified as ulcerative colitis and 1 as salmonellosis. Cases without symptoms of infection were somewhat more common in the control group. Thus, judging from the clinical data *Y. enterocolitica* which is known to cause intestinal symptoms may also give rise to erythema nodosum.

Table 9
INFECTIOUS SYMPTOMS CONNECTED WITH ERYTHEMA NODOSUM

Group of symptoms	Stimulated by *Y. enterocolitica*	Control group
Acute tonsillitis Otitis media beta-Hemolytic streptococci in the throat	1/73 (1.4)	69/260 (26.5)
Rhino-pharyngitis Sinusitis Bronchitis Bronchopneumonia Pneumonia	8/73 (11.0)	51/260 (19.6)
Diarrhea Abdominal pains Syndrome of appendicitis	43/73 (58.9)	18/260 (6.9)
Various infectious symptoms	1/73 (1.4)	8/260 (3.1)
No symptoms of infection	22/73 (30.1)	132/260 (50.8)

Note: Percentages in parentheses.

Table 10
CASES OF YERSINIOSIS DUE TO *Y. ENTEROCOLITICA* OBSERVED IN MALMO LABORATORY DURING 1966—1976 DIVIDED BY SYMPTOMOLOGY

Acute noncomplicated cases of yersiniosis due to *Y. enterocolitica*		Complicated cases of yersiniosis due to *Y. enterocolitica*	
Acute terminal ileitis	92	Arthritis	231
Pseudoappendicitis	100	Erythema nodosum	312
Acute abdominal pains	87	Fever without other etiology	97
Acute gastroenteritis in children (until 4 yrs)	106	Acute glomerulonephritis	4
Acute enteritis	274	Hypersedimentation with other etiology	19
Total	660	Total	663

A. Joint Pain in Connection with Erythema Nodosum

As known, erythema nodosum is often accompanied by pain of the joints, especially those of the legs and arms; 52% of the cases of erythema nodosum stimulated by *Y. enterocolitica* and 2.3% of the controls had such symptoms. The difference was not significant. Such joint pain seems to be a component of the syndrome regardless of the causal factor.

In contrast to the observation that arthritis subsequent to infection with *Y. enterocolitica* is predominantly (75%) associated with the genetic group HLA-B27,[28] this is not the case with erythema nodosum.

The answer to the question of how often erythema nodosum is a complication of yersiniosis due to *Y. enterocolitica* may reside in the registration of such cases in a district in South Sweden where the incidence is high (Table 10). Nearly half of the complicated cases in this material were patients with erythema nodosum.

It may be stated that a patient, especially a woman, with erythema nodosum or erythema multiforme developing after intestinal disorders must be suspected to have had yersiniosis. This suspicion is particularly heightened if the patient resides in a Scandinavian country.

REFERENCES

1. **Kelley, M. L., Jr.**, Skin lesions associated with chronic ulcerative colitis, *Am. J. Dig.*, 7, 255, 1962.
2. **Girard, M., Bel, A., Coulet, M., Vallon, C., and Bocher, M.**, Les lésions cutaniés à cause de la rectocolite hémorragique, *Lyon. Med.*, 210, 609, 1963.
3. **Morger, F.**, Zur "appendizitischen" Form der *Pasteurella-pseudotuberculosis*-Infektion beim Kind, *Praxis*, 51, 142, 1962.
4. **Mollaret, H. H.**, Une nouvelle cause possible d'Erythème noueux; le bacille de Malassez et Vignal ("*Pasteurella pseudotuberculosis*"), *Presse Med.*, 70, 1923, 1962.
5. **Undenstock, R. and Jeantaud-Bonnelie, R.**, Adénitie mésentérique et Erythème noueux due au bacille de Malassez et Vignal "*Pasteurella pseudotuberculosis*", *J. Med. Bordeaux*, 141, 1023, 1964.
6. **Kerzoncuf, A.**, L'érythème noueux à bacille de Malassez et Vignal, Ph.D. thesis Fac. Med., Copedith, Paris, 1967.
7. **Winblad, S.**, Erythema nodosum associated with infection with *Yersinia enterocolitica*, *Scand. J. Infect. Dis.*, 1, 11, 1969.
8. **Hannuksela, M. and Ahvonen, P.**, Erythema nodosum due to *Yersinia enterocolitica*, *Scand. J. Infect. Dis.*, 1, 17, 1969.
9. **Helander, I., Olkkonen, L., and Hopsu-Havu, V. K.**, Yersinia Infection as a cause of Erythema Nodosum, *Z. Hout. Geschlechtskr.*, 48, 399, 1973.
10. **Hannuksela, M. and Ahvonen, P.**, Skin manifestations in human yersiniosis, *Ann. Clin. Res.*, 7, 368, 1975.
11. **Niemi, K.-M., Hannuksela, M., and Salo, O. P.**, Skin lesions in human yersiniosis, *Br. J. Dermatol.*, 54, 155, 1976.
12. **Debois, J., Vandepitte, J., and Degreef, H.**, *Yersinia enterocolitica* as a cause of erythema nodosum, *Dermatologica*, 156, 65, 1978.
13. **Myging, N. and Thulin, H.**, Erythème noueux dans l'infection par "*Yersinia enterocolitica*", *Presse Med.*, 77, 1660, 1969.
14. **Berard, B. P.**, Premier cas francais d'erytheme noueux a *Yersinia enterocolitica* avec isolement de la souche, Ph.D. thesis Fac. Med., Paris, 1970.
15. **Assis, J.**, Erythema nodosum a *Yersinia enterocolitica*, *Ned. Tijdschr. Geneeskd.*, 115, 2156, 1971.
16. **Bagger, O. V. and Fischer, A. B.**, *Yersinia enterocolitica* som årsag til erythema nodosum, *Ugeskr. Laege.*, 134, 212, 1972.
17. **Zaremba, M. and Bolinska, J.**, Erythema nodosum as a sequel of infection due to *Yersinia enterocolitica* bacille (the first report in Poland), *Przegl. Dermatol.*, 63, 75, 1976.
18. **de Muylder, E., Thulliez, A., and Wauters, G.**, Erytheme noueux per *Yersinia enterocolitica*, *Arch. Belg. Dermatol. Syphiligr.*, 27, 25, 1978.
19. **Manigand, G., Pointud, Ph., Mallet, D., and Deparis, M.**, L'érythème noueux du aux infections à *Yersinia enterocolitica*, *Ann. Med. Intern.*, 129, 211, 1978.
20. **Wauters, G.**, Diagnostic biologique des infection à *Yersinia enterocolitica*, *Med. Malad. Infect.*, 3, 437, 1973.
21. **Wauters, G.**, Improved methods for the isolation and the recognition of *Yersinia enterocolitica*, *Contrib. Microbiol. Immunol.*, 2, 68, 1973.
22. **Winblad, S.**, Studies on serological typing of *Yersinia enterocolitica*, *Acta Pathol. Microbiol. Scand. Suppl.*, 187, 115, 1967.
23. **Winblad, S.**, Studies on O-antigen factors of *Yersinia enterocolitica*, *Symp. Ser. Immunbiol. Standard. Int. Symp. Pseudotub.*, 2, 337, 1968.
24. **Winblad, S.**, Studies on the O-serotypes of *Yersinia enterocolitica*, *Contrib. Microbiol. Immunol.*, 2, 27, 1973.
25. **Niléhn, B.**, Studies on *Yersinia enterocolitica* with special reference to bacterial diagnosis and occurrence in human acute enteric disease, *Acta Pathol. Microbiol. Scand. Suppl.*, 206, 1, 1969.

26. **Wauters, G.,** Contribution à l'étude de *Yersinia enterocolitica,* Ph.D. thesis Vander, Louvain, 1970.
27. **Wauters, G., LeMinor, L., Chalon, A., and Lassen, J.,** Supplement au schema antigènique de *"Yersinia enterocolitica",* Ann. Inst. Pasteur, 122, 951, 1972.
28. **Aho, K., Ahvonen, P., Lassus, A., Sievers, K., and Tillikainen, A.,** HLA 27 in reactive arthritis. A study of Yersinia arthritis and Reiter's disease, *Arthritis Rheum.,* 17, 521, 1974.
29. **Juhlin, I. and Ericson, C.,** A new medium for the bacteriologic examination of stools (LSU-agar), *Acta Pathol. Microbiol. Scand.,* 12, 185, 1961.
30. **Rappaport, F., Konforti, N., and Navon, B.,** A new enrichment medium for certain salmonellae, *J. Clin. Pathol.,* 9, 261, 1956.

Chapter 13

THE OCCURRENCE OF ANTIBODIES TO *YERSINIA ENTEROCOLITICA* IN THYROID DISEASES

Louis Shenkman and Edward J. Bottone

TABLE OF CONTENTS

I.	Introduction	136
II.	Antibodies to Yersinia in Patients with Thyroid Disease	137
	A. Scandinavia	137
	B. United States	138
	C. Israel	139
	D. Great Britain	140
III.	Methodology	141
IV.	Discussion	142
References		143

I. INTRODUCTION

With the exception of several distinct thyroid disorders whose pathogenesis have been defined, the etiology of most forms of thyroid disease is still unclear. The greatest attention has been directed towards understanding the pathogenetic mechanisms of Graves' disease and Hashimoto's thyroiditis, two disorders which share many pathologic and clinical similarities. Many causal factors such as (1) emotional stress, (2) genetic predisposition, (3) alterations in humoral and cell-mediated immunity, and (4) infections have been proposed for Graves' disease and Hashimoto's thyroiditis. The strongest evidence marshaled to date appears to implicate an immunologic pathway as the pathogenetic mechanism of these two disorders.

A substantial body of evidence has accumulated which indicates that Graves' disease and Hashimoto's thyroiditis are probably closely related immunologic disorders. Circulating thyroid antibodies are virtually always present in the sera of these patients and in 50% of their relatives. Thymic enlargement is present in both conditions and lymphocytosis, lymphadenopathy, and splenomegaly are commonly seen in Graves' disease. Circulating immunoglobulins capable of stimulating thyroid function occur in virtually all patients with Graves' disease and in a small number of those with Hashimoto's thyroiditis. Both disorders are associated with an increased incidence of other diseases which may have an immunologic basis, such as pernicious anemia and rheumatoid arthritis.

Evidence has been obtained that cell-mediated immunity plays a role in Hashimoto's thyroiditis and Graves' disease. Several investigators have demonstrated that if lymphocytes from animals with experimental thyroiditis are transferred to recipient animals, the recipients will develop thyroiditis.[6,7] It has been shown that lymphocytes from patients with Hashimoto's thyroiditis are cytotoxic to thyroid cells in tissue culture.[8] In patients with Graves' disease the presence of thymus-derived (T) lymphocytes sensitized against thyroid antigens have been demonstrated by migration-inhibition factor assay.[9,10]

Volpé[11] has incorporated many of these observations into an interesting theory that Graves' disease and Hashimoto's thyroiditis are due primarily to an inherited defect of immune surveillance, permitting specific thyroid-directed clones of T-lymphocytes to survive, interact with antigens on thyroid cells, and establish a cell-mediated immune reaction. Such T-lymphocytes may also direct and cooperate with groups of bone marrow-derived (B) lymphocytes which in turn produce immunoglobulin. In Graves' disease, the main immunoglobulins have thyroid-stimulating properties (such as long acting thyroid stimulated (LATS) and human thyroid stimulator). In Hashimoto's thyroiditis, the immunoglobulins produced are the well-known antithyroid antibodies.

The suggestion that patients with thyroid disease may have a defect in immune surveillance has been supported by recent studies which show that the depletion of T-cells by irradiating thymectomized animals leads to the development of chronic thyroiditis and antithyroid antibodies.[12] If the animals are reconstituted with normal syngeneic lymphoid cells, the autoimmune response to thyroid components is abrogated, thus providing evidence that the thyroiditis was related to a defect in immune surveillance.

From the available evidence at hand, it does appear that cellular and humoral immune mechanisms are operative in the pathogenesis of Graves' disease and Hashimoto's thyroiditis. While Volpe postulates that the primary defect is one of immune surveillance which allows a clone of thyroid-reactive T lymphocytes to expand,[5] others including Werner and Fierer[13] have suggested that an "occult" thyroid antigenic change, perhaps induced by viral or bacterial infection, is at fault. According to this view, an infectious agent causes an alteration in a thyroid antigen, and this antigenic change then serves as

Table 1
GEOGRAPHIC DISTRIBUTION AND RESULTS OF REPORTED STUDIES FOR ANTIBODIES TO *Y. ENTEROCOLITICA* IN THE SERUM OF PATIENTS WITH THYROID DISEASES

Author	Year	Country	Nature of *Y. enterocolitica* O:3 antigen	Serologic-immunological modality	Results
Bech et al.[14]	1974	Denmark	NS[b]	Agglutination titer ≥ 10[a]	Antibodies to serotype O:3 in 50% of patients with Graves' disease, 47% with nontoxic goiter, and 22% of controls
Lidman et al.[15]	1974	Sweden	NS	Agglutination titer 40 IFL[c]	74/100 sera with O:3 antibody by IFL with membrane region of thyrotoxic thyroid epithelial cells; absorption with formalin-killed O:3 cells abolished reaction
von Bonsdorf and Friman[18]	1974	Finland	NS	Agglutination titer ≥ 40	3/6 patients with thyroiditis had anti-*Yersinia* O:3, and O:9 antibodies
Horne et al.[19]	1975	England	NS	Agglutination IFL	2/3 sera with O:3 antibodies reacted X-1 IFL with thyroid cell membrane
Lidman et al.[16]	1976	Sweden	Heat-killed grown at 37°C	Agglutination IFL	79/163 sera with O:3 or O:9 agglutinins gave marginal staining of membrane region of thyroid epithelial cells; absorption with sonicated O:3 and O:9 antigens abolished IFL
Shenkman and Bottone[20]	1976	U.S.	Live, harvested after growth at 25°C	Agglutination titer ≥ 8	48/67 (75%) patients with a variety of thyroid disorders had $\geq 1:8$ titer; less than 8% prevalence in controls
Keddie et al.[23]	1977	England	NS	Agglutination	No antibodies to *Y. enterocolitica* O:3 noted with 26 patients with thyroid diseases
Bech et al.[17]	1978	Denmark	NS	Agglutination Leucocyte migration inhibition	Delayed type hypersensitivity towards *Y. enterocolitica* in patients with thyroid disease in contrast to controls
Reynolds et al.[24]	1978	Ireland	NS	Agglutination titer 160	Antibodies to O:3 in serum of a patient with thyrotoxicosis but absent in family members and 19 other patients with thyroid disease
Weiss et al.[22]	1979	Israel	Live, harvested after growth at 25°C	Agglutination	Antibodies to serotypes O:3, O:8, O:9 in 42% of 36 thyroid patients—77 controls negative

[a] Reciprocal of titer.
[b] Not specified.
[c] Indirect fluorescence.

the stimulus for the development of cell-mediated immunity, antithyroid antibodies, or thyroid-stimulating immunoglobulins.

Until recently, there has been little support for the role of an infectious agent in the pathogenesis of thyroid disease. However, several reports demonstrating the presence of anti-*Yersinia* antibodies in the sera of many patients with thyroid disease have rekindled interest in the possible relationship between infectious agents and thyroid disease.

II. ANTIBODIES TO YERSINIA IN PATIENTS WITH THYROID DISEASE

Retrospectively, as depicted in Table 1, a number of investigators working in diverse geographical locales have reported on the occurrence of anti-*Yersinia* antibodies in patients with various thyroid diseases.

A. Scandinavia

Initially in 1974, Bech et al.[14] in Denmark reported that patients with thyroid disease have a high prevalence of antibodies to *Y. enterocolitica*, serotype O:3.[14] Fifty percent of their patients with Graves' disease, 53% with nontoxic goiter (39% with diffuse goiter and 67% with nodular goiter), had antibody titers of 1:10 or greater. In contrast, only 22% of controls (17% healthy subjects and 28% of hospitalized patients with nonthyroidal disease) had similar titers.

Lidman et al.,[15] working in Sweden with cryostat tissue sections, reported that of 100 sera demonstrating *Yersinia* serotype O:3 agglutinins to a titer of 1:40, 74 reacted by indirect immunofluorescence with the membrane region of thyrotoxic thyroid epithelial cells. Repeated absorption of serum with formalin-killed *Y. enterocolitica* cultivated at 37°C abolished the cross-reactivity to the membrane tissue. In contrast, similar treatment of sera with an extract of pig stomach to remove potential smooth muscle antibody which would also cross-react with membrane tissue did not abolish membrane fluorescence.

By 1976, Lidman et al.[16] had extended their earlier observations to include 96 additional sera which also showed marginal staining of the membrane region of thyroid epithelial cells, but did not react with any other tissue. Of these sera, 26 were obtained from patients with acute *Y. enterocolitica* serotype O:3 infection while three out of four sera demonstrating serotype O:9 agglutinins also reacted with the cryostat thyroid tissue sections. These investigators showed, that (in contrast to their initial study) absorption with sonicated *Y. enterocolitica* O:3 and O:9 antigen, but not heat-killed whole cells, abolished thyroid fluorescence.

On the basis of these two studies, Lidman et al. suggested the presence of cross-reacting antigens in *Y. enterocolitica* and antigens in thyroid epithelial cells.

Utilizing a leucocyte-migration inhibition assay, Bech et al.[17] in 1978 amplified their previous studies (1974) and further substantiated the association between agglutinating antibodies to *Y. enterocolitica* and thyroid disease. These investigators showed a delayed type of hypersensitivity toward *Y. enterocolitica* O:3 in 64 patients with thyroid disease in contrast to 25 control subjects and two patients with known *Y. enterocolitica* infection. These authors attributed their findings to a cross-reaction between antigenic determinants of *Y. enterocolitica* and the thyroid cell.

Von Bondsdorff and Friman[18] reported on a 59-year-old patient with subacute thyroiditis whose illness was heralded by high fever, nausea, and diarrhea, symptoms compatible with *Yersinia* gastroenteritis. Although *Yersinia* could not be recovered from the patient's stool culture, antibody titers to serotype O:3 rose from 1:80 to 1:320 and subsequently fell again to 1:80 over a 6-week period. Because of this occurrence, these investigators examined the sera of five additional patients treated for thyroiditis, two of which also had antibody titers of at least 1:40 against *Y. enterocolitica*.

These reports suggesting a relationship between *Y. enterocolitica* serotypes O:3 and O:9 and thyroid disease in Scandinavia were certainly very intriguing. However, inapparent infections with *Yersinia* are common in Scandinavia as evidenced by the high prevalence of antibodies in healthy controls.

In Great Britain, where yersiniosis is uncommon, Horne et al.[19] tested 30 sera (10 positive for thyroid cell membrane antibody, 10 positive for smooth muscle antibody, and 10 negative for all autoantibodies) for *Y. enterocolitica* antibodies, all of which were negative. Nevertheless, two of three sera derived from patients with antibodies to *Y. enterocolitica* serotype O:3 did react by indirect immunofluorescence with thyroid-cell membrane.

B. United States

In order to determine whether the presence of antibodies to *Yersinia* in patients with thyroid disease is unique to Scandinavia or also occurs in the U.S. where serotype O:3 infection is still uncommon, Shenkman and Bottone[20] undertook a study to determine

Table 2
NUMBER OF SUBJECTS WITH SIGNIFICANT TITERS (≥ 1:8) AGAINST *Y. ENTEROCOLITICA* SEROTYPES O:3, O:8, AND O:9

	No.	Serotype O:3	Serotype O:8	Serotype O:9
Normal controls	110	9	2	0
"Febrile agglutinin" group	151	6	6	6
Rheumatoid arthritis @ SLE	43	3	0	0
Thyroid disease	67	35[a]	19[a]	14[a]

[a] The number of subjects with significant titers in each control group differed significantly at the level of $p < .001$ by Chi Square analysis for each serotype when compared with subjects with thyroid disease.

the frequency of antibodies to this particular *Yersinia* serotype as well as to serotypes O:8 and O:9 in patients with a variety of thyroid disorders, in normal subjects, and in patients with nonthyroidal disease.

Sixty-seven patients with a variety of thyroid disorders were studied: 36 had Graves' disease, 11 nontoxic nodular goiter, 6 autonomous adenoma, 5 idiopathic primary hypothyroidism, 7 Hashimoto's thyroiditis, and 2 carcinoma of the thyroid. Three patient populations were selected as control groups. The first consisted of 110 healthy subjects whose sera were submitted as part of a routine syphilis screening program. The second group consisted of 151 sera submitted for assay for febrile agglutinins. The third control group consisted of sera from 43 patients with rheumatoid or systemic lupus erythematosus.

In contrast to the low prevalence (less than 8%) of antibody titers to *Y. enterocolitica* serotype 0:3 in normal controls and in patients with nonthyroidal disease, patients with thyroid disease had a markedly increased frequency (75%) of significant titers to this organism (Table 2). Of the entire group of 67 patients, 35 or 52% had titers of 1:8 or greater against serotype O:3, 20 (29%) had significant titers against serotype O:8, and 14 (22%) had significant titers against serotype O:9. Of the entire group of 67, 48 (75%) had significant titers to one or more of the 3 serotypes studied. Antibody titers were demonstrated with the sera of 24 of 36 patients with Graves' disease, 5 of 6 with autonomous adenoma, 7 of 7 with Hashimoto's thyroiditis, 3 of 5 with idiopathic primary hypothyroidism, 4 of 11 with nontoxic nodular goiter, and 1 of 2 with thyroid carcinoma. Antibodies to serotype O:3 were the most prevalent, occurred in the highest titers, and were found particularly in patients with Graves' disease. None of these patients had clinical evidence of a *Yersinia* infection.

In this study, thyroid status did not correlate with the presence of antibody titers against *Yersinia*. There was no significant difference in titers between the patients with Graves' disease who were thyrotoxic and those who had been rendered euthyroid at the time of study. Similarly, in patients with idiopathic primary hypothyroidism and Hashimoto's thyroiditis, there was no difference in titers between the hypothyroid subjects and those who had been rendered euthyroid with thyroid hormone replacement.

C. Israel

The prevalence of *Y. enterocolitica* infections as determined culturally and serologically is low in Israel and in a recent study was 0.1% in symptomatic patients.[21] Thus, it was considered worthwhile to examine the sera of patients with thyroid disease for antibodies to *Y. enterocolitica* serotypes O:3, O:8, and O:9 and compare the findings with the results

Table 3
PREVALENCE OF *Y. ENTEROCOLITICA* IN SERA OF THYROID PATIENTS RESIDING IN ISRAEL

Disease	No. of patients	Antibody to serotype			No. of sera reactive against a minimum of one serotype
		O:3	O:8	O:9	
Control	77	0	0	0	0
Graves' disease	9	3	2	0	4
Hashimoto	9	2	4	1	5
Primary myxedema	4	1	2	1	2
Autonomous nodule	2	0	1	0	1
Adenoma	3	0	0	0	0
MN goiter	7	1	0	0	1
S. A. thyroiditis	2	1	1	0	2

noted in Scandinavia and in the U.S. Weiss et al.,[22] therefore, studied 36 patients with various thyroid disorders. Nine patients had Graves' disease, nine Hashimoto's thyroiditis, three thyroid adenoma, two autonomous nodules, seven multinodular goiter, four primary hypothyroidism, and two subacute thyroiditis. The control group consisted of 77 patients referred to the thyroid clinic, but who were free of thyroid disease.

The prevalence of positive *Yersinia* antibodies can be seen in Table 3. Significant anti-*Yersinia* titers (1:8 to 1:64) were obtained with the sera of 42% of 36 thyroid patients, 4 of the 9 with Graves' disease, 5 of 9 with Hashimoto's thyroiditis, 2 of 4 with primary hypothyroidism, 1 of 7 with multinodular goiter, 2 with subacute thyroiditis, and 0 of 3 with an adenoma. None of the sera, even undiluted, from the 77 subjects comprising the control group were reactive against the *Yersinia* antigens tested. The differences in prevalence of *Yersinia* antibodies between the controls and the thyroid patients as a whole or subgrouped according to their disease was statistically significant $p(X^2) \leq 0.0001$ as determined by the Chi square test.

No correlation was found between thyroid patients with or without antithyroglobulin antibodies and with *Yersinia* antibodies. These findings indicate that the prevalence of significant titers in patients with thyroid disorders is elevated (42%) in Israel and in this regard correlates with that reported on in both Scandinavia (89%) and in the U.S. (75%) irrespective of the prevalence of positive titers in the general population. Indeed, in the study by Shmilovitz and Kretzer,[21] undertaken in Israel, none of 129 sera obtained from adults comprising a control group were reactive against *Y. enterocolitica* serotype O:3. These observations strengthen the view that thyroid patients behave differently with regard to *Y. enterocolitica* antibodies than do subjects in the general population.

D. Great Britain

In contrast to the high prevalence of anti-*Yersinia* antibodies in thyroid disease patients noted in Scandinavia, the U.S., and in Israel, several investigators from Great Britain have not been able to find a similar association.

Keddie et al.[23] from Manchester, England, studied 26 consecutive patients admitted for thyroid surgery. Seven had Hashimoto's thyroiditis, eleven Graves' disease, and nine colloid goiter. None were found to have agglutinating antibodies to serotype O:3. Sera were not tested for antibodies to serotypes O:8 and O:9.

Reynolds et al.[24] examined 19 patients with thyroid disease in Ireland. Fourteen had Graves' disease; two a single autonomous nodule, two multinodular toxic goiter, one

Hashimoto's thyroiditis, and one idiopathic hypothyroidism. One patient, a 13-year-old girl with Graves' disease and exophthalmos, was found to have a 1:160 antibody titer to serotype O:3 *Y. enterocolitica*. The patient underwent subtotal thyroidectomy 9 months later, and of note was the finding that at the time of surgery antibody titers were 1:80. As in the previously noted study, antibodies to serotype O:8 and O:9 were not examined.

The seeming lack of correlation in prevalence of anti-*Yersinia* antibodies in thyroid disease patients noted in these two studies, as compared to the previously quoted studies from Scandinavia, the U.S., and Israel, may be related to methodologic factors, or perhaps to slight antigenic differences in the strains of serotype O:3 *Y. enterocolitica* occurring in the various geographic regions and, as detailed below, used for testing.

III. METHODOLOGY

The methods used to demonstrate cross-reacting *Yersinia* antibodies in the sera of patients with thyroid diseases have varied according to investigators. In most instances the serological approach has been one of agglutination, but little information has been presented as to the nature of the *Y. enterocolitica* antigen used in these studies (Table 1).

The phenotypic expression of surface antigens of *Y. enterocolitica* is associated with temperature of growth. Examples of this phenomenon include the synthesis of flagella at 25°C but not at 37°C, loss of phage susceptibility after growth at 37°C, and increased resistance to the bactericidal effect of normal serum by *Y. enterocolitica* cells grown at 25°C.[25] According to Niléhn,[25] these findings may reflect differences in cell wall structure due to a temperature-dependent metabolism. It is, therefore, critical that a degree of antigen uniformity be maintained among investigators searching for anti-*Yersinia* antibodies in the sera of thyroid patients. The studies performed in our laboratory made use of live suspensions of *Y. enterocolitica* O:3, O:8, and O:9 harvested after overnight growth at 25°C on trypticase soy agar containing 5% sheep blood. The phage type (VIII, IXb) of the serotype O:3 isolate did not influence the outcome.

The nature of the *Y. enterocolitica* antigen used for demonstrating serum agglutinins is critical. Our earlier experiences with the serologic diagnosis of yersiniosis have highlighted the subtlety inherent in this approach. In one instance, the serum of a patient with *Y. enterocolitica* bacteremia reacted faintly in agglutination tests with the blood isolate after it was grown at 37°C. In contrast, the same serum sample rendered distinct clumping of the same isolate to a titer of 1:1024 after growth at 25°C. A similar phenomenon was noted by Shmilovitz and Kretzer.[21] These investigators showed agglutinins in the sera of patients with known *Y. enterocolitica* infection only after growth at 25°C of the locally derived test organism. Using a serotype O:3 strain from their stock collection, however, posed no such problem as these same sera agglutinated this strain irrespective of whether the strain was grown at 25 or 37°C.

Additionally, in our laboratory we have encountered a "prozone" phenomenon with several sera containing high-titered (1:1024) anti-*Yersinia* antibodies. The variable results observed by different authors regarding the presence of agglutinins in the sera of thyroid patients may, in part, be explained by one or all of the aforementioned factors. The instances wherein agglutinins were observed with the sera of thyroid patients against test organisms grown at 37°C may be analogous to the experience of Shmilovitz and Kretzer with their collection strain. Repeated subculture of test O:3 or O:9 strains at 37°C may well have resulted in loss of surface antigens and exposure of recessed antigens against which serum from thyroid patients reacted.

Of interest, and possibly related, are the findings of Lidman et al.[16] who, in their second study, while unable to absorb anti-*Yersinia* agglutinins from the sera of thyroid patients with a heat-killed suspension of *Y. enterocolitica* prepared after growth at 37°C, could remove the cross-reacting antibody using a sonicated *Y. enterocolitica* preparations. Sonication may have revealed subsurface antigens normally expressed at 25°C which were then available for reacting with sera derived from the thyroid patients. It is also conceivable that the complete absorption of *Yersinia* agglutinins from sera with whole *Y. enterocolitica* cells as noted in their initial study, and its absence in their second study, was related to using a different strain or to additional surface alterations of the original strain used as an antigen source.

IV. DISCUSSION

Several possible mechanisms may be considered to account for the high prevalence of antibodies to *Y. enterocolitica* in the sera of patients with thyroid disorders. An intriguing suggestion is that actual infection with *Yersinia* may, in some fashion, be responsible for the development of thyroid disease. Clinical infections with *Yersinia* have been followed by the development of a variety of autoantibodies, including antibodies directed against smooth muscle, connective tissue, renal tubular epithelium, and the basement membrane of thyroid epithelial cells. It is also interesting to note that *Yersinia* infection may be followed by postinfective sequelae (such as erythema nodosum, myocarditis, and polyarthritis) similar to those observed after Group A streptococcal infection.[26] The propensity of *Yersinia* infections to be followed by the development of a variety of autoantibodies suggests that this microorganism shares antigens with mammalian tissue.

One might then postulate that *Yersinia* infection results in the proliferation of a clone of T cells directed at a cross-reacting thyroid antigen, which then mount a cell-mediated autoimmune response against the thyroid. Similarly, B cells may be activated with the resultant production of a variety of autoantibodies.

Arguing against a direct pathogenetic role of *Yersinia* in the development of thyroid disease is the fact that clinical yersiniosis is seen at a much lower frequency than would be expected by the high prevalence of titers in patients with thyroid disease. In the U.S. and Israel, for example, infections with serotypes O:3 are still uncommon, while a large number of patients with thyroid disease were found to have antibodies to this organism. Additionally, of the published[21,22] studies originating from Israel, *Y. enterocolitica* agglutinins have only been demonstrated with the sera of thyroid patients and those with known *Y. enterocolitica* infections. In these two studies, none of 206 control sera contained antibody to *Y. enterocolitica*. Conceivably, patients with thyroid disease may be more susceptible to infection with *Yersinia* than the general population and the demonstration of a high prevalence of anti-*Yersinia* antibodies in these patients is a reflection of their increased susceptibility. However, one would have to postulate inapparent or subclinical infection since none of the reported patients with thyroid disease gave a history of illness compatible with yersiniosis. The possibility exists that thyroid patients may be asymptomatic gastrointestinal carriers of *Y. enterocolitica*, which could then explain their serum agglutinins. To date, however, stool cultures have been examined for 20 thyroid patients. At least two specimens were obtained from all subjects, while eight had three specimens examined. Stools were plated directly onto Endo and MacConkey agar plates, and were also subjected to cold enrichment in phosphate buffered saline (pH. 7.2) for 28 days with weekly plating to Endo and MacConkey agar. In no instance was *Yersinia* isolated. These results indicate that inapparent infection with *Yersinia* probably does not occur in patients with thyroid disease, and that the high prevalence of antibodies to *Yersinia* is unlikely to be a result of exposure to this bacterium.

The association of anti-*Yersinia* antibodies in patients with thyroid disease may be explained on the basis of a cross-reaction between some thyroid antigen(s) and antigen(s) of *Yersinia*. That bacterial antigens of both Gram-positive and Gram-negative organisms may cross-react with various components of mammalian tissues has been recognized for some time.[27] Notable examples of this phenomenon include the discovery of common antigens shared by group A streptococci and human myocardium[28] and the observations that many Gram-negative bacteria have antigens common to blood group substances.[29]

In our laboratory a number of experiments have been undertaken in an attempt to determine whether antigens of *Y. enterocolitica* cross-react with constituents of thyroid tissue. First, rabbits were inoculated either with human or rabbit thyroid tissue preparations, and their sera subsequently examined for the development of antibodies to *Yersinia*. The inoculum consisted of minced homogenized thyroid tissue emulsified in Freund's complete adjuvant, and injected subcutaneously into the footpads of New Zealand white rabbits twice monthly. The rabbits were bled periodically and serum studied serologically for anti-*Yersinia* antibodies. By agglutination and counter-immunoelectrophoresis, anti-*Yersinia* antibodies could not be detected. Anti-*Yersinia* antibodies have been detected, however, by indirect immunofluorescence. Serum from immunized rabbits was exposed to heat-killed *Yersinia* grown at 25°C and reacted with fluorescein labelled goat anti-rabbit IgG. At titers of 1:64, fluorescence was seen against *Yersinia* serotypes O:3, O:8, and O:9, but not against *Escherichia coli* or B-hemolytic streptococci. Fluorescence was abolished by adsorbing serum with dense suspensions of the homologous *Yersinia*, but not with *E. coli* or B-hemolytic streptococci. Further studies are in progress in an attempt to learn more about the putative cross-reacting antigens.

REFERENCES

1. **Gibson, J. G.**, Emotions and the thyroid gland: a critical appraisal, *J. Psychosom. Res.*, 6, 99, 1962.
2. **Martin, L.**, The hereditary and familial aspects of exophthalmic goiter and nodular goiter (with a genetical note by R. A. Fisher), *Q. J. Med.*, 14, 207, 1945.
3. **Farid, N. R., Munro, R., Row, V. V., and Volpé, R.**, Peripheral thymus-dependent (T) lymphocytes in Graves disease and Hashimoto's thyroiditis, *N. Engl. J. Med.*, 288, 1313, 1973.
4. **Iverson, K.**, An epidemic of thyrotoxicosis in Denmark during World War II, *Am. J. Med. Sci.*, 217, 121, 1949.
5. **Volpé, R., Farid, N. R., Von Westarp, C., and Row, V. V.**, the pathogenesis of Graves' disease and Hashimoto's thyroiditis, *J. Clin. Endocrinol. Metab.*, 3, 239, 1974.
6. **Felix-Davies, D. and Waksman, B. H.**, Passive transfer of experimental immune thyroiditis in the guinea pig, *Arthritis Rheum.*, 4, 416, 1961.
7. **Twarog, F. J. and Rose, N. R.**, Transfer of autoimmune thyroiditis of the rat with lymph node cells, *J. Immunol.*, 104, 1467, 1970.
8. **Laryea, E., Row, V. V., and Volpé, R.**, The effect of blood leucocytes from patients with Hashimoto's Disease on human thyroid cells in monolayer culture, *Clin. Endocrinol.*, 2, 23, 1973.
9. **Lamki, J., Row, V. V., and Volpé, R.**, Cell-mediated immunity in Graves' Disease and in Hashimoto's thyroiditis as shown by the demonstration of migration inhibition factor (MIF), *J. Clin. Endocrinol. Metab.*, 36, 358, 1973.
10. **Brostoff, J.**, Migration inhibition studies in human disease, *Proc. R. Soc. Med.*, 63, 905, 1970.
11. **Volpé, R.**, The role of autoimmunity in hypoendocrine and hyperendocrine function with special emphasis on autoimmune thyroid disease, *Ann. Intern. Med.*, 87, 86, 1977.
12. **Penhale, W. J., Irvine, W. J., Inglis, J. R., and Farmer, A.**, Thyroiditis in T cell-depleted rats: suppression of the autoallergic response by reconstitution with normal lymphoid cells, *Clin. Exp. Immunol.*, 25, 6, 1976.
14. **Bech, K., Larsen, J. H., Hansen, J. M., and Nerup, J.**, *Yersinia enterocolitica* infection and thyroid disorders (letter), *Lancet*, 2, 951, 1974.
15. **Lidman, K., Eriksson, U., Fagraeus, A., and Norberg, R.**, Antibodies against thyroid cells in *Yersinia enterocolitica* infection, (letter), *Lancet*, 2, 1449, 1974.

16. **Lidman, K., Eriksson, U., Norberg, R., and Fagraeus, A.**, Indirect immunofluorescence staining of human thyroid by antibodies occurring in *Yersinia enterocolitica* infections, *Clin. Exp. Immunol.*, 23, 429, 1976.
17. **Bech, K., Clemmensen, D., Larsen, J. H., and Nerup, J.**, Cell-mediated immunity to *Yersinia enterocolitica* serotype 3 in patients with thyroid disease, *Allergy*, 33, 82, 1978.
18. **Von Bonsdorff, M. and Friman, C.**, *Yersinia enterocolitica* infection and thyroid disorders (letter), *Lancet*, 2, 1565, 1974.
19. **Horne, C. H. W., Briggs, R. D., Nicol, A. G., and Hopkins, L. M.**, Thyroid-cell-membrane antibodies (letter), *Lancet*, 2, 411, 1975.
20. **Shenkman, L. and Bottone, E. J.**, Antibodies to *Yersinia enterocolitica* in thyroid disease, *Ann. Intern. Med.*, 85, 735, 1976.
21. **Shmilovitz, M. and Kretzer, B.**, Isolates of *Yersinia enterocolitica* from clinical cases in Northern Israel, *Isr. J. Med. Sci.*, 14, 1048, 1978.
22. **Weiss, M., Rubinstein, E., Bottone, E. J., Shenkman, L., and Bank, H.**, *Yersinia enterocolitica* antibodies in thyroid disorders, *Isr. J. Med. Sci.* 15, 553, 1979.
23. **Keddie, N., Metcalfe-Gibson, C., and Tooth, J. A.**, *Yersinia* and thyroid disease (letter), *Lancet*, 1, 368, 1977.
24. **Reynolds, M. T., Keane, C. T., Tomkin, G. H., Roberts, J. C., Lenehan, T. J., and Mair, N. S.**, Antibodies to *Yersinia enterocolitica* serotype 3 in thyroid disease, *Br. Med. J.*, 2, 400, 1978.
25. **Niléhn, B.**, The relationship of incubation temperature to serum bactericidal effect, pathogenicity and in vivo survival of *Yersinia enterocolitica*, in *Contributions to Microbiology and Immunology*, Vol. 2, Winblad, S., Ed., S. Karger, Basel, 1973, 85.
26. **Leino, R. and Kalliomaki, J. L.**, Yersiniosis as an internal disease, *Ann. Intern. Med.*, 81, 458, 1975.
27. **Lyampert, I. N. and Danilova, T. A.**, Immunological phenomena associated with cross-reactive antigens of micro-organisms and mammalian tissues, *Prog. Allergy*, 18, 423, 1975.
28. **Kaplan, M. H. and Meyeserian, M.**, An immunological cross-reaction between group-A streptococcal cells and human heart tissue, *Lancet*, 1, 706, 1962.
29. **Springer, G. F.**, Blood-group and Forssman antigenic determinants shared between microbes and mammalian cells, *Prog. Allergy*, 15, 9, 1971.

Chapter 14
ZOONOTIC *YERSINIA ENTEROCOLITICA* INFECTION: HOST RANGE, CLINICAL MANIFESTATIONS, AND TRANSMISSION BETWEEN ANIMALS AND MAN

Bengt Hurvell

TABLE OF CONTENTS

I.	Introduction	146
II.	Isolations of *Y. enterocolitica* in Animals	146
III.	Clinical Manifestations	149
IV.	Transmission of *Y. enterocolitica* Between Animals and Man	152
References		155

I. INTRODUCTION

Yersinia enterocolitica was originally considered to be a variant of *Pasteurella pseudotuberculosis*.[1,2] Other names that have been proposed are *Bacterium enterocolitica*,[3] *Pasteurella X*, and *Pasteurella Y*.[4,5] The species name *Y. enterocolitica* was proposed by Fredriksen in 1964.[6]

The International Subcommittee on *Yersinia* accepted five reference strains of *Y. enterocolitica* at a symposium on *Yersinia, Pasteurella,* and *Francisella* held in Malmö, Sweden in 1972.[7] At this symposium it was also proposed that *Y. enterocolitica* should belong to the family *Enterobacteriaceae*, and in the latest edition of *Bergey's Manual of Determinative Bacteriology Y. enterocolitica* is placed under this family.[8]

The frequency of performed isolations of *Y. enterocolitica* has risen dramatically in recent years. The increase is noted not only in the number of patients with *Y. enterocolitica* infection, but also in the number of countries in which *Y. enterocolitica* has been recovered from man and animals; this applies, in particular, to various countries in Europe. Up to the present, reports have been presented from 30 different countries, but presumably, this infection occurs in several other countries, notably in Africa and Asia. There are great variations in the frequency of performed isolation, but the distribution of the different serotypes of *Y. enterocolitica* also varies from country to country. The bacterium seems to be distributed over the greater part of the world and isolations indicate that it occurs frequently in nature.[9-11] Epidemiological studies have shown that *Y. enterocolitica* occurs in water, fruits, and vegetables and in a large number of wild and domesticated animals. In spite of increased watchfulness and awareness as regarding infection in man, there are many unknown factors relating to the ecology, routes of spread, and other details of the bacterium.

This paper will deal with zoonotic aspects of *Y. enterocolitica* infection with special reference to the prevalence of *Y. enterocolitica* in animals and any clinical manifestations in the different animal species. The tentative role of animals as the transmitters of *Y. enterocolitica* bacteria to humans will also be discussed.

II. ISOLATIONS OF *Y. ENTEROCOLITICA* IN ANIMALS

Recovery of *Y. enterocolitica* from animals was described for the first time in the early 1960s from swine, chinchillas, and hares.[2,12,13] Later, isolations from swine,[14-37] chinchilla,[14-16,21,25,38-42] and hares[16,21,23,25,43,44] have been reported from several countries, notably in Europe.

Isolates from farm animals other than pigs have also been reported, namely from cattle,[16,21,25,27,45-47] chicken,[27,47] goats,[48] horses,[16,47] and sheep.[16,21,42] *Y. enterocolitica* has been isolated from the following pet animals: cat,[16,18,23,33,49] dog,[14-16,18,21,23,29,31,33,47,49-53] guinea-pig,[16,21,54] and rabbit.[21,23,54] Besides the isolates from chinchilla, *Y. enterocolitica* has been recovered from another commercially-reared fur-bearer, namely the mink.[42,54] *Y. enterocolitica* has been isolated from other free-living wild mammals besides hares, namely the beaver,[49,55] deer,[39,56] racoon,[55] red fox,[57] and small rodents.[23,26,47,50,57-62] There are even a few isolations from free flying birds, such as the buzzard,[63] Canada goose,[49,55] and pigeon.[42,54,64] *Y. enterocolitica* has also been recovered from cage birds.[16,40,47,55]

Isolations of *Y. enterocolitica* have been obtained from zoo animals, such as the camel,[54,55] monkey,[56,65] ocelot,[42] and Pekin robin.[55] Futher, *Y. enterocolitica* has been isolated from fish,[23,66] frog,[67] snail,[67] and oyster.[49]

Most of the isolated strains from animals differ, generally, both biochemically and serologically from the strains found as causes of disease in man. The distribution of the different serotypes in animals has not yet been made fully clear. Table 1 summarizes the

Table 1
DISTRIBUTION OF *Y. ENTEROCOLITICA* SEROTYPES ISOLATED FROM DIFFERENT ANIMALS

Animals	Serotypes	Year	Authors
Canada goose	4/33	1973	Toma[49]
	4/33	1974	Hacking and Sileo[55]
Camel	5/27	1974	Hacking and Sileo[55]
Canard	6	1976	Alonso et al.[47]
Cat	9	1973	Ahvonen et al.[18]
	3	1973	Rakovsky et al.[23]
	3	1973	Toma[49]
Cattle	5, 6	1973	Vandepitte et al.[25]
	4, 5, 7, 10, 12, 14, 22, 27, 31	1975	Inoue and Kurose[46]
	17	1976	Alonso et al.[47]
Chicken	4/10	1976	Alonso et al.[47]
Chinchilla	1	1972	Mollaret[16]
	6	1973	Vandepitte et al.[25]
Deer	12	1973	Otsuki et al.[56]
Dog	3, 9	1973	Ahvonen et al.[18]
	3	1973	Rakovsky et al.[23]
	5/27	1973	Toma[49]
	3, 4	1975	Tsubokura et al.[50]
	7, 10, 13	1976	Alonso et al.[47]
	6	1976	Befekadu et al.[51]
	3	1976	Farstad et al.[52]
	3, 5, 7	1976	Pedersen[29]
	20	1976	Wilson[53]
	3, 5a, 5b, 6, 9	1977	Kaneko et al.[31]
Fox (red)	6	1977	Kapperud[57]
Goat	2	1972	Krogstad et al.[48]
Hare	2	1972	Mollaret[16]
	2a/2b/3	1973	Rakovsky et al.[23]
	2	1973	Vandepitte et al.[25]
Horse	10	1976	Alonso et al.[47]
Monkey	5	1970	Mair et al.[65]
	4, 5, 6, 12, 14	1973	Otsuki et al.[56]
Oyster	5, 6/30, 7/13	1973	Toma[49]
Pekin robin	6/30	1974	Hacking and Sileo[55]
Rabbit	7/8	1973	Rakovsky et al.[23]
Racoon	5/27	1974	Hacking and Sileo[55]
Small rodents	6	1973	Rakovsky et al.[23]
	1, 2a/2b/3, 4, 8	1974	Aldova and Lim[58]
	5a, 6, 7, 9, 14, 16, 25	1974	Zen-Yoji et al.[26]
	4, 5, 7, 12, 20	1975	Tsubokura et al.[50]
	1/2a, 4, 5, 6, 7/8, 10	1976	Alonso et al.[47]
	6, 7	1976	Głośnicka et al.[60]
	3	1977	Aldova et al.[61]
	1, 3, 4, 6, 7, 12, 16	1977	Kapperud[57]
	3	1977	Pokorna and Aldova[62]
Swine	3	1972	Ahvonen[17]
	3	1972	Mollaret[16]
	3	1973	Ahvonen et al.[18]
	3, 9	1973	Akkermann and Hill[19]
	3, 9	1973	Esseveld and Goudzwaard[20]
	3	1973	Rabson and Koornhof[22]
	3	1973	Rakovsky et al.[23]
	3, 5, 10, 12	1973	Tsubokura et al.[24]
	3, 4, 5, 6	1973	Vandepitte et al.[25]

Table 1 (continued)
DISTRIBUTION OF *Y. ENTEROCOLITICA* SEROTYPES ISOLATED FROM DIFFERENT ANIMALS

Animals	Serotypes	Year	Authors
	3, 5a, 5b, 6, 9, 12, 14, 16, 25	1974	Zen-Yoji et al.[26]
	3, 5, 5/27, 6/30, 7/13, 8/14	1975	Toma and Deidrick[28]
	10	1976	Alonso et al.[47]
	3, 5, 6, 7, 12, 17, 19, 26b, 26c	1976	Pedersen[29]
	3, 5, 6, 7, 8, 10, 11, 12, 13, 14, 16, 19	1976	Tsubokura et al.[30]
	3	1978	Hurvell et al.[35]
	2b, 3, 5, 5/27, 6/30, 7/8, 9	1977	Narucka and Westendoorp[32]
	9	1978	Bockemühl and Roth[34]
	3, 4, 4/13, 4/15, 5, 5a, 6, 7, 7/13, 13, 13/15, 15	1978	Hurvell[63]
Trout (brown)	4/33, 17, 18, 21	1976	Kapperud and Jonsson[66]

different serotypes for isolates obtained in animals. It is based on the data available in the papers which are referred to in Table 1.

It has long been suspected that the animal kingdom would be a reservoir for *Y. enterocolitica* and hence, the source of human infections.[28,29,31,41,57-59,61,66] It will be seen from Table 1 that the different isolates represent a variety of serotypes. Many animal species, notably chinchillas and most wild animals, have serotypes that do not correspond to those usually found in man. Some *Y. enterocolitica* strains, e.g., O:2 (biotype 5), which have been isolated mainly from hares, and O:1 (biotype 3), isolated mainly from chinchillas, have never been isolated from man.[28] Yet, strains from domestic animals, such as cattle, goats, and horses, have also serological patterns that deviate from those of human isolates. This holds true for strains isolated from cage birds and zoo animals as well.

Regarding pigs, dogs, and cats, the conditions are different. In studies reported from Europe, Japan, Canada, and South Africa *Y. enterocolitica* serotype O:3 has in many cases been isolated from pigs (see references in Table 1). This serotype had also been isolated from cats and dogs. Serotype O:9 has been recovered from pigs as well as from dogs and cats. As these serotypes are common disease-producers in man and the most common human isolates in Europe,[17,68,69] pigs in particular have been suspected to be reservoirs of infection in this part of the world. Several workers in Japan, Canada, and South Africa have also unequivocally suspected pigs as a main reservoir for human *Y. enterocolitica* infections.[20,22-24,26,28-30,32,35]

Serotype O:5 has also been frequently isolated from pigs (see references in Table 1). This serotype is common in human infections in Japan.[26] Recovery of this serotype in swine strongly suggests that these animals are of great importance as a source of *Y. enterocolitica* to humans.[24,26,30]

The cause of this geographical distribution of the different serotypes in man as well as animals remains an unsolved mystery.[11] The occurrence and the distribution of the serotypes differ not only between the continents, but also between geographically adjacent countries. Of the Nordic countries, Finland has a much higher frequency of serotype O:9 than, for instance, Denmark, Norway, and Sweden. Moreover, in Sweden serotype O:9 has been isolated exclusively in the northern part of the country. Serotype O:3 has often been isolated both from man and from animals in Canada; this serotype is extremely unusual in the U.S.[11,70] Another epidemiologically mystifying fact is that isolates of serotype O:3 in Canada belong to phage type 9b. This phage type, referred to as the Canadian phage type, has indeed never been isolated in Europe, Africa, Asia, or elsewhere.[28,71] The reverse is the case for serotype O:8, which is the predominating disease-producing serotype in man in the U.S.[11,72,73]

A fact of special epidemiological interest is that this American serotype O:8 has not yet been isolated from animals in the U.S. Deer have, however, been suspected to be a reservoir for human infections with serotype O:8.[74]

III. CLINICAL MANIFESTATIONS

In the early 1960s, several enzootics in chinchillas were reported to occur in Europe where great numbers of animals died in septicemia and a bacterium described as *Pasteurella pseudotuberculosis*-like could be isolated.[12,14,38] Similar outbreaks of disease in chinchillas in December 1964 and in February 1965 have been reported from California in the U.S. as well as from Mexico, where cases were discovered in February 1964 on a farm with 2000 chinchillas.[39] The bacterial isolates from all these outbreaks have later been identified as *Y. enterocolitica*.[39,68]

Diseased chinchillas showed sluggishness, anorexia, weightloss, profuse salivation, and diarrhea. In general, the outbreaks were not explosive, but the animals fell ill one after another. Many of the animals died in septicemia and the bacteria were isolated from different organs. Autopsy revealed exudative inflammation of the nasal passages, necrotic lesions in spleens, livers, mucosa and submucosa of colon, caecum, and ileum as well as peritonitis, perihepatitis, perisplenitis, and catarrhal enteritis.[12,38,39] Langford also described similar clinical manifestations and pathoanatomical changes in chinchillas from which *Y. enterocolitica* had been isolated.

Mollaret and Lucas[43] report a number of cases in hares which during the years 1961 to 1964 had died with septicemia due to *Y. enterocolitica*. They describe the following pathoanatomical changes: pleuritis and peritonitis with well-defined areas of exudation, enlarged spleens, and extensive enteritis.

Although, as will be seen from Table 1, several reports have been presented on isolation of *Y. enterocolitica* from a variety of other free-living wild animals, descriptions of observed clinical manifestations or pathoanatomical changes are extremely sparse. Hacking and Sileo[55] describe abscess formation in the liver and intestine of beavers and in the liver of racoons. They concluded that these abscesses were caused by *Y. enterocolitica* and that yersiniosis was the cause of death. The same authors reported the recovery of *Y. enterocolitica* in a Canada goose, which, however, lacked both overt gross pathoanatomical lesions and histological organic changes.

Langford[42] reported that *Y. enterocolitica* has been isolated from diseased or dead pigeons and a mink in which "the frequency of observed lesions in tissues from the greatest to the least was as follows: liver, spleen, intestine, lymph node, kidney, and lung."

Isolation of *Y. enterocolitica* from deer has been reported from the U.S.[39] and Japan.[56] No clinical symptoms were noted, however, and the isolations were made in healthy, apparently normal, deer.

As regards small wild rodents, clinical data are sparse and few reports of findings at autopsy and histological examinations have been presented. Besides enlarged spleens noted with voles and field mice trapped in Czechoslovakia, no gross pathoanatomical changes were noted in animals from which *Y. enterocolitica* was isolated.[58] On the other hand, the authors state that the frequency of positive *Y. enterocolitica* isolations was not higher in animals with enlarged spleens. Totally, samples were taken from 446 animals and 60 *Y. enterocolitica* strains were isolated. The frequency of positive findings in *Microtus arvatis* and *Apodemus sylvaticus* ranged from 3.4% to 26.7%.

Kapperud[57,59] has also made a comprehensive detailed study of the frequency of *Y. enterocolitica* in small wild living rodents. He found *Y. enterocolitica* in 9% of his material from widespread localities in Fennoscandia. Autopsy of 38 naturally infected small rodents showed no apparent gross pathoanatomical changes. Histopathological examination of spleen and liver from six of the small rodents also gave negative results.

Three cases of yersiniosis in monkeys caused by *Y. enterocolitica* have been described by McClure et al.[75] The following pathoanatomical changes were noted: necrotic abscesses in the liver, spleen, and several different lymph nodes and inflammatory lesions in the intestine; in addition, coalescent mucosal ulcers were seen in the proximal two thirds of the colon of one of the monkeys.

Mollaret et al.[76] report six cases of infections with *Y. enterocolitica* in a zoo in Sao Paolo in 1968. All the monkeys died within a few days of the onset of symptoms. The only clinical symptoms that appeared were rapid emaciation and, in one of the animals, profuse diarrhea.

Mair et al.[65] reported an outbreak of gastroenteritis among bushbabies (Galago) in a private zoo in Yorkshire, England. Of the animals, five fell ill suddenly with sluggishness and diarrhea as apparent clinical symptoms. The diseased animals died within 5 hr. The clinical symptomatology resembled that seen in earlier cases of *Y. pseudotuberculosis* infections in the zoo. The diarrhea seemed to be more profuse than was usual in the cases of infections with *Y. pseudotuberculosis*. The liver and spleen were congested and multiple pale necrotic foci about 1 mm in diameter were present in the spleen. At serological typing the isolated *Y. enterocolitica* was found to belong to serotype O:5.

As regards clinical symptoms and/or pathoanatomical changes in diseases due to *Y. enterocolitica* in other zoo animals, reports are sparse. Hacking and Sileo[55] describe a case in a male Pekin robin in Canada which died suddenly with *Y. enterocolitica* infection. Clinical examination showed that the intestinal content was a frothy yellow fluid. At the postmortem examination however, no apparent organic changes were noted. In another Canadian zoo, *Y. enterocolitica* serotype 5/27 was recovered from a camel with diarrhea.[55]

As to *Y. enterocolitica* in domestic animals, it is evident from what has been previously mentioned and noted in Table 1 that numerous reports on isolations of these bacteria from different animal species have been presented, whereas descriptions of clinical symptoms and pathoanatomical changes are few. This hold true for studies in pigs, in particular. In most cases, *Y. enterocolitica* was isolated from seemingly quite healthy pigs and sampling was mostly carried out in connection with normal slaughter for human consumption.[19,22-24,26,28-30,32,35]

In a few studies[17-20] samples were collected from animals that had had clinical symptoms of diarrhea. Ahvonen[17] describes an outbreak of disease in which serotype O:3 was isolated from six stool specimen of pigs which had had diarrhea for 2 weeks.

Esseveld and Goudzwaard[20] report the results from a study in which a group of pigs with diarrhea and a group of seemingly healthy pigs without diarrhea were compared. *Y. enterocolitica* serotype O:3 was isolated in 6.1% and serotype O:9 in 2.2% of the animals in the first group. The second group comprised 138 animals and *Y. enterocolitica* serotype O:9 was isolated from 19 of them. The animals of the latter group came from different farms where some pigs showed sero-reactions against *Brucella*. This cross-reaction between *Brucella* and *Y. enterocolitica* and serotype O:9 was observed and described for the first time by Ahvonen et al.[77] This serological cross-reaction is remarkable because the homologous and the heterologous titers are of the same order of magnitude.[78-81] The problems associated with this extraordinary cross-reaction have been tackled by several workers.[82] Immunological and immunochemical investigations have shown that two monosaccharides (glucose and galactose) are the most important common components of the respective lipopolysaccharide molecules of *Y. enterocolitica* serotype O:9 and *B.*

abortus, responsible for the serological cross-reaction.[83,84] The diagnosis of brucellosis and of yersiniosis in both man and animals is complicated by the manifestation of this cross-reaction.[14,20,34,78,82] This cross-reaction makes it impossible to perform a differential serological diagnosis using common routine test procedures, such as agglutination test, complement-fixation test, and immuno-diffusion test. In view of this diagnostic dilemma, two test systems, namely ELISA (enzyme-linked immunosorbent assay)[85,86] and electroimmunoassay,[87] have been developed and adapted in order to allow serological differential diagnosis of brucellosis and yersiniosis caused by *Y. enterocolitica* serotype O:9.

Pedersen[29] reported a study of autopsy material from 22 diseased pigs for routine diagnostic purpose for the presence of *Y. enterocolitica*. Selection by postmortem diagnosis was not made. Different serotypes of *Y. enterocolitica* were isolated in 5.4%, serotypes O:3 and O:5 predominating. For comparison, 100 apparently healthy pigs were also examined. Nine different serotypes, with predominance for serotypes O:3, O:5, and O:6, were isolated in 17% of this group. This study showed that *Y. enterocolitica* can be isolated both from pigs with various diseases and from apparently healthy pigs at slaughter.

Regarding other domestic animals, *Y. enterocolitica* has been isolated from the cardiac valves of a 2½-year-old charollais bull. The bull had had a temperature of 41°C and autopsy showed changes indicative of endocarditis.[45]

Isolation of *Y. enterocolitica* serotype O:2 from an enzootic in a goat herd in Norway has been reported.[48,88,89] The herd consisted of 161 animals and during a 3-month period in the winter and spring of 1972, 49 of 100 animals in one section of a stable fell ill; 32 of them were kids. A total of 19 animals died, including 14 kids. Some died suddenly without any apparent clinical symptoms, whereas others had diarrhea, usually for a short period. Pathoanatomical examination carried out in 16 animals showed signs of acute catarrhal enteritis in 13. Agglutinating antibodies against the isolated *Y. enterocolitica* were found in 9 of 10 kids that had recovered from a period of diarrhea. The serum titers varied from 1/40 to 1/640. Samples from the environment of the infected herd were examined for the presence of *Y. enterocolitica* with a negative result.

On account of this enzootic, two goats were experimentally infected with living cultures of *Y. enterocolitica* serotype O:2.[88] They received orally about 10^9 smooth (S)-form cells daily for 7 days. Three months later the same animals were given 2×10^9 cells intraperitoneally. The feedings did not produce any clinical symptoms, but in one of the goats clinical signs were observed 3 days after i.p. injection. The goat was depressed and had a slightly raised temperature and some fluid mucous feces. *Y. enterocolitica* was isolated from the feces. Serological examinations during the enzootic showed that in the first 2 weeks after the acute phase of the disease there were rapid titer rises (up to 1/640) followed by falls to 1/80 to 1/20 in the next 3 to 6 months. This agrees well with the antibody kinetics in man in infection with *Y. enterocolitica*.[17,68,69]

Among 190 healthy goats from 11 different parts of Norway 3.1% had titers of 1/160 and 16.3% titers of 1/80. These results suggest that the reacting animals must have been exposed to *Y. enterocolitica* during the previous 3 to 6 months. Yet, the organism could not be isolated from any of the examined fecal samples of these goats. The results of this study indicate that *Y. enterocolitica* is ubiquitously present in the environment of goats in the different parts of the country[88] and that it does not produce symptoms in exposed goats. The bacteria can cause diarrhea and death among goats,[48] if predisposing factors are present, such as overcrowding in the stable, abrupt changes of food, poor quality of the food, or other pathological conditions or deficiency diseases.[88]

Because of their close contact with humans, pet animals, such as cats and dogs, have long been suspected to be reservoir for human infections with *Y. enterocolitica*. These suspicions have gained further support by the fact that the human-pathogenic serotypes O:3 and O:9 have been isolated from dogs and from cats on several occasions (Table 1).

A case of chronic enteritis in a dog from which *Y. enterocolitica* serotype O:3 was isolated has been reported from Norway.[52] The dog, a 9-month-old Whippet, had had moderate diarrhea and anorexia since the age of 6 to 7 weeks with short intermittent periods of normal stools. The dog was killed, since repeated treatment with antibiotics was ineffective. Pathoanatomical examination showed that the dog was emaciated and dehydrated. Numerous bald patches and dry scaly crusts were seen on the skin over the thorax and abdomen. The intestinal contents were mucous and scanty throughout the intestine. The mucosa of the jejunum and ileum was hyperemic and thickened. The mesenteric lymph glands were slightly enlarged; histologically, marked plasma-cell proliferation was noted in the medullary cords of these glands. *Y. enterocolitica* was isolated from intestinal content but this does not necessarily indicate that the organism was the cause of the disease. Nor did the pathological picture provide a definite answer.

Although *Y. enterocolitica* has been isolated at a high rate from dogs[29,31] (in some cases in excess of 10%), data on clinical symptoms are lacking and the pathogenicity of *Y. enterocolitica* in dogs is still not known.[29,31,51]

As regards cats, reports from Finland[18] and Canada[49] of isolations of *Y. enterocolitica* from cats with diarrhea have been presented. No further data are available. In the case of cats, too, the pathogenic role of *Y. enterocolitica* remains unknown.

IV. TRANSMISSION OF *Y. ENTEROCOLITICA* BETWEEN ANIMALS AND MAN

The paths of *Y. enterocolitica* infection are to a large extent unknown. Clinical and epidemiological data indicate, however, that the most likely portal of entry for an infection both in animals and in man would be the digestive tract. Since *Y. enterocolitica* has been isolated from the intestinal contents of a great number of animals, as previously mentioned, and occurs abundantly in different environments, it has been presumed that this organism would follow the same epidemiological pattern as other enterobacteria, e.g., various *Salmonella* and *Escherichia* species. The proposed modes of infection with *Y. enterocolitica* to man are intake of contaminated food, contact with infected animals, or person-to-person infection. Although many outbreaks have been reported from hospitals and schools, e.g., in Finland, Japan, and Czechoslovakia, the sources and the paths of infection have not been made clear and extensive work on sampling and epidemiological investigation has resulted solely in the presumption that the infection would be transmitted by food or water.[10,90-93]

However, the first known and described outbreak in which food stuff infected with *Y. enterocolitica* was proved to be the cause of infection in man has recently been reported from the U.S. In September and October 1976, a total of 222 children at five schools in New York State fell ill. The cause of the epidemic was chocolate milk from which *Y. enterocolitica* O:8 could be isolated.[94]

A characteristic feature of *Y. enterocolitica*, as distinct from other bacteria belonging to the family *Enterobacteriaceae*, is its ability to multiply rapidly at temperatures around +4°C. Figure 1 shows the difference in the growth curve for *Salmonella typhimurium* and for *Y. enterocolitica*.

This ability of *Y. enterocolitica* to grow at low temperatures and its distribution in nature have led to the suspicion that drinking water would be a reservoir for man. Workers from several European countries[16,23,25,95-97] have reported a seasonal variation of *Y. enterocolitica* outbreaks, the frequencies of infections in man reaching peaks in the

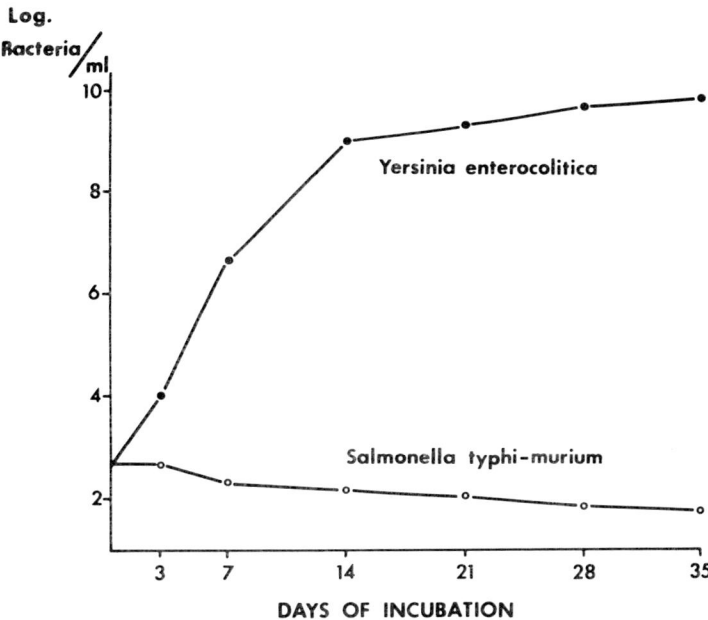

FIGURE 1. Growth-parameters of *Salmonella typhimurium* and *Y. enterocolitica* at +4°C. (From Leistner, L., Heckelmann, H., Kashiwazaki and Albertz, R., *Fleischwirtschaft*, 55, 1599, 1975. With permission.).

autumn and winter (Figure 2). The seasonal variation has been regarded as supporting evidence for water-borne infection. The bacteria would be spread into the waterways by humans and animals in the summer months, and when the temperature of the water remains around +4°C in the autumn and winter, *Y. enterocolitica* would multiply profusely, with the risk of infection via the drinking water. *Y. enterocolitica* has been isolated from drinking water in Europe as well as in Canada and the U.S.[49,74,98-103]

Several authors claim that the nonavian wildlife reservoir for *Y. enterocolitica* is by far much more widespread than what the relatively few reports of findings of *Y. enterocolitica* in the terrestrial ecosystems indicate.[47,55,57] Antigenically related *Y. enterocolitica* strains have been isolated in high frequencies both from terrestrial animals and from adjacent fresh water. But the question whether the transmission occurs from water to animals or vice versa remains unanswered.[57]

Virtually all the isolated *Y. enterocolitica* strains from water and from small rodents, such as voles, field mice, and shrews, differ both biochemically and serologically from strains isolated from man and domestic animals in pathological conditions. With our present knowledge about the pathogenicity of *Y. enterocolitica* it seems, therefore, less probable that these small rodents would play a decisive role as reservoirs of infection or be involved in the spread of infection with *Y. enterocolitica* to man and domestic animals.

Keet[74] has proposed that deer would be a reservoir of *Y. enterocolitica* infection to man. He describes a case of septicemia caused by *Y. enterocolitica* serotype O:8 in a man who had been on a hunting trip in the Adirondack Mountains in the state of New York. The same serotype of *Y. enterocolitica* was recovered from mountain stream water used for drinking water at the patient's campsite. Keet claims that as the patient had been eating nothing but canned food, the infection could possibly have occurred via the stream water used. This would, in turn, have been contaminated by deer or other animal feces.[11,74]

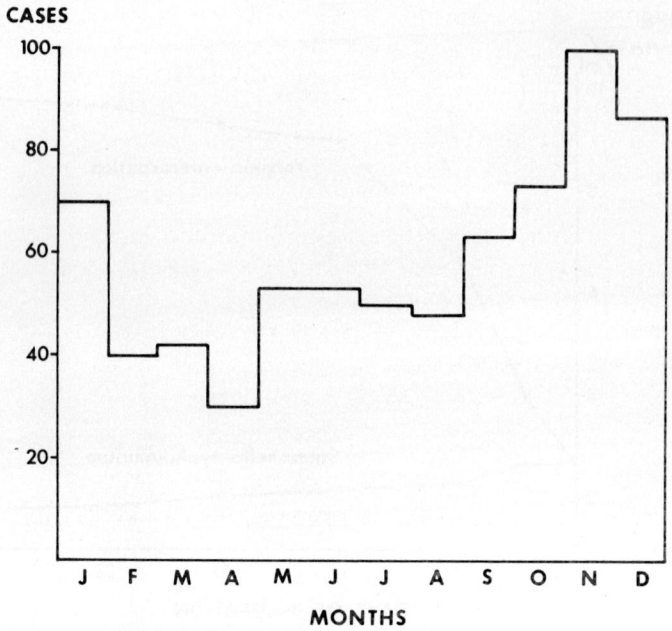

FIGURE 2. Seasonal variations of *Yersinia* infections in humans in Sweden. (From Winblad, S., *Läkartidningen*, 10, 270, 1973. With permission.).

So far, only a few outbreaks of *Y. enterocolitica* infection have been reported in which there seems to be a direct culture-proved association between humans on one side and pet animals or domestic animals on the other. Several studies have been presented, however, in which various factors indicated such a relationship.

In the U.S., a case of *Y. enterocolitica* serotype O:20 infection in a 4-month-old infant was traced directly to three puppies from which the same serotype was isolated.[53] One month before the onset of the child's illness the family's pet dog whelped 11 puppies. Eight of them died but showed no evidence of diarrhea. The isolation of *Y. enterocolitica* serotype O:20 from the three puppies, the death of their litter mates before the child fell ill, and the isolation of the same organism in dried stool from the basement of the house suggest earlier infection in the dogs. The question whether the dogs and the child were infected by one and the same source or the child was infected by the dogs remains unsettled, however.

Befekadu et al.[51] also discuss a case of a child who could possibly have been infected with *Y. enterocolitica* from an asymptomatic dog.

In another outbreak, 16 of 21 persons with two fatalities in four related and neighboring families were involved.[104] In these cases, too, puppies were epidemiologically implicated as a source of infection, but attempt at isolating *Y. enterocolitica* bacteria from the dogs failed.

Ahvonen[17] describes an outbreak of infection with *Y. enterocolitica* serotype O:9 in a family of six persons. The source of infection was presumed to be a kitten that had had diarrhea for a week before one of the children in the family developed severe diarrhea. *Y. enterocolitica* serotype O:9 was isolated from the kitten and from the infant.

Although conclusive evidence is still lacking and it has not been possible to establish a direct link between *Y. enterocolitica* infection in man and in domestic animals, strong evidence exists for the possibility of such a transmission. Suspicion has been directed towards abattoir workers[105] or persons who are in charge of domestic animals and in daily contact with animals, such as farmers and keepers.[106]

Krogstad et al.[48] noted that some persons who had been in contact with goats from which *Y. enterocolitica* serotype O:2 was isolated, also suffered from diarrhea. In one of these persons who had had diarrhea and abdominal pain for a week, antibodies against *Y. enterocolitica* serotype O:2 was demonstrated, the titer being 1/1250. However, the organism could not be isolated from fecal specimens from this patient.

Earlier in this paper it was mentioned that swine have been considered to be a source of infection in man. The histories of many patients with yersiniosis contain data on pork ingestion[11] or direct contact with pigs or pig keepers.[17,23,106] Rakovsky et al.[23] propose that one cause of the increased frequency of yersiniosis in man in Czechoslovakia over the period December to March would be that it coincides with the period of domestic pig slaughter.

REFERENCES

1. **Hässig, A., Karrer, J., and Pusterla, F.**, Über Pseudotuberculose beim Menschen, *Schweiz. Med. Wochenschr.*, 79, 971, 1949.
2. **Dickinson, A. B. and Mocquot, G.**, Studies on the alimentary tract of pigs. I. *Enterobacteriaceae* and other Gram-negative bacteria, *J. Appl. Bacteriol.*, 24, 252, 1961.
3. **Schleifstein, J. and Coleman, M. B.**, *Bacterium enterocolitica*, N. Y. State Dep. Health Annu. Rep. Div. Lab. Res., 56, 1943.
4. **Daniels, J. J. H. M.**, Untersuchungen an als *Pasteurella pseudotuberculosis* diagnostizierten Stammen von Chinchillas, *Zentralbl. Veterinaermed.*, 10, 413, 1963.
5. **Mollaret, H. H. and Destombes, P.**, Les germes "X" en pathologie humaine, *Presse Med.*, 72, 2913, 1964.
6. **Frederiksen, W.**, A study of some *Yersinia pseudotuberculosis*-like bacteria ("*Bacterium enterocoliticum*" and "*Pasteurella* X"), Proc. 14th Scand. Congr. Pathology Microbiology, Universiteitsforlaget Tryckningssentral, Oslo, 1964, 103.
7. **Mollaret, H. H. and Knapp W.**, International Committee on Systematic Bacteriology, Subcommittee on the Taxonomy on *Pasteurella*, *Yersinia*, and *Francisella*. Minutes of the meeting, 13 April 1972, *Int. J. Syst. Bacteriol.*, 22, 401, 1972.
8. **Mollaret, H. H. and Thal, E.**, *Yersinia* in *Bergey's Manual of Determinative Bacteriology*, 8th ed., Williams & Wilkins, Baltimore, 1974, 330.
9. **WHO Collaborating Center for** *Yersinia Pseudotuberculosis*, worldwide spread of infections with *Yersinia enterocolitica*, *WHO Chron.*, 30, 494, 1976.
10. **Morris, G. K. and Feeley, J. C.**, *Yersinia enterocolitica*: a review of its role in food hygiene, *Bull. W.H.O.*, 54, 79, 1976.
11. **Bottone, E. J.**, *Yersinia enterocolitica*: a panoramic view of a charismatic microorganism, *CRC Crit. Rev. Microbiol.*, 5, 1977, 211.
12. **Becht, H.**, Untersuchungen über die Pseudotuberculose beim Chinchilla, *Dtsch. Tieraerztl. Wochenschr.*, 69, 626, 1962.
13. **Daniëls, J. J. H. M.**, Untersuchungen an als *Pasteurella pseudotuberculosis* diagnostizierten Stämmen von Chinchillas, *Zentralbl. Veterinaermed.*, 10, 413, 1963.
14. **Mollaret, H. H. and Chevalier, A.**, Contribution a l'etude d'un nouveau groupe de germes proches du bacille de malassez et vignal. I. Caracteres culturaux et biochimique, *Ann. Inst. Pasteur Paris*, 107, 121, 1964.
15. **Mollaret, H. H. and Guillon, J. C.**, Contribution at l'etude d'un nouveau groupe de germes *Yersinia enterocolitica* proches du bacille de malassez et vignal. II. Pouvoir pathogène experimental, *Ann. Inst. Pasteur Paris*, 109, 608, 1965.
16. **Mollaret, H. H.**, Un domaine pathologique nouveau: l'infection a *Yersinia enterocolitica*, *Ann. Biol. Clin.*, 30, 4, 1972.
17. **Ahvonen, P.**, Studies on Human Yersiniosis in Finland, Ph.D. thesis, University of Helsinki, Finland, 1972.
18. **Ahvonen, P., Thal, E., and Vasenius, H.**, Occurrence of *Yersinia enterocolitica* in animals in Finland and Sweden, in *Contributions to Microbiology and Immunology*, Vol. 2, Winblad, S., Ed., S. Karger, Basel, 1973, 2.
19. **Akkermans, J. P. W. M. and Hill, W. K. W.**, *Yersinia enterocolitica* serotype 9 infection as a factor interfering with serodiagnosis of *Brucella* infections in swine, *Neth. J. Vet. Sci.*, 5, 73, 1972.

20. **Esseveld, H. and Goudzwaard, C.,** On the epidemiology of *Y. enterocolitica* infections: pigs as the source of infections in man, in *Contributions to Microbiology and Immunology,* Vol. 2, Winblad, S., Ed., S. Karger, Basel, 1973, 99.
21. **Mair, N. S.,** Yersiniosis in wildlife and its public health implications, *J. Wildl. Dis.,* 9, 64, 1973.
22. **Rabson, A. R. and Koornhof, H. J.,** *Yersinia enterocolitica* infections in South Africa, in *Contributions to Microbiology and Immunology,* Vol. 2, Winblad, S., Ed., S. Karger, Basel, 1973, 102.
23. **Rakovsky, J., Pauekova, V., and Aldová E.,** Human *Yersinia enterocolitica* infections in Czechoslovakia, in *Contributions to Microbiology and Immunology,* Vol. 2, Winblad, S., Ed., S. Karger, Basel, 1973, 93.
24. **Tsubokura, M., Otsuki, K., and Itagaki, K.,** Isolation of *Y. enterocolitica* from swine, *Jpn. J. Vet. Sci.,* 35, 419, 1973.
25. **Vandepitte, J., Wauters, G., and Isebaert, A.,** Epidemiology of *Yersinia enterocolitica* infections in Belgium, in *Contributions to Microbiology and Immunology,* Vol. 2, Winblad, S., Ed., S. Karger, Basel, 1973, 111.
26. **Zen-Yoji, H., Sakai, S., Maruyama, T., and Yanagawa, Y.,** Isolation of *Yersinia enterocolitica* and *Yersinia pseudotuberculosis* from swine, cattle, and rats at an abattoir, *Jpn. J. Microbiol.,* 18, 103, 1974.
27. **Leistner, L., Hechelmann, H., Kashiwazaki, M., and Albertz, R.,** Nachweis von *Yersinia enterocolitica* in Faeces und Fleisch von Schweinen, Rindern und Geflügel. (Detection of *Yersinia enterocolitica* in the faeces and meat of pigs, bovines and poultry), *Fleischwirtschaft,* 55, 1599, 1975.
28. **Toma, S. and Deidrick, V. R.,** Isolation of *Yersinia enterocolitica* from swine, *J. Clin. Microbiol.,* 2, 478, 1975.
29. **Pedersen, K. B.,** Isolation of *Yersinia enterocolitica* from danish swine and dogs, *Acta Pathol. Microbiol. Scand. Sect. B,* 84, 317, 1976.
30. **Tsubokura, M., Fukuda, T., Otsuki, K., Kubota, M., Itagaki, K., Yamaoki, K., and Wakatsuki, M.,** Studies on *Yersinia enterocolitica.* II. Relationship between detection from swine and seasonal incidence and regional distribution of the organism., *Jpn. J. Vet. Sci.,* 38, 1, 1976.
31. **Kaneko, K., Hamada, S., and Kato, F.,** Occurrence of *Yersinia enterocolitica* in dogs, *J. Vet. Sci.,* 39, 407, 1977.
32. **Narucka, U. and Westendoorp, Y. F.,** Een onderzoek noar het voorhomen van *Yersinia enterocolitica* en *Yersinia pseudotuberculosis* bij klinisch normale varkens. (Studies for the presence of *Yersinia enterocolitica* and *Yersinia pseudotuberculosis* in clinically normal pigs), *Tijdschr. Diergeneeskd.,* 102, 299, 1977.
33. **Vaschenok, G. I., Volokhov, S. V., Sergeeva, N. A., Punko, T. A., Bukulina, L., II, Denisova, N. F., Chapkevich, O. B., and Balashova, L. M.,** Characteristics of the strains of pseudotuberculosis and intestinal yersiniosis isolated from some species of domestic animals, *Zh. Microbiol. Epidemiol. Immunbiol.,* 1, 104, 1977.
34. **Bockemühl, J., and Roth, J.,** *Brucella* titers in subclinical infections due to *Yersinia enterocolitica* serotype O:9 in a pig-breeding farm, Zentralbl. Bakteriol. Hyg. Abt. Orig. Reihe A, 240, 86, 1978.
35. **Hurvell, B., Glatthard, V., and Thal, E.,** Isolation of *Yersinia enterocolitica* from swine at an abattoir in Sweden in *Contributions to Microbiology and Immunology,* Vol. 5, S. Karger, Basel, 1979, 243.
36. **Pedersen, K. B.,** Occurrence of *Yersinia enterocolitica* in the throat of swine, in *Contributions to Microbiology and Immunology,* Vol. 5, S. Karger, Basel, 1979, 253.
37. **Wauters, G.,** Carriage of *Yersinia enterocolitica* serotype 3 by pigs as a source of human infection, in *Contributions to Microbiology and Immunology,* Vol. 5, S. Karger, Basel, 1979, 249.
38. **Akkermans, J. P. W. M. and Terpstra, J. I.,** Pseudotuberculosis in chinchillas caused by a special species, *Tijdschr. Diergeneesk.,* 2, 91, 1963.
39. **Wetzler, T. F. and Hubbert, W. T.,** *Yersinia enterocolitica* in North America, Int. Symp. Pseudotuberculosis, Paris, 1967; *Symp. Ser. Immunobiol. Stand.,* Vol. 9, S. Karger, Basel, 1968, 343.
40. **Halen, P. and Meulemans, G.,** Compte rendu de la troisième réunion dé groupe belge d'étude de *Yersinia entérocolita, Méd. Malad. Infect.,* 1, 226, 1971.
41. **Hubbert, W. T.,** Yersiniosis in mammals and birds in the United States. Case reports and review, *Am. J. Trop. Med. Hyg.,* 21, 458, 1972.
42. **Langford, E. V.,** *Yersinia enterocolitica* isolated from animals in the Fraser valley of British Columbia, *Can. Vet. J.,* 13, 109, 1972.
43. **Mollaret, H. H. and Lucas, A.,** Sur le particularités biochimiques des souches de Yersinia enterocolitica isolées chez lièvres, *Ann. Inst. Pasteur Paris,* 108, 121, 1965.
44. **Dubois, P., Fievez, L., and Granville, A.,** Le liévre porteur asymptomatique de *Yersinia enterocolitica, Ann. Med. Vet.,* 116, 447, 1972.
45. **Goyon, M.,** Endocardite végétante á *Yersinia enterocolitica* chez un bovin, *Rec. Med. Vet.,* 45, 61, 1969.
46. **Inoue, M. and Kurose, M.,** Isolation of *Yersinia enterocolitica* from cow's intestinal contents and beef meat, *Jpn. J. Vet. Sci.,* 37, 91, 1975.

47. Alonso, J. M., Bercovier, H., Servan, J., Bourdin, M., and Mollaret, H. H. Contribution á l'etude épidémiologique des infections á *Yersinia enterocolitica*. II. Enquéte écologique, *Méd. Malad. Infect.*, 6(10), 434, 1976.
48. Krogstad, O., Teige, J., Jr., and Lassen, J., *Yersinia enterocolitica* type 2 associated with disease in goat, *Acta Vet. Scand.*, 13, 594, 1972.
49. Toma, S., Survey on the incidence of *Yersinia enterocolitica* in the province of Ontario, *Can. J. Public Health*, 64, 477, 1973.
50. Tsubokura, M., Fukuda, T., Otsuki, K., Kubota, M., Itagaki, K., and Tanamachi, S., Isolation of *Yersinia enterocolitica* from some animals and meats, *Jpn. J. Vet. Sci.*, 37, 213, 1975.
51. Befekadu, E., Bercovier, H., Alonso, J. M., and Mollaret, H. H., Röle épidémiologique éventuel des animaux domestiques dans l'infection humaine à *Yersinia enterocolitica*. A propos d'une observation, *Méd. Malad. Infect.* 6, no. 1, 29, 1976.
52. Farstad, L., Landsverk, T., and Lassen, J., Isolation of *Yersinia enterocolitica* from a dog with chronic enteritis. A case report, *Acta Vet. Scand.*, 17, 261, 1976.
53. Wilson, H. O., McCormick, J. B., and Feeley, J. C., *Yersinia enterocolitica* infection in a 4-month-old infant associated with infection in household dogs, *J. Pediatr.*, 89, no. 5, 767, 1976.
54. Mollaret, H. H., personal communication, 1977.
55. Hacking, M. A. and Sileo, L., *Yersinia enterocolitica* and *Yersinia pseudotuberculosis* from wildlife in Ontario, *J. Wildl. Dis.*, 10, 452, 1974.
56. Otsuki, K., Tsubokura, M., Itagaki, K., Hirai, K., and Nigi, H., Isolation of *Yersinia enterocolitica* from monkeys and deers, *Jpn. J. Vet Sci.*, 35, 447, 1973.
57. Kapperud, G., *Yersinia enterocolitica* and *Yersinia* like microbes isolated from mammals and water in Norway and Denmark, *Acta Pathol. Microbiol. Scand. Sect. B*, 85, 129, 1977.
58. Aldová, E. and Lim, D., *Yersinia enterocolitica* in small rodents communication 1: pilot study in two wildlife areas, *Zentralbl. Bakteriol. Abt. Orig. Reihe A*, 226, 491, 1974.
59. Kapperud, G., *Yersinia enterocolitica* in small rodents from Norway, Sweden and Finland, *Acta Pathol. Microbiol. Scand. Sect. B*, 83, 335, 1975.
60. Głośnicka, R., Kunikowska, D., Dominowska, C., Wegner, Z., Przyborowski, R., and Malottke, R., Description of Yersinia reservoires in synanthropic rodents in ports and in the Gdansk coastal area, *Bull. Inst. Mar. Trop. Med. Med. Acad. Gdansk*, 27(2), 247, 1976.
61. Aldová, E., Cerny, J., and Chmela, J., Findings of *Yersinia* in rats and sewer rats, *Zentralbl. Bakteriol. Parasitenk. Infektionskr. Hyg. Abt. Orig. Reihe A*, 239, 208, 1977.
62. Pokorna, B. and Aldová, E., Finding of *Yersinia enterocolitica* in Rattus rattus, *J. Hyg. Epidemiol. Microbiol. Immunol.*, 21(1), 104, 1977.
63. Hurvell, B., unpublished data, 1978.
64. McDiarmid, A., Some disease of free-living wildlife. X. *Yersinia*, in *Advances in Veterinary Science and Comparative Medicine*, Brandly, C. A. and Cornelis, C. E., Eds., Academic Press, New York, 1975, 118.
65. Mair, N. S., White, G. D., Schubert, F. K., and Harbourne, J. F., *Yersinia enterocolitica* infection in the bushbaby (Galago), *Vet. Rec.*, 86, 69, 1970.
66. Kapperud, G. and Jonsson, B., *Yersinia enterocolitica* in brown trout (Salmon trutta L) from Norway, *Acta Pathol. Microbiol. Scand. Sect. B*, 84, 66, 1976.
67. Botzler, R. G., Wetzler, T. F., and Cowan, A. B., *Yersinia enterocolitica* and Yersinia-like organisms isolated from frogs and snails, *Bull. Wildl. Dis. Assoc.*, 4, 110, 1968.
68. Niléhn, B., Studies on *Yersinia enterocolitica* with special reference to bacterial diagnosis and occurrence in human acute enteric disease. Thesis, *Acta Pathol. Microbiol. Scand.*, Suppl. 206, 1, 1969.
69. Wauters, G., Contribution à l'étude de *Yersinia enterocolitica*, Ph.D. thesis, Vander, Louvain, 1970.
70. Bissett, M. L., *Yersinia enterocolitica* isolates from humans in California, 1968–1975, *J. Clin. Microbiol.*, 4, 137, 1976.
71. Nicolle, P., Mollaret, H., and Broult, J., Recherches sur la lysogénie, la lysosensibilité, et la lysotypie et la serologie de *Yersinia enterocolitica*, in *Contributions to Microbiology and Immunology*, Vol. 2, Winblad, S., Ed., S. Karger, Basel, 1973, 54.
72. Weaver, R. E. and Jordan, J. G., Recent human isolates of *Yersinia enterocolitica* in the United States, in *Contributions to Microbiology and Immunology*, Vol. 2, Winblad, S., Ed., S. Karger, Basel, 1973, 120.
73. Highsmith, A. K., Feeley, J. C., and Morris, G. K., *Yersinia enterocolitica*. A review of the bacterium and recommended laboratory methodology, *Health Lab. Sci.*, 14(4), 253, 1977.
74. Keet, E. E., *Yersinia enterocolitica* septicemia. Source of infection and incubation period identified, *N. Y. State J. Med.*, 74, 2226, 1974.
75. McClure, H. M., Weaver, R. E., and Kaufmann, A. F., Pseudotuberculosis in nonhuman primates: infection with organisms of the *Yersinia enterocolitica* group, *Lab. Anim. Sci.*, 21(3), 376, 1971.

76. **Mollaret, H. H., Giorgi, W., Matera, A., Pestana de Castro, A. F., and Guillon, J. C.**, Isolement de *Yersinia enterocolitica* chez le Signe callitriche au Brésil, *Rec. Med. Vet.*, 146, 919, 1970.
77. **Ahvonen, P., Jansson, E., and Aho, K.**, Marked cross-agglutination between Brucellae and a subtype of *Yersinia enterocolitica*, *Acta Pathol. Microbiol. Scand.*, 75, 291, 1969.
78. **Corbel, M. J. and Cullen, G. A.**, Differentiation of the serological response to *Yersinia enterocolitica* and *Brucella abortus* in cattle, *J. Hyg.*, 68, 519, 1970.
79. **Diaz, R., Lacalle, R., Medrano, M. P., and Leong, D.**, Immunobiological activities of the endotoxin from *Yersinia enterocolitica* strain M. Y., 79, Proc. 5th Int. Congr. Infectious Diseases, Vol. 2, Spitzer, K. H., Ed., Vienna, 1970, 11.
80. **Hurvell, B., Ahvonen, P., and Thal, E.**, Serological cross-reactions between different *Brucella* species and *Yersinia enterocolitica*, Proc. 11th Nord. Vet. Congr., Bergen, 1970, 282.
81. **Szita, J., Káli, M., and Rédey, B.**, Serological diagnosis of *Yersinia enterocolitica*, *Acta Microbiol. Acad. Sci. Hung.*, 18, 113, 1971.
82. **Hurvell, B.**, Serological Cross-Reactions between Different *Brucella* Species and *Yersinia enterocolitica*. An Immunological and Immunochemical Study, Ph.D. thesis, Royal Veterinary College, Stockholm, Sweden, 1973.
83. **Hurvell, B.**, Serological cross-reactions between different *Brucella* species and *Yersinia enterocolitica*. Biological and chemical investigations of lipopolysaccharides from *Brucella abortus* and *Yersinia enterocolitica* type IX, *Acta Pathol. Microbiol. Scand. Sect. B*, 81, 105, 1973.
84. **Hurvell, B. and Lindberg, A. A.**, Serological cross-reactions between different *Brucella* species and *Yersinia enterocolitica*. Immunochemical studies on phenol-water extracted lipopolysaccharides from *Brucella abortus* and *Yersinia enterocolitica* type IX, *Acta Pathol. Microbiol. Scand. Sect. B*, 81, 113, 1973.
85. **Carlsson, H. E., Hurvell, B., and Lindberg, A. A.**, Enzyme-linked immunosorbent assay (ELISA) for titration of antibodies against *Brucella abortus* and *Yersinia enterocolitica*, *Acta Pathol. Microbiol. Scand. Sect. C*, 84, 168, 1976.
86. **Hurvell, B., Lindberg, A. A., and Carlsson, H. E.**, Differentiation of antibodies against *Brucella abortus* and *Yersinia enterocolitica* by enzyme-linked immunosorbent assay in *Contributions to Microbiology and Immunology*, Vol. 5, S. Karger, Basel, 1979, 73.
87. **Hurvell, B.**, Differentiation of cross-reacting antibodies against *Brucella abortus* and *Yersinia enterocolitica* by electroimmuno assay, *Acta Vet. Scand.*, 16, 318, 1975.
88. **Krogstad, O.**, *Yersinia enterocolitica* infection in goat, a serological and bacteriological investigation, *Acta Vet. Scand.*, 15, 597, 1974.
89. **Krogstad, O.**, Yersinia enterocolitica in goat—properties and diagnosis, *Nord. Veterinaermed.*, 27, 31, 1975.
90. **Asakawa, Y., Akahane, S., Kagata, N., Noguchi, M., Sakazaki, R., and Tamura, K.**, Two community outbreaks of human infection with *Yersinia enterocolitica*, *J. Hyg. Camb.*, 71, 715, 1973.
91. **Toivanen, P., Toivanen, A., Olkkonen, L., and Aantaa, S.**, Hospital outbreak of *Yersinia enterocolitica* infection, *Lancet*, 801, 1973.
92. **Zen-Yoji, H., Sakai, S., Mauyama, T., and Yanagawa, T.**, Isolation of *Yersinia enterocolitica* and *Yersinia pseudotuberculosis* from swine, cattle and rats at an abattoir, *Jpn. J. Microbiol.*, 18, 103, 1974.
93. **Olsovsky, Z., Olsakova, S., Chobot, S., and Sviridov, V.**, Mass occurrence of *Yersinia enterocolitica* in two establishments of collective care of children, *J. Hyg. Epidemiol. Microbiol. Immunol.*, 19, 22, 1975.
94. **Black, R. E., Jackson, R. J., Tsal, T., Medvesky, M., Shayegani, M., Feeley, J. C., McLeod, K. I. E., and Wakelee, A. M.**, Epidemic *Yersinia enterocolitica* infection due to contaminated chocolate milk, *N. Engl. J. Med.*, 298, 76, 1978.
95. **Arvastson, B., Damgaard, K., and Winblad, S.**, Clinical symptoms of infection with *Yersinia enterocolitica*, *Scand. J. Infect. Dis.*, 3, 37, 1971.
96. **Rusu, V., Lucinescu, S., Stanescu, C., Totescu, E., and Muscan, A.**, Clinical and epidemiological aspects of the human *Yersinia enterocolitica* infections in Romania, in *Contributions to Microbiology and Immunology*, Vol. 2, Winblad, S., Ed., S. Karger, Basel, 1973, 126.
97. **Winblad, S.**, *Yersinia enterocolitica*, Läkartidningen, 70, 270, 1973.
98. **Lassen, J.**, *Yersinia enterocolitica* in drinking-water, *Scand. J. Infect. Dis.*, 4, 125, 1972.
99. **Wauters, G.**, Souches de *Yersinia enterocolitica* isolées de l'eau, *Rev. Ferment. Ind. Aliment*, 7, 18, 1972.
100. **Harvey, S., Greenwood, J. R., Pickett, M. J., and Mah, R., A.**, Recovery of *Yersinia enterocolitica* from streams and lakes of California, *Appl. and Environ. Microbiol.*, 32, 352, 1976.
101. **Saari, T. N. and Quan, T. J.**, Waterborne *Yersinia enterocolitica* in Colorado, in *Abstracts of the Annual Meeting of the American Society for Microbiology*, American Society for Microbiology, Washington, D.C., 1976, 45.

102. **Eden, K. V., Rosenberg, M. L., Stoopler, M., Wood, B. T., Highsmith, A. K., Skaliy, P., and Wells, J. G.,** Waterborne gastrointestinal illness at a ski resort. Isolation of *Yersinia enterocolitica* from drinking water, *Public Health Rep.,* 92, 245, 1977.
103. **Highsmith, A. K., Feeley, J. C., Skaliy, P., Wells, J. G., and Wood, B. T.,** Isolation of *Yersinia enterocolitica* from well water and growth in distilled water, *Appl. Environ. Microbiol.,* 34, 745, 1977.
104. **Gutman, L. T., Ottesen, E. A., Quan, T. J., Noce, P. S., and Katz, S. L.,** An inter-familial outbreak of *Yersinia enterocolitica* enteritis, *N. Engl. J. Med.,* 288, 1372, 1973.
105. **Finlayson, M. H., Coldrey, N. A., Street, B., and Brede, H. D.,** *Yersinia enterocolitica* in the Western Cape., *S. Afr. Med. J.,* 111, 1973.
106. **Szita, J., Káli, M. and Rédey, B.,** Incidence of *Yersinia enterocolitica* infection in Hungary, in *Contributions to Microbiology and Immunology,* Vol. 2, Winblad, S., Ed., S. Karger, Basel, 1973, 106.

Chapter 15

THE OCCURRENCE OF *YERSINIA ENTEROCOLITICA* IN FOODS

Wei Hwa Lee, Carl Vanderzant, and Norman Stern

TABLE OF CONTENTS

I.	Introduction 162	
II.	Occurrence of *Y. enterocolitica* in Foods 162	
	A. Dairy Products 162	
	B. Meats 163	
	1. Beef and Lamb 163	
	2. Pork 164	
	3. Poultry 165	
	4. Seafood 165	
	C. Vegetables 165	
III.	Isolation Procedures for *Y. enterocolitica* in Foods 165	
IV.	Significance of *Y. enterocolitica* in Foods 166	
V.	Survival and Control of *Y. enterocolitica* 167	
References	 169	

I. INTRODUCTION

Y. enterocolitica can cause diseases in humans and animals, but the source of infections is often not known. Several recent large outbreaks of *Y. enterocolitica* infections in Canada,[1] Japan,[2,3] and the U.S.[4] have led to a worldwide investigation of foods as a vehicle of *Y. enterocolitica* infections. An additional public health hazard is the ability of *Y. enterocolitica* to grow at refrigeration temperatures (0 to 5°C) used to store perishable foods. *Y. enterocolitica* infections have been documented in animals;[5-9] dogs and pigs are known carriers of this bacterium.[10-12] The close association of *Y. enterocolitica* with domestic livestock has led to a widespread investigation of meats as a possible source of *Y. enterocolitica* infection.

Worldwide studies indicate that *Y. enterocolitica* is fairly common in raw milk, raw meats, raw oysters, and water.[13] Yet only one large documented outbreak in the U.S. (New York State) was actually traced to *Y. enterocolitica* contamination of chocolate milk.[4] In addition, one single human infection, also in New York State, was attributed to drinking contaminated stream water.[14]

This discrepancy in the seemingly common occurrence of *Y. enterocolitica* in foods and the seldom documented food-borne infection may be partially explained by the following reasons. First, current enrichment procedures for the recovery of *Y. enterocolitica* from foods are poor as compared to, for example, enrichment methods for *Salmonella*, and for this reason low levels of clinically important strains of *Y. enterocolitica* may fail to be detected in foods. Some clinical strains of *Y. enterocolitica* are especially sensitive to inhibitory chemicals used in enrichment media and will not grow in selenite cystine with novobiocin or in the modified enrichment broths. Recovery of clinical strains of *Y. enterocolitica* from foods, inoculated with several hundred cells per gram, is still a difficult and uncertain process.[15] Thus, low levels of *Y. enterocolitica* in foods may escape detection and still cause food-borne illness.

On the other hand, *Y. enterocolitica* is commonly recovered from foods. Many of the *Y. enterocolitica* strains isolated from foods are environmental biotypes which do not usually cause the typical gastrointestinal symptoms attributed to typical *Y. enterocolitica* infections.[16,17] The species *Y. enterocolitica* is composed of several groups of moderately related bacteria as explained by Brenner in Chapter 1. The presence of so many environmental biotypes in foods may actually mask the presence and recovery of the clinically important strains from foods. To compound the confusion, environmental and clinical biotypes can have the same O serotypes.[18,19]

II. OCCURRENCE OF *Y. ENTEROCOLITICA* IN FOODS

A. Dairy Products

Y. enterocolitica was isolated from ice cream[20] and pasteurized milk[4,21] as early as 1970. In 1975, *Y. enterocolitica* was recovered from pasteurized milk and eggnog in a Canadian Hospital.[22] The organism also was recovered from a milking machine, from raw milk and cream in Czechoslovakia,[23] and from raw milk in Australia.[24] Two large outbreaks of *Y. enterocolitica* infection involving 137 children and 1 adult (out of 831 persons) who had gone on outings were reported in Quebec, Canada.[1] Two samples of raw milk from the site of the outbreak were found to be contaminated with *Y. enterocolitica*. However, the serotype of the isolates from milk were not the same as that recovered from children.[1,25]

Recent outbreaks of *Y. enterocolitica* infections in Canada have led to an extensive survey of raw milk from Ontario. Schiemann and Toma[22] recovered 42 strains of *Y. enterocolitica* from 29 samples of raw milk out of a total of 131 samples tested. About one half of the strains were rhamnose positive. These strains are environmental biotypes and are biochemically different from the predominant clinical strains serotype O:3, biotype 4 usually recovered from patients in Ontario, Canada.[26] The authors tested (1) cold enrichment in Butterfields phosphate buffer (PB) of pH 7.2, (2) enrichment in modified Rappaport broth containing magnesium chloride, malachite green, and carbenicillin (RMC)[27] at 23°C, (3) RMC inoculated from a pre-enrichment in PB and then incubated for 5 days at 23°C, and (4) enrichment in cooked meat broth for 28 days at 23°C followed by incubation in RMC for 5 days at 23°C. They found that a combination of cold enrichment in PB followed by inoculation into RMC was the best procedure for the recovery of environmental biotypes from raw milk.[22]

An outbreak of intestinal illness with an unusual number of appendectomies in Oneida County, New York led the Center for Disease Control (CDC) to uncover a severe outbreak of food-borne illness caused by *Y. enterocolitica*.[4] About 220 persons in 5 area schools were believed to have been infected. *Y. enterocolitica* with the same serotype (O:8) and biotype (Nilehn 2) was recovered from one carton of chocolate milk and from patients suffering from gastroenteritis and pseudoappendicitis. The strains isolated from milk and humans produced a heat-stable enterotoxin as determined by the infant mouse assay.[28] Cells grown at 25°C showed enterotoxin activity, but not at 36°C. These strains also caused a positive Sereny test (keratoconjunctivitis in guinea pigs). Most strains showed a positive Sereny test only when cells were grown at 25°C and not at 36°C. This is the only documented case of a food-borne outbreak of *Y. enterocolitica* involving refrigerated foods. Unfortunately, the source and the number of *Y. enterocolitica* in the chocolate milk that caused this severe and widespread infection is not known. Zink et al.[29] reported that the invasive property (as measured by the Sereny test) of some strains associated with human illness in New York was correlated with the presence of a 41 Mdal plasmid. Several strains which were cured of this plasmid showed a negative Sereny test.

Y. enterocolitica was recovered from one cheese sample in Quebec, Canada.[25] Commercial Grade A pasteurized milk, cheese made from pasteurized milk, and milk products such as ice cream prepared from pasteurized mixes should ordinarily not pose a health problem unless recontaminated with *Y. enterocolitica*, because the pasteurization process is adequate to destroy *Y. enterocolitica*. Consumption of raw milk or dairy products made from raw milk may pose potential public health problems.

B. Meats

In many areas of the world, healthy dogs and pigs have shown to be carriers of *Y. enterocolitica* with the same biotype (4) and serotype (O:3) as the predominant clinical strains isolated from humans. For this reason, Mollaret[30] considered pigs an important zoonotic source of *Y. enterocolitica* infections. The presence of *Y. enterocolitica* in animals suggests that it also may be present in meats. Extensive surveys have been conducted to recover *Y. enterocolitica* from meats. Many researchers used some form of cold enrichment in phosphate buffer solutions (PBS) for the recovery of *Y. enterocolitica* in meats.

1. Beef and Lamb

Cattle and sheep are generally not considered to be carriers of the clinical biotypes of *Y. enterocolitica*. However, beef sometimes may be consumed without adequate heating and there has been some concern about *Y. enterocolitica* contamination in beef. In an extensive survey conducted in Germany by Leistner et al.,[31] 4 strains of *Y. enterocolitica* and 6 related bacteria were isolated from 37 samples of beef. Two of the beef isolates belonged to biotypes that can cause typical human illness, but their virulence was not tested.[31] Environmental biotypes of *Y. enterocolitica* were isolated from 15 of 61 samples of beef in Japan.[32] Selenite medium with 40 mg novobiocin per liter and a 1/15 M phosphate buffer solution (PBS) of pH 7.6 were used as cold enrichments. The authors concluded that cold enrichment in the 1/15 M PBS was superior for the recovery of environmental biotypes from beef.[32] In the U.S., environmental biotypes of *Y. enterocolitica* were recovered from 10 of 98 samples of vacuum packaged beef and from 2 of 18 samples of vacuum packaged lamb. Isolation was carried out by direct plating of surface swabs of the meat samples on a nonselective medium.[33] Also, environmental biotypes of *Y. enterocolitica* were easily recovered from samples of retail ground beef in the U.S.[63] In Quebec, Canada, *Y. enterocolitica* was recovered from raw beef samples with selenite broth enrichment.[25]

2. Pork

The predominant human serotype O:3 has consistently been recovered from the cecal contents and mesenteric lymph nodes of healthy pigs from different parts of the world,[34-36] except in the U.S. where such recovery has not been reported. Recently, several European investigators reported a high incidence of *Y. enterocolitica* serotype O:3 in the throat of pigs where it appears to be a part of the normal microflora. In Belgium, Wauters et al.[12] found that 47 of 426 throat swabs from pigs contained *Y. enterocolitica* serotype O:3. They used as enrichment the RMC broth developed by Wauters.[27] In the Netherlands, Narucka and Westendoorp[37] recovered 22 *Y. enterocolitica* isolates with diverse serotypes of O:3, O:5, O:5,27, O:6,30, O:7,8, and O:9 as well as 7 isolates of *Yersinia pseudotuberculosis* from the throat swabs (tonsils) of 163 healthy pigs. Isolation was carried out by direct plating on desoxycholate agar (DCA) with plate incubation for 2 to 3 days at 24°C, by enrichment in selenite broth plus 40 mg novobiocin per liter for 2 days at 24°C and subsequent plating on DCA, or by enrichment in phosphate buffer at pH 7.6 for 21 days at 4°C and subsequent plating on DCA. In Denmark, Pederson[38] recovered 84 strains of *Y. enterocolitica* serotype O:3 from throat swabs of 282 normal pigs at the time of slaughter. Isolation was carried out with cold enrichment in phosphate buffered saline and nutrient broth with subsequent plating on MacConkey and *Salmonella-Shigella* (SS) agar. Pederson also examined the 21-day-old enrichment broths for the presence of serotype O:3 bacteria by applying fluorescent O:3 antibody stain. Enrichment broths which showed positive fluorescent antibody reactions were restreaked on plating agars if no *Y. enterocolitica* were recovered from the initial isolation attempt.

Narucka and Westendoorp[37] did not recover *Y. enterocolitica* from the muscle tissues of 163 pigs, nor from 60 samples of raw ground pork. One strain, serotype O:9, was recovered from one of the pork livers. In a related study, Wauters and Janssens,[39] using enrichment in RMC, recovered *Y. enterocolitica* serotype O:3 from 75 pork tongues out of 142 samples obtained from 34 different butcher shops in 19 Belgian communities. This figure represents a remarkable recovery (53% of samples) of potentially pathogenic strains of *Y. enterocolitica* from a food source. In a very extensive survey of Japanese foods, Asakawa et al.[35] recovered two potentially pathogenic strains of serotype O:3 and one strain of serotype O:5 from 300 samples of pork. Serotype O:3 also was recovered from the feces of 2 of 571 butchers and from 2 of 50 chopping blocks. Inoue and Nagao[19] recovered 27 environmental biotypes of *Y. enterocolitica* from 40 samples of pork in Okayama, Japan. In Germany, Leistner et al.[31] isolated 10 strains of *Y. enterocolitica* and 10 related bacteria from 29 samples of pork; one strain was similar to the clinical biotypes. *Y. enterocolitica* was recovered from one ham and two sausage samples in Quebec, Canada.[25] Environmental biotypes of *Y. enterocolitica* have been easily recovered from retail pork samples in the U.S.[40] There is, however, no published report of the distribution of *Y. enterocolitica* in the pig population or pork products from the U.S.

From the reports cited above, it appears that pigs are definitely carriers of strains of *Y.*

enterocolitica associated with human illness. They appear more concentrated in the area of the tongue and tonsils than in the mesenteric lymph nodes or cecal contents. The presence of clinical strains of *Y. enterocolitica* in cooked pork or other meat products has not been adequately studied and warrants further investigation. The public health significance of environmental biotypes of *Y. enterocolitica* found in meats will be discussed later.

3. Poultry

Poultry has not been proven to be a carrier of clinical biotypes of *Y. enterocolitica*. Leistner et al.[31] isolated 35 strains of *Y. enterocolitica* and 56 related bacteria from 121 samples of chicken meat in Germany. Inoue and Nagao[19] recovered 16 strains of *Y. enterocolitica*-related bacteria from 40 samples of chicken in Japan. *Y. enterocolitica* was also recovered from 3 of 75 turkey meat samples in the U.S.[41] and from 2 poultry samples in Quebec, Canada.[25] *Y. enterocolitica* invasive to HeLa cells was recovered from chicken in the U.S.[64]

4. Seafood

Some seafoods may be consumed raw and thus the presence of virulent strains of *Y. enterocolitica* could cause food-borne illness. The presence of environmental biotypes of *Y. enterocolitica* has been reported in fish,[16,42,43] mussels,[44] and oysters in Canada[45] and in the U.S.[40] Also, about 100 *Y. enterocolitica* isolates were recovered from 240 oyster samples tested by the Food and Drug Administration Minneapolis Center for Microbiological Investigation Laboratory.[65] Recovery from clams has not been reported. Recently, Peixotto et al.[66] isolated *Y. enterocolitica*-like bacteria from shrimp and crab of the Gulf of Mexico.

C. Vegetables

Vegetables are also stored under refrigeration and may be eaten raw. In France, 3 strains of *Y. enterocolitica* were recovered from 66 samples of carrots, tomatoes, and green salads from hospital foods.[46] Some isolations from vegetables in Czechoslovakia have been observed.[23] *Y. enterocolitica* invasive to HeLa cells was recovered from celery in the U.S.[64] It is of interest to note that *Y. pseudotuberculosis* was recovered from 3 to 16% of stored fresh onions, carrots, beets, and potatoes, as well as from 3 to 9% of the sauerkraut, pickles, and pickled tomatoes tested in the cold Primorsk region of the Soviet Union.[47]

III. ISOLATION PROCEDURES FOR *Y. ENTEROCOLITICA* IN FOODS

Many problems related to the recovery of low numbers of *Y. enterocolitica* in foods need to be resolved. The ability of many current enrichment procedures to recover clinical strains of *Y. enterocolitica* from different types of food has not been very effective. In addition, recovery of *Y. enterocolitica* from many types of foods has not been evaluated by studies involving inoculation and subsequent recovery. Also, the species *Y. enterocolitica* is a heterogenous collection of bacteria and different strains may require different conditions for recovery.

In general, one group of clinical strains is nearly as resistant as *Salmonella* and can grow in enrichment media used for *Salmonella* such as selenite-cystine broth with 40 mg novobiocin per liter added[48] or in RMC broth.[27] Typical resistant strains are serotypes O:3 and O:9.

In one study,[15] *Y. enterocolitica* (O:3, IP 134) was added to foods and enrichment was carried out in RMC broth. Under these conditions, *Y. enterocolitica* (at a level of about

100 per gram) was recovered from oysters in only one of three tests, and from ground pork in five of six tests. Recovery of strain IP 134 by cold enrichment in phosphate buffer was also poor. *Y. enterocolitica* (at a level of 100 per gram) was recovered from oysters in one of three tests, and from ground pork in only one of six tests using the phosphate buffer cold enrichment technique. Enrichment in selenite broth with 40 mg novobiocin per liter and subsequent plating on desoxycholate citrate agar was effective in the isolation of *Y. enterocolitica* from the tonsils of pigs.[37] The RMC and selenite enrichment methods are fairly rapid (2 to 3 days at 25°C) for the recovery of the resistant O:3 and O:9 serotypes, but often will not recover the more sensitive clinical strains in foods.

Another group of clinical strains of *Y. enterocolitica*, like *Shigella*, are sensitive to chemicals used in enrichment and are very difficult to recover from foods. Typical sensitive strains are some of the O:8 clinical strains isolated in the U.S. In studies with inoculated foods, recovery of *Y. enterocolitica* (IP 107, about 100 per gram) was accomplished from fresh oysters in 1 of 3 tests and with ground pork in 3 of 6 tests by the use of phosphate buffer cold enrichment.[15] In the same study, the use of RMC broth completely inhibited growth and recovery of this sensitive serotype (O:8) as well as that of most of the native environmental biotypes in pork and oysters. These results indicate that current enrichment methods may not detect low levels of some of the clinically important strains in foods.

Environmental biotypes of *Y. enterocolitica* have intermediate to low resistance to selective enrichment broths and are easily recovered from foods. Their presence in foods makes it much more difficult to recover the important clinical strains. Recovery procedures developed for environmental biotypes in uninoculated foods are probably not applicable for the isolation of typical human clinical biotypes in foods.

Various enteric plating media such as desoxycholate citrate lysine sucrose urea,[49] MacConkey, *Salmonella-Shigella* (SS), and SS with 20 additional grams of sodium desoxycholate per liter[27] have been used for the isolation of *Y. enterocolitica*. Some of these plating media are inhibitory to the sensitive clinical strains of *Y. enterocolitica*. The following plating media were also developed or found to be helpful for the isolation of lactose-negative *Y. enterocolitica* colonies. *Y. enterocolitica* forms small black colonies on bismuth sulfite agar at 25°C within 2 to 3 days.[50] Addition of a fermentable substrate such as sucrose in DHL agar[31] (Eiken Ltd., Japan), sorbitol to DNAse agar with Tween 80 (DST),[51] mannitol to a modified deoxycholate-citrate-mannitol agar (YM),[52] or Tween 80 to MacConkey agar (MT)[51] makes it possible to recognize *Y. enterocolitica* colonies on these plates with some practice. For example, *Y. enterocolitica* forms distinctive translucent colonies on DST and MT agars.[51] It is probably advisable to use two or three different plating media for the recovery of *Y. enterocolitica* since enrichment methods are unpredictable or ineffective.

IV. SIGNIFICANCE OF *Y. ENTEROCOLITICA* IN FOODS

This section will discuss the evidence that was presented in this chapter from the food sanitarian's standpoint. On one hand there is substantial evidence that environmental biotypes of *Y. enterocolitica* are fairly common in drinking water and foods such as raw milk, meats, oysters, fish, and vegetables. Unchlorinated drinking water may be consumed without boiling. Also, rare or raw meats, fresh vegetables, raw clams and oysters, and even raw fish are consumed by a substantial portion of the general population. The consumption of these foods and water would expose these people to a regular intake of environmental biotypes of *Y. enterocolitica*.

The rhamnose-positive environmental biotypes may be associated with human illness, but the illness is usually mild, or has completely different foci of infections and symptoms than those caused by typical clinical biotypes of *Y. enterocolitica*.[16,17,53,54] The virulence and infectivity of salicin and esculin-positive strains and other environmental biotypes with atypical fermentation patterns are not clear at present. The exposure of humans to environmental biotypes of *Y. enterocolitica* suggests that in most cases, the environmental biotypes do not cause human illness that is noted by health authorities. However, confirmed cases of human infections caused by environmental biotypes of *Y. enterocolitica* suggest that a few strains may be virulent and can cause human disease under specific conditions.[16,17] For this reason, the virulence of environmental isolates of *Y. enterocolitica* from foods should be tested. Zink et al.[29] reported a possible plasmid-mediated virulence in *Y. enterocolitica* strains associated with illness in New York State. The possibility exists then, though not proven, that virulence may be acquired by some of the environmental biotypes by acquiring a plasmid that mediates virulence.

On the other hand, human clinical biotypes have so far been recovered from only few food and water samples. Perhaps this is because current enrichment and plating procedures are not effective in detecting low levels of typical human clinical biotypes from foods. While *Y. enterocolitica* recovered from acute human cases and foods involved in the illness are likely to be virulent regardless of the serotype and biotype, the virulence of a food isolate not related to a food-borne illness cannot be assumed and should be tested. Although it is theoretically possible for virulent *Y. enterocolitica* to contaminate and then grow in many types of refrigerated foods, actual food-borne outbreaks caused by *Y. enterocolitica* are rare. Based on the limited number of studies cited in this chapter, the recovery of typical clinical biotypes from foods is infrequent, with the exception of raw pork tongues and occasionally from raw meats and liver.

V. SURVIVAL AND CONTROL OF *Y. ENTEROCOLITICA*

Since *Y. enterocolitica* can grow at refrigeration temperatures used to store foods, investigations have been carried out to test the effect of some physical or chemical food preservation methods on the survival or inhibition of *Y. enterocolitica*. Unfortunately the biotypes or serotypes of strains used in some of these experiments are not reported. There is no information on the variation of resistance of the different *Y. enterocolitica* biotypes to the various physical and chemical agents.

Refrigeration — Hanna et al.[33] followed the growth of microorganisms in vacuum packaged beef and lamb during refrigerated storage for 35 days. A higher incidence of *Y. enterocolitica* was observed as storage progressed. In a subsequent study, Hanna et al.[55] tested the growth of 2 clinical and 3 *Y. enterocolitica*-like strains on raw and cooked beef and pork at 0, 1, 5, 7, and 25°C. Good growth of one clinical strain (ATCC 23715) was observed at 0, 1, and 5°C in 10 to 14 days, at 7°C within 5 days, and at 25°C in 1 day. Leistner et al.[31] tested the growth of combined European and also Japanese clinical reference strains and environmental biotypes on sterile chicken meat and pork at 0 to 1, 4, and 15°C. Good growth was observed at 0 to 1°C in 14 days, at 4°C in 7 days, and at 15°C in 2 days. *Y. pseudotuberculosis* also grew well at 4°C.[31] In contrast, Asakawa et al.[35] showed good growth of serotype O:3 inoculated into sliced ham and stored at 5 or 10°C, but no growth was noted at 0°C. Stern[56] found that a combination of 4 strains of *Y. enterocolitica* (2 clinical and 2 environmental biotypes) inoculated into sterile milk at levels of 250/mℓ increased to 4.6×10^7 cells per milliliter within 20 days at 3°C. The ability of *Y. enterocolitica* to compete with other psychotropic organisms normally present in milk was also demonstrated as being poor.[56]

Freezing — Under certain conditions freezing may have adverse effects on the survival of *Y. enterocolitica*. When beef roasts, inoculated at the surface with clinical and *Y. enterocolitica*-like strains, were stored at −23°C for about 1 month, large reductions in counts were observed.[57] These reductions were greater when the survivors were plated on bismuth sulfite agar than on a nonselective tryptic soy agar. In this study, cell numbers decreased from approximately 3×10^6 to 10,000 per gram within 28 days of storage at the same temperature.[57] Asakawa et al.[35] showed that the *Y. enterocolitica* count of sliced ham stored at $-10°C$ decreased from 3000 to less than 50 per gram within 7 days. On the other hand, Leistner et al.[31] reported only slight decreases in the *Y. enterocolitica* count (300 per gram) of frozen chicken broilers stored for 90 days at $-18°C$. Differences in the survival of *Y. enterocolitica* during freezing may have been caused by differences in sensitivity between strains and by differences in the characteristics of the foods.

Heating — Hanna et al.[57] reported that *Y. enterocolitica* populations in beef roasts of 3×10^6 per gram were destroyed when the final internal temperatures were 60 to 62°C. At 51°C some survivors remained. The heat resistance of several strains of *Y. enterocolitica* and similar organisms also was determined in skim milk. Considerable variation existed in heat resistance of the cultures tested.[58] Heating for 3 to 10 min at 55°C caused large reductions in counts, and at 60°C no survivors were detected after 1 to 3 min.[58] Pruitt and Johnson[69] reported injury of stationary phase cells of *Y. enterocolitica* when heated for 30 min at 51.5°C in phosphate buffer at pH 7.2. Recovery was tested on 1.5% trypticase and 0.5% phytone agar with or without 0.15% bile salts or 3% NaCl. The heat resistance of *Y. enterocolitica* is not greater than the heat resistance of other *Enterobacteriaceae*.

Vacuum packaging — Hanna et al.[60] reported on the development of *Y. enterocolitica* on beef steaks placed in packages of different oxygen permeability and stored at 1, 2.5, and 5°C for 21 to 35 days. *Y. enterocolitica* counts of steaks inoculated with *Y. enterocolitica* were consistently higher on samples packaged in the more oxygen-permeable film than on those stored in vacuum packages with very low oxygen tension. In addition to *Y. enterocolitica*, *Lactobacillus* spp. were predominant on the vacuum packaged steaks at the end of the storage period, whereas *Pseudomonas* spp. were predominant on the beef stored in oxygen-permeable film.

Radiation — El-Zawary and Rowley[61] reported on the effects of radiation on and recovery of *Y. enterocolitica*. D values (one log reduction of cells) for 3 virulent strains (IP 107, IP 134, and WA) were in the range of 10 to 12 krad. The radiation resistance of strain IP 107 increased with decreasing temperature. Over a temperature range of -30 to $25°C$ greater radiation injury occurred at the higher temperatures. *Y. enterocolitica* appeared quite radiation sensitive and should not create a problem in foods irradiated with at least 300 krad.

pH — The effect of pH on growth of *Y. enterocolitica* was determined in brain heart infusion at pH levels ranging from 5 to 9.[57] Growth of *Y. enterocolitica* was most rapid between pH 7 and 8. Little or no growth of ATCC 23715 occurred over a 24-hr period at pH 5. Stern et al.[62] observed growth of 2 strains of *Y. enterocolitica* associated with gastroenteritis in New York (CDC A2635 and A2611) and 2 environmental biotypes (IP 867 and 955) within 24 hr at 25°C in brain heart infusion broth at a pH of 4.6. The recovery of *Y. enterocolitica* in tartar sauce in Czechoslovakia[23] and the presence of *Y. enterocolitica* in pickled products[47] suggests survival of *Yersinia* in acidic environments and warrants further investigation.

Chemical preservation — The effects of the commonly used antimicrobial agents in foods such as sulfur dioxide, sorbate, propionate, acetate, benzoate, and parabenzoate on *Y. enterocolitica* have not been reported in the literature. No unusual salt tolerance was observed in the growth of four strains of *Y. enterocolitica*. Stern et al.[62] reported that strains CDC A2611, CDC A2635, IP 867, and IP 955 were inhibited in brain heart infusion broth with 5 to 7% NaCl, which is similar to that experienced with other *Enterobacteriaceae*.

REFERENCES

1. **Health and Welfare Canada,** *Yersinia enterocolitica* gastroenteritis outbreaks-Montreal, *Can. Dis. Weekly Rep.,* 2(Suppl. 11), 41, 1976 (see also correction, 2(19), 73, 1976.
2. **Asakawa, Y., Akahane, S., Kagata, N., Noguchi, M., Sakazaki, R., and Tamura, K.,** Two community outbreaks of human infection with *Yersinia enterocolitica, J. Hyg.,* 71, 75, 1973.
3. **Zen-Yoji, H., Maruyama, T., Sakai, S., Kimura, S., Mizuno, T., and Momose, T.,** An outbreak of enteritis due to *Yersinia enterocolitica* occurring at a junior high school, *Jpn. J. Microbiol.,* 17, 220, 1973.
4. **Black, R. E., Jackson, R. J., Tasai, T., Medvesky, M., Shayegani, M., Feeley, J. C., MacLeod, K. I. E., and Wakelee, A. M.,** Epidemic *Yersinia enterocolitica* infection due to contaminated chocolate milk, *N. Engl. J. Med.,* 298, 76, 1978.
5. **Gutman, L. T., Ottesen, E. A., Quan, T. I., Noce, P. S., and Katz, S. L.,** An interfamilial outbreak of *Yersinia enterocolitica* enteritis, *N. Engl. J. Med.,* 288, 1372, 1973.
6. **Hacking, M. A. and Sileo, L.,** *Yersinia enterocolitica* and *Yersinia pseudotuberculosis* from wildlife in Ontario, *J. Wildl. Dis.,* 10, 452, 1974.
7. **Wilson, H. D., McCormick, J. B., and Feeley, J. C.,** *Yersinia enterocolitica* infection in a 4-month-old infant associated with infection in household dogs, *J. Pediatr.,* 89, 767, 1976.
8. **McClure, H. M., Weaver, R. E., and Kaufmann, A. F.,** Pseudotuberculosis in nonhuman primates: infection with organisms of the *Yersinia enterocolitica* group, *Lab. Anim. Sci.,* 21, 376, 1971.
9. **Farstad, L., Ladsverk, T., and Lassen, J.,** Isolation of *Yersinia enterocolitica* from a dog with chronic enteritis, *Acta Vet. Scand.,* 17, 261, 1976.
10. **Kaneko, K., Hamada, S., and Kato, E.,** Occurrence of *Yersinia enterocolitica* in dogs, *Jpn. J. Vet. Sci.,* 39, 407, 1977.
11. **Pedersen, K. B.,** Isolation of *Yersinia enterocolitica* from Danish swine and dogs, *Acta Pathol. Microbiol. Scand.,* 84B, 317, 1976.
12. **Wauters, G., Pohl, P., and Stevens, J.,** Portage de *Yersinia enterocolitica* par le porc de boucherie, *Med. Mal. Infect.,* 6, 484, 1976.
13. **Lee, W. H.,** An assessment of *Yersinia enterocolitica* and its presence in foods, *J. Food Prot.,* 40, 486, 1977.
14. **Keet, E. E.,** *Yersinia enterocolitica* septicemia. Source of infection and incubation period identified, *N.Y. State J. Med.,* 74, 2226, 1974.
15. **Lee, W. H.,** Testing for the recovery of *Yersinia enterocolitica* in foods and their ability to invade HeLa cells, *Contrib. Microbiol. Immunol.,* 5, 228, 1979.
16. **Alonso, J. M., Bejot, J., Bercovier, H., and Mollaret, H. H.,** Sur un groupe de souches de *Yersinia enterocolitica* fermentant le rhamnose. Intérêt diagnostique et particularités écologiques, *Med. Mal. Infect.,* 5, 490, 1975.
17. **Bottone, E. J.,** Atypical *Yersinia enterocolitica:* clinical and epidemiological parameters, *J. Clin. Microbiol.,* 7, 562, 1978.
18. **Highsmith, A. K., Feeley, J. C., Skaliy, P., Wells, J. G., and Wood, B. T.,** Isolation of *Yersinia enterocolitica* from well water and growth in distilled water, *Appl. Environ. Microbiol.,* 34, 745, 1977.
19. **Inoue, M. and Nagao, H.,** Isolation of *Yersinia enterocolitica* from bovine cecal contents, commercial meat, and meat shops, *J. Jpn. Vet. Med. Assoc.,* 29, 612, 1976.
20. **Mollaret, H., Nicolle, P., Brault, J., and Nicolas, R.,** Importance actuelle des infections a *Yersinia enterocolitica, Bull. Acad. Nat. Med. Paris,* 156, 704, 1972.
21. **Sarrouy, J.,** Isolement d'une *Yersinia enterocolitica* a partir du lait, *Med. Mal. Infect.,* 2, 67, 1972.
22. **Schiemann, D. A. and Toma, S.,** Isolation of *Yersinia enterocolitica* from raw milk, *Appl. Environ. Microbiol.,* 35, 54, 1978.
23. **Aldova, E., Cerna, J., Janeckova, M., and Pegrimkova, J.,** *Yersinia enterocolitica* and its demonstration in foods, *Czech. Hyg.,* 20, 395, 1975.
24. **Mollaret, H. H., Bercovier, H., and Alonso, J. M.,** Summary of the data received at the WHO reference center for *Yersinia enterocolitica, Contrib. Microbiol. Immunol.,* 5, 174, 1979.
25. **Capriolli, T., Drapeau, A. J., and Kasatiya, S.,** *Yersinia enterocolitica:* serotypes and biotypes isolated from humans and the environment in Quebec, Canada, *J. Clin. Microbiol.,* 8, 7, 1978.
26. **Toma, S. and Lafleur, L.,** Survey on the incidence of *Yersinia enterocolitica* infection in Canada, *Appl. Microbiol.,* 28, 469, 1974.
27. **Wauters, G.,** Improved methods for the isolation and the recognition of *Yersinia enterocolitica, Contrib. Microbiol. Immunol.,* 2, 68, 1973.
28. **Feeley, J. C., Wells, J. G., Tsai, T. F., and Puhr, N. D.,** Detection of enterotoxigenic and invasive strains of *Yersinia enterocolitica, Contrib. Microbiol., Immunol.,* 5, 329, 1979.

29. Zink, D. L., Feeley, J. C., Wells, J. G., Vanderzant, C., Vickery, J. C., Roof, W. D., and O'Donovan, G. A., Plasmid-mediated tissue invasiveness in *Yersinia enterocolitica*, *Nature (London)*, 283, 224, 1980.
30. Mollaret, H. H., L'infection humaine à *"Yersinia enterocolitica"* en 1970, à la lumière de 642 cas recents, *Pathol. Biol.*, 19, 189, 1971.
31. Leistner, L., Hechelmann, H., Kashiwazaki, M., and Albertz, R., Nachweis von *Yersinia enterocolitica* in Faeces und Fleisch von Schweinen, Rindern und Geflügel, *Fleischwirtschaft*, 55, 1599, 1975.
32. Inoue, M. and Kurose, M., Isolation of *Yersinia enterocolitica* from cow's intestinal contents and beef meat, *Jpn. J. Vet. Sci.*, 37, 91, 1975.
33. Hanna, M. O., Zink, D. L., Carpenter, Z. L., and Vanderzant, C., *Yersinia enterocolitica*-like organisms from vacuum-packaged beef and lamb, *J. Food Sci.*, 41, 1254, 1976.
34. Toma, S. and Deidrick, V. R., Isolation of *Yersinia enterocolitica* from swine, *J. Clin. Microbiol.*, 2, 478, 1975.
35. Asakawa, Y., Akahane, S., Shiozawa, K., and Honma, T., Investigations of the source and route of *Yersinia enterocolitica* infection, *Contrib. Microbiol. Immunol.*, 5, 115, 1979.
36. Zen-Yoji, H., Sakai, S., Maruyama, T., and Yanagawa, Y., Isolation of *Yersinia enterocolitica* and *Yersinia pseudotuberculosis* from swine, cattle and rats at an abattoir, *Jpn. J. Microbiol.*, 18, 103, 1974.
37. Narucka, U. and Westendoorp, J. F., Studies for the presence of *Yersinia enterocolitica* and *Yersinia pseudotuberculosis* in clinically normal pigs, *Tijdschr. Diergeneeskd.*, 102, 299, 1977.
38. Pedersen, K. B., The occurrence of *Yersinia enterocolitica* in the throats of swine, *Contrib. Microbiol. Immunol.*, 5, 253, 1979.
39. Wauters, G. and Janssens, M., Portage de *Yersinia enterocolitica* par le porc de boucherie. II. Recherche de *Yersinia enterocolitica* sur des langues de porc achetèes en boucherie, *Med. Mal. Infect.*, 6, 517, 1976.
40. Lee, W. H., McGrath, P. P., Carter, P. H., and Eide, E. L., The ability of some *Yersinia enterocolitica* strains to invade HeLa cells, *Can. J. Microbiol.*, 23, 1714, 1977.
41. Guthertz, L. S., Fruin J. T., Spicer, D., and Fowler, J. L., Microbiology of fresh comminuted turkey meat, *J. Milk Food Technol.*, 39, 823, 1976.
42. Kapperud, G. and Jonsson, B., *Yersinia enterocolitica* in brown trout (*Salmo trutta* L.) from Norway, *Acta Pathol. Microbiol. Scand.*, 84B, 66, 1976.
43. Rakovsky, J., Paučkova, V., and Aldova, E., Human *Yersinia enterocolitica* infections in Czechoslovakia, *Contrib. Microbiol. Immunol.*, 2, 93, 1973.
44. Spadaro, M. and Infortuna, M., Isolamenta di *Yersinia enterocolitica* in *Mitilus galloprovincialis* lamk, *Boll. Soc. Ital. Biol. Sper.*, 44, 1896, 1968.
45. Toma, S., Survey on the incidence of *Yersinia enterocolitica* in the province of Ontario, *Can. J. Publ. Health*, 64, 477, 1973.
46. Louiseau-Marolleau, M. L. and Alonso, J. M., Isolement de *Yersinia enterocolitica* lors d'une étude systematique des aliments en milieu hospitalier, *Med. Mal. Infect.*, 6, 373, 1976.
47. Kuznetsov, V. G., Rakovsky, V. I., Grebenshchikov, L. A., Loskutov, A. N., Dagayeva, A. R., and Somov, G. P., Study of the epidemiology of the fareastern scarlatina like fever in the Primorsk region. Report I. Contamination of vegetables, roots and pickled provisions with *Yersinia pseudotuberculosis* in the foci of scarlatina-like fever, *Zh. Mikrobiol. Epidemiol. Immunobiol.*, 10, 34, 1975.
48. Van Noyen, R. and Vandepitte, J., L'isolement de *Yersinia enterocolitica* par une technique usuelle de coproculture. *Ann. Inst. Pasteur (Paris)*, 4, 463, 1968.
49. Nilehn, B., Studies on *Yersinia enterocolitica* with special reference to bacterial diagnoses and occurrence in human acute enteric disease, *Acta Pathol. Microbiol. Scand. Suppl.*, 206, 1, 1969.
50. Hanna, M. O., Stewart, J. C., Carpenter, Z. L., and Vanderzant, C., Development of *Yersinia enterocolitica*-like organisms in pure and mixed cultures on different bismuth sulfite agars, *J. Food Prot.*, 40, 676, 1977.
51. Lee, W. H., Two plating media modified with Tween 80 for isolating *Yersinia enterocolitica*, *Appl. Environ. Microbiol.*, 33, 215, 1977.
52. Saari, T. N. and Jansen, G. P., Waterborne *Yersinia enterocolitica* in the Midwest United States, *Contrib. Microbiol. Immunol.*, 5, 185, 1979.
53. Bottone, E. J., Chester, B., Malowany, M. S., and Allerhand, J., Unusual *Yersinia enterocolitica* isolates not associated with mesenteric lymphadenitis, *Appl. Microbiol.*, 27, 858, 1974.
54. Dabernat, H. J., Bauriaud, R., Lemozy, J., LeFevre, J. C., and Lareng, M. B., *Yersinia enterocolitica* fermentant le rhamnose. A propos de 15 souches isolées chez des enfants, *Med. Mal. Infect.*, 4, 156, 1978.
55. Hanna, M. O., Stewart, J. C., Zink, D. L., Carpenter, Z. L., and Vanderzant, C., Development of *Yersinia enterocolitica* on raw and cooked beef and pork at different temperatures, *J. Food Sci.*, 42, 1180, 1977.

56. **Stern, N. J., Pierson, M. O., and Kotula, A. W.**, Growth and competitive nature of *Yersinia enterocolitica* in whole milk, *J. Food Sci.*, 45, 1, 1980.
57. **Hanna, M. O., Stewart, J. C., Carpenter, Z. L., and Vanderzant, C.**, Effect of heating, freezing and pH on *Yersinia enterocolitica*-like organisms from meat, *J. Food Prot.*, 40, 689, 1977.
58. **Hanna, M. O., Stewart, J. C., Carpenter, Z. L., and Vanderzant, C.**, Heat resistance of *Yersinia enterocolitica* in skim milk, *J. Food. Sci.*, 42, 1134, 1977.
59. **Pruitt, D. V. and Johnson, M. G.**, Recovery of tolerance to NaCl or bile salts no. 3 by heat-stressed cells of *Yersinia enterocolitica* 107 exposed to 5°C or an overlay. Abstr. Annu. Meet. Am. Soc. Microbiol., 1978, 187.
60. **Hanna, M. O., Stewart, J. C., Carpenter, Z. L., and Vanderzant, C.**, Effect of packaging methods on the development of *Yersinia enterocolitica* on beef steaks, *J. Food Saf.*, 1, 29, 1977.
61. **El-Zawahry, Y. A. and Rowley, D. B.**, Radiation resistance and injury of *Yersinia enterocolitica*, *Appl. Environ. Microbiol.*, 37, 50, 1979.
62. **Stern, N. J., Pierson, M. D., and Kotula, A. W.**, Effects of pH and sodium chloride on *Yersinia enterocolitica* growth at room and refrigeration temperatures, *J. Food Sci.*, 45, 64, 1980.
63. **Lee, W. H.**, unpublished results, 1978.
64. **Mehlman, I.**, personal communication, 1980.
65. **Anderson, D.**, personal communication, 1980.
66. **Peixotto, S. S., Finne, G., Hanna, M. O., and Vanderzant, C.**, Presence, growth, and survival of *Yersinia enterocolitica* in oysters, shrimp, and crab, *J. Food Prot.*, 42, 972, 1980.

Chapter 16

EPIDEMIOLOGICAL ASPECTS OF *YERSINIA ENTEROCOLITICA* IN NEW YORK STATE WITH EMPHASIS ON A RECENT FOOD-BORNE OUTBREAK

Mehdi Shayegani

TABLE OF CONTENTS

I. Introduction 174

II. *Y. enterocolitica* Outbreak in Oneida County, New York 174
    A. Method of Isolation and Identification 174
    B. Serotypes 175
    C. Biotyping 175
    D. Phage Typing 175

III. Isolation of *Y. enterocolitica* from Humans in Oneida County
    (October 1976 to May 1977) 175

IV. Isolation of *Y. enterocolitica* from Animals and the Environment in
    Oneida County 177

V. Statewide Survey of *Y. enterocolitica* in New York State Not
    Related to the Outbreak 177
    A. *Y. enterocolitica* Isolated During or Shortly after the Outbreak
       (October 1976 to December 1977) 177
    B. Surveillance of *Y. enterocolitica* 15 Months after the Outbreak
       (January to December 1978) 178

VI. Serotypes, Biotypes, and DNA Relatedness Groups of *Y. enterocolitica*
    Isolates 179

VII. Phage Types of Selected Isolates of *Y. enterocolitica* 180

VIII. Discussion 181

References 182

I. INTRODUCTION

Yersinia enterocolitica was first recognized in the U.S. in 1933[1] in the Bacteriology Laboratory of the New York State Department of Health's Division of Laboratories and Research. In 1939 Schleifstein and Coleman[2] of this laboratory supplied the first detailed description of a new species then called *Bacterium enterocoliticum*. The same laboratory became involved in the investigation of the first documented food-borne outbreak of *Y. enterocolitica* which occurred in Oneida County, New York in October 1976. Following the outbreak the prevalence of the organism was further studied in humans, animals, and the environment in the state of New York. The following subjects are discussed in this report:

1. *Y. enterocolitica* outbreak and follow-up study in Oneida County (October 1976 to May 1977)
2. Statewide survey of *Y. enterocolitica* in New York State, not related to the outbreak
   a. *Y. enterocolitica* isolated during or shortly after the outbreak (October 1976 to December 1977)
   b. Surveillance of *Y. enterocolitica*, 15 months after the outbreak (January to December 1978)
3. Classification of *Y. enterocolitica* isolates by serotype, biotype, and DNA relatedness group
4. Phage types of selected isolates of *Y. enterocolitica*
5. Discussion

II. *Y. ENTEROCOLITICA* OUTBREAK IN ONEIDA COUNTY, NEW YORK

In late September 1976 an outbreak of illness with symptoms of abdominal pain and fever occurred among school children in Oneida County. Appendectomies were performed on several of the children. Histological examination did not confirm the diagnoses of appendicitis, and *Y. enterocolitica* was isolated at the Rome Hospital Laboratory from culture of the surgical incision of one of the patients. The identification was confirmed by our laboratory. A joint epidemiologic investigation by New York State and the Center for Disease Control, Atlanta, revealed that 217 children attending five Oneida County schools had symptoms of yersiniosis, 36 were hospitalized and 16 had appendectomies. Chocolate milk was epidemiologically incriminated in the outbreak when *Y. enterocolitica* with the same serotype (O:8) and biotype as that found in the patients was isolated from one out of four unopened chocolate milk containers.[3-5]

A. Method of Isolation and Identification[5,6]

Fecal specimens or rectal swabs were placed in 10 mℓ of sterile phosphate-buffered saline pH 7.6 and immediately hand-carried or mailed to our laboratory. If delivery was delayed, specimens were refrigerated until they could be delivered. On receipt, part of each specimen was plated on Endo, *Salmonella-Shigella*, and deoxycholate-citrate agars; the remainder was refrigerated at 4°C and replated after 1 and 3 weeks.

Milk samples were placed in cooked-meat medium and refrigerated at 4°C. Water samples were filtered through a 0.45-μm Millipore filter and the filter was removed and placed in cooked-meat medium and refrigerated. After 1 and 3 weeks of cold enrichment both milk and water samples were inoculated on the same plating media as the fecal specimens. All plates were incubated at 35 to 37°C and read at 48 hr.

Colonies with typical *Y. enterocolitica* morphology were transferred to triple sugar iron (TSI) and urea agar. Cultures which were acid in the butt and slant of the TSI and were urease positive were examined by further biochemical tests. Isolates with typical colonial morphology but giving a different reaction in TSI and urea were further studied for identification as strains with an atypical biochemical reaction. Each isolate was tested with 36 biochemicals. The methyl red, Voges-Proskauer, ONPG, lecithinase, and OF (oxidation-fermentation) lactose tests were performed at room temperature (approximately 25°C) only. The rest of the biochemical tests were performed at 35 to 37°C except for the motility test which was incubated at both room temperature and 35 to 37°C.[5] For biotyping purposes of representative atypical strains, some biochemicals were tested at both 25°C and 35 to 37°C.

B. Serotypes

Isolates were serotyped in our laboratory by slide agglutination with antisera O:1 to O:21 prepared by Hudson and Quan and supplied to us by Carter of the Trudeau Institute at Saranac Lake, New York, and antisera O:22 to O:34 provided us by Toma of the National Reference Center for Yersinia, Public Health Laboratory, Ontario Ministry of Health, Toronto, Canada. In an earlier study of the outbreak some isolates were serotyped by Bissett of the California Department of Health. Later serotyping of all isolates from the outbreak was done by James C. Feeley of the Bureau of Epidemiology of the Center for Disease Control using a serial dilution method. Certain selected strains were serotyped by Mollaret of the Institut Pasteur, Paris.

C. Biotyping

The isolates were divided into biotypes according to the schemes of Niléhn,[7] Wauters,[8] and Knapp and Thal[9] and into DNA relatedness groups of Brenner et al.[10,11]

D. Phage Typing

Representative isolates of *Y. enterocolitica* (various serotypes and biotypes) from different sources were phage typed by Mollaret at the Yersinia Center, Institut Pasteur, Paris, France. Later several isolates of serotype O:3 were phage-typed by Toma, Canadian National Reference Center for Yersinia, Toronto, Canada.

III. ISOLATION OF *Y. ENTEROCOLITICA* FROM HUMANS IN ONEIDA COUNTY (OCTOBER 1976 TO MAY 1977)

During the investigation of *Y. enterocolitica* in Oneida County October 1976 to May 1977, a total of 371 human specimens from 306 persons were tested in our laboratory: the source of the specimens studied were feces (365), throat (4), pustule (1), and sputum (1). Also received were five human isolates for identification or confirmation: three from feces, one from a surgical incision, and one from vomitus. Of the 306 persons from whom specimens were received, 154 had typical symptoms of yersiniosis, 95 were asymptomatic, and no clinical information was available to us on 57. Of the 49 *Y. enterocolitica* isolated from the persons, 35 serotype O:8 strains were isolated from symptomatic persons (in one case two strains from one person), two from asymptomatic persons, and none from persons with no clinical information (Table 1). Also isolated were 12 *Y. enterocolitica* of other than serotype O:8 — two serotype O:5, one O:3, one O:6,31, and eight nontypable.[5]

Follow-up specimens on 40 of the 41 symptomatic patients, from whom *Y. enterocolitica* was originally isolated, were tested several months later. *Y. enterocolitica*

Table 1
TOTAL *Y. ENTEROCOLITICA* AND SEROTYPE O:8 ISOLATED FROM HUMANS IN THE ONEIDA COUNTY OUTBREAK

Clinical classification	No. of persons	No. of isolates (%)	No. of serotype O:8 (%)
Symptomatic	154	42 (27.3)	35 (22.7)
Asymptomatic	95	2 (2.1)	2 (2.1)
No information	57	5 (8.8)	0 (0)
Total	306	49 (16.0)	37 (12.1)

Table 2
Y. ENTEROCOLITICA ISOLATES FROM ANIMALS IN ONEIDA COUNTY DURING THE OUTBREAK

Animals	No.	Positives No.	Positives %	Serotypes Type	Serotypes No. of isolates
Cow	148	17	11.5	4,33	13
				8	2
				7,8	1
				6,31	1
Pig	40	8	20.0	NT[a]	3
				$10K_1$	2
				16,29	2
				6,31	1
Mouse	6	1	16.7	NT	1
Rat	6	1	16.7	NT	1
Dog	2	0	0		
Cat	1	0	0		
Horse	1	0	0		
Mole	1	0	0		
Total	205	27	13.2		

[a] Nontypable with antisera O:1 through O:34.

was reisolated from only three of them and in each case the strain recovered was a different serotype from the original isolate. A nontypable strain was isolated from a patient who originally had serotype O:8 and serotype O:6,31 was isolated from two sisters who originally had nontypable strains.

The majority of *Y. enterocolitica* serotype O:8 isolated in Oneida County were encountered at the time of the outbreak (October and November). This serotype was isolated from 23% of the symptomatic persons tested. The isolation of this serotype declined sharply in December and in January, and serotype O:8 was not isolated from February through June. However, this particular serotype comprised the majority (77%) of all *Y. enterocolitica* strains isolated from Oneida County residents during October 1976 to May 1977.

Table 3
Y. ENTEROCOLITICA ISOLATES FROM ENVIRONMENT IN ONEIDA COUNTY DURING THE OUTBREAK

Sources	No.	Positive No.	Positive %	Serotypes Type	Serotypes No. of isolates
Milk					
Raw	21	4	19.0	6,31	2
				8	2
Pasteurized	18	0	0		
Water	31	8	25.8	NT[a]	5
				3	1
				6,31	1
				7	1
Total	70	12	17.1		

[a] Nontypable with antisera O:1 through O:34.

IV. ISOLATION OF Y. ENTEROCOLITICA FROM ANIMALS AND THE ENVIRONMENT IN ONEIDA COUNTY

Table 2 shows the results of this investigation. Y. enterocolitica was isolated from 13.2% of the animals tested. The sources of the cow specimens tested were feces (133), cervix (8), placenta (3), lung (1), liver (1), discharge (1), and stomach contents (1). Specimens from pigs consisted of feces or rectal swabs. Either feces or portions of organs obtained during autopsy were cultured from the other animals. Two strains of Y. enterocolitica which were serotype O:8 and one strain which was serotype O:7,8 were isolated from cows. Neither of these serotypes, however, were isolated from any of the other animals.

As shown in Table 3, Y. enterocolitica was isolated from four of the raw milk samples, two of the isolates being serotype O:8. Yersinia was not isolated from pasteurized milk. In testing local brooks and drinking water, 26% of the samples contained Yersinia, however, serotype O:8 was not isolated and most of the strains were nontypable.

V. STATEWIDE SURVEY OF Y. ENTEROCOLITICA IN NEW YORK STATE, NOT RELATED TO THE OUTBREAK

A. Y. enterocolitica Isolated During or Shortly after the Outbreak (October 1976 to December 1977)

Three hundred sequentially selected human fecal specimens (100 in the Winter 1976 to 1977 and 200 in the Spring 1977), submitted to our laboratory for *Salmonella* and *Shigella* screening from laboratories in various parts of upstate New York were studied for *Yersinia*. The specimens were plated both fresh and after cold enrichment. Animal and environmental specimens also were studied. In addition, human specimens (feces or isolates) received in our laboratory for isolation, identification or confirmation of Y. enterocolitica were included. To our knowledge, none of the above specimens was related to the outbreak.

Table 4
Y. ENTEROCOLITICA ISOLATED FROM HUMANS IN NEW YORK STATE DURING OR SHORTLY AFTER THE OUTBREAK, AND NOT RELATED TO THE OUTBREAK

Source	No. Specimens	No. Isolates	Serotype — No. of isolates per type
Selected routine fecal	300	7	3 of O:25 and one each of O:7,8, O:16, O:30,31, NT[a]
Fecal requested for Yersinia	3	3	One each of O:5, O:8, NT
Isolates for identification			
Fecal	4	4	3 of O:8, one of O:3
Throat	3	3	One each of O:7,8, O:16,34 NT
Appendix	2	2	One each of O:7,8, O:8
Sputum	1	1	O:4,33

[a] Nontypable with antisera O:1 through O:34.

Of the 300 selected human fecal specimens, *Y. enterocolitica* was isolated from seven (2.3%). Three of these isolates were from members of a Dutchess County family (aged 12, 14, and 43 years) and were serotype O:25; four other isolates were from unrelated children age 15 to 22 months in Greene and Fulton Counties and were serotype O:7,8, O:16, O:30,31, and nontypable.

Tables 4 and 5 demonstrate the results of this study. Serotype O:3, O:5, O:7,8 and O:8 were isolated from humans but not from animals or the environment. *Y. enterocolitica* with the same serotype and biotype responsible for the Oneida County outbreak were isolated from five human specimens originating in five counties including one from Oneida County.

B. Surveillance of *Y. enterocolitica* 15 Months After the Outbreak (January to December 1978)

The prevalence of *Y. enterocolitica* in New York state was investigated in a 1-year study from January to December 1978. The procedure for isolation and identification was the same as described in this report except the specimens were tested only after 3 weeks of cold enrichment. The following materials were studied:

1. Fecal specimens from patients with signs or symptoms of gastroenteritis. These specimens were collected specifically for this study in phosphate buffer pH 7.6 at six laboratories located in different geographical locations in New York State.
2. Human fecal specimens routinely received in our laboratory in buffered glycerol for *Salmonella* and *Shigella* screening.
3. Surface water collected from rivers and streams in the State.
4. Fecal specimens from animals and samples of raw milk collected primarily from Oneida County.

The final results of this study, which consisted of examining about 2500 specimens, are not available at the time of this writing. Preliminary findings indicate that *Y.*

Table 5
Y. ENTEROCOLITICA ISOLATED FROM ANIMALS AND WATER IN NEW YORK STATE DURING OR SHORTLY AFTER THE OUTBREAK, AND NOT RELATED TO THE OUTBREAK

Source	No. Specimens	No. Isolates	Serotype
Pig	57	5	Two each of O:11,24, NT[a], One of O:16,29
Cow	20	1	NT
Turkey	4	0	—
Dog	1	1	NT
Water	1	1	NT

[a] Nontypable with antisera O:1 through O:34.

enterocolitica is not as common in humans as it is in surface water. Occasional typical isolates of the serotypes most often associated with illness (O:3, O:5, and O:8) were isolated from humans. The majority of the human and environmental isolates belonged to various other serotypes or were nonserotypable with antisera O:1 through O:34.

VI. SEROTYPES, BIOTYPES, AND DNA RELATEDNESS GROUPS OF Y. ENTEROCOLITICA ISOLATES

The strains of *Y. enterocolitica* isolated in our studies were first grouped according to their serotypes and each serotype according to biotypes (Niléhn "N", Wauters "W", Knapp and Thal "KT") and finally according to DNA relatedness groups of Brenner et al.

Forty-six isolates serotype O:8 or O:7,8 were biotyped as N2, W1, KT2. Forty-one human isolates of these biotypes were serotype O:8, thirty-seven from the Oneida County outbreak. Four other human isolates (one from the outbreak) and one isolate from a cow's cervix in Oneida County were serotype O:7,8. Five other isolates also serotype O:8, differed in their biotypes. Two isolates from cows and two from milk were biotyped as N1, W1, KT3. Another isolate was from a human source and was biotyped N3, W3, KT1.

There were 11 isolates from various sources (human, water, milk, cow, and pig) and locations in the state, which were serotype O:6,31 and were biotyped as N1, W1, KT3.

Fifteen isolates were serotype O:4,33. Thirteen of them from cows were biotype N1, W1, KT3, and two others, one from human and one from water, were in two other different biotypes.

There were six human isolates of serotype O:5. Four (three from Oneida County) were biotype N1, W1, KT3, and two others were biotypes N3, W3, KT1.

Seven isolates, six from humans, and one from water, were serotype O:3. The water isolate was biotype N2, W1, KT2 (similar in biotype to the 46 isolates of serotype O:7,8 or O:8). One human isolate was biotype N3, W3, KT1, while the five other human isolates were biotype N4, W4, KT1 and phage type IXb.

Other strains of *Y. enterocolitica* isolated in New York from various sources up to this time belonged to various serotypes: O;2,3, O:7, O:10, O:10K$_1$, O:11,24, O:14, O:16, O:16,29, O:16,34, O:18, O:21, O:21,28, O:25, O:30,31, O:31, and O:34. The isolates nontypable with serotypes O:1 through O:34 belonged to a variety of biotypes. Twenty-

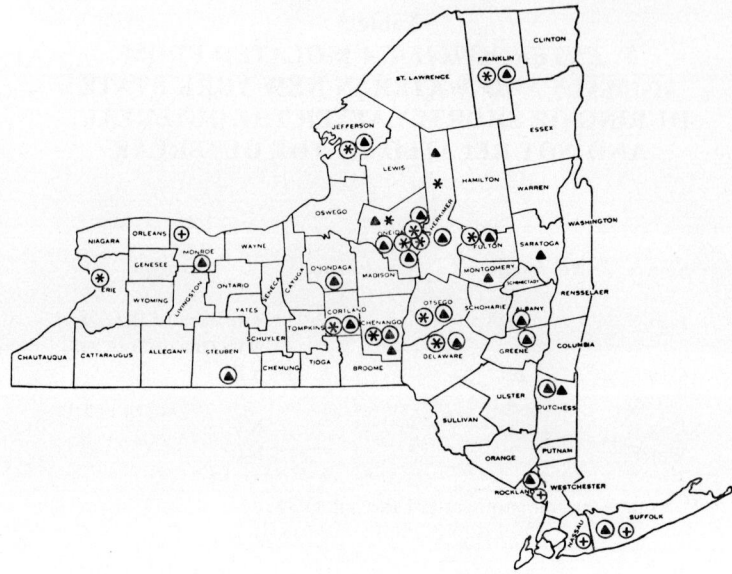

FIGURE 1. *Y. enterocolitica* serotype O:8 (or O:7,8) and DNA relatedness group of typical *Y. enterocolitica* in humans, animals, and environment in New York State. ⊛: *Y. enterocolitica* serotype O:8 (or O:7,8) from humans (one isolate in Buffalo is from early 1979). ★: *Y. enterocolitica* serotype O:8 (or O:7,8) from animals and environment. ⊕: *Y. enterocolitica* serotype O:3 (phage type IXb) from humans. Ⓐ: Typical *Y. enterocolitica* of Brenner et al. DNA relatedness group from humans. ▲: Typical *Y. enterocolitica* of Brenner et al. DNA relatedness group from animals and environment.

three of the nonserotypable isolates were from humans, 27 from water, and several were from various animals.

Brenner et al.[10,11] DNA relatedness group one corresponds to biochemically typical *Y. enterocolitica*. In our studies up to this time, nearly 80% of human isolates, 46% of pig isolates, and 28% of water isolates (including some nontypable strains), all isolates from cows and milk, one isolate from a dog, and one from a mouse were typical *Y. enterocolitica* (group one). DNA relatedness group two, and three (rhamnose-positive) and group four (sucrose-negative), which remain in the genus *Yersinia* with their species yet to be named, were found by us in human isolates as well as environmental sources. The distribution of *Y. enterocolitica* serotypes O:3 and O:8 and typical *Y. enterocolitica* (DNA relatedness group one) in New York State is shown in Figure 1.

VII. PHAGE TYPES OF SELECTED ISOLATES OF *Y. ENTEROCOLITICA*

Representatives isolates of *Y. enterocolitica* from various serotypes and biotypes were phage typed by Institut Pasteur, Paris. Most representative isolates from human, water, milk, cow, pig with serotype O:3 (biotype N2, W1, KT2 or N3, W3, KT1), O:4,33, O:5, O:6,31, O:7, O:7,8, O:8, O:10, O:10K$_1$, O:25, and nontypable were phage type X$_z$. Isolates which were serotype O:7,8, O:8, O:10K$_1$ from humans and cows were phage X$_0$ although they were in the same biotypes as those in phage type X$_z$. An isolate serotyped as O:16,29 (from a pig) was phage type X$_0$ Five isolates serotyped as O:3 biotype N4, W4, KT1 were phage type IXb.

VIII. DISCUSSION

Y. enterocolitica serotype O:8, biotype Niléhn 2, Wauters 1, and Knapp and Thal 2, the most common strain isolated from human infections in the U.S. and the same strain as was first reported from our laboratory in the 1930s, proved to be the causative agent of the extensive food-borne outbreak in Oneida County in 1976. It was isolated from patients (school children) and from one carton of chocolate milk made by a local dairy. The reservoir for contamination of chocolate milk remains unresolved. Attempts to isolate the organism from chocolate syrup, raw milk, dairy equipment, water, and employees, and cows, pigs rodents, and other animals at the dairy, were unsuccessful.[4] In a further study[5] we did isolate from the cervix of a cow from Oneida County *Y. enterocolitica*, serotype O:8(O:7,8) which had the same biotype pattern as the isolates from the patients. Four other isolates from Oneida County, two from cow and two from raw milk, were the same serotype, but differed in biotype. Although the isolation of these strains from these sources is not sufficient to reveal the origin of the organism in the outbreak, it hints at the possibility of contamination through a cow or milk. It is possible that either the lot of raw milk used to make chocolate milk was not properly pasteurized because of a temporary malfunction of the pasteurization process or that the contamination occurred when the chocolate and the milk were combined.

Isolation of Y. *enterocolitica* serotype O:8 was reported[12] from a stream in the Adirondack Mountains in upstate New York a few weeks after the isolation of the organism with the same serotype and biotype from a case of septicemia in a man who was on a hunting trip in the mountains and used mountain stream water for drinking. Deer were suggested as the reservoir for the organism.

While *Y. enterocolitica* has worldwide geographical distribution, each country seems to have its own locally predominant strain. In Canada, *Y. enterocolitica* is more prevalent than in the U.S., with the most common human isolate serotype O:3,[13,14] a type rarely isolated in the U.S. In the Montreal area the second most common cause of bacterial gastroenteritis is *Y. enterocolitica* serotype O:3,[15] while across the St. Lawrence River in upstate New York, and in other parts of the U.S., enteritis caused by *Y. enterocolitica* O:3 is not common in the human population.[16]

Serotype O:3 is also a common pathogen in Europe, Japan, and South Africa. The strains from these three geographic areas and Canada share a common biotype, but differ in phage type.[16] "European" and "Japanese" *Y. enterocolitica* serotype O:3 is phage type VIII, "South African" is phage type IXa, and "Canadian" is phage type IXb. In the U.S., isolation of two strains of serotype O:3 from humans was reported in California[17] and New York.[18] Of 7 *Y. enterocolitica* serotype O:3 isolated in our studies, the 5 human isolates had typical biotypes (N4, W4, KT1) and were phage-typed as "Canadian" phage IXb. Two patients were from Monroe County (near the New York-Canada border) and one each from Suffolk, Nassau, and Rockland Counties (near New York City).

Y. enterocolitica serotypes O:5 and O:6,30 are the second and third most frequent human isolates in Canada.[14] In the present study we report six human isolates serotype O:5 (two biotypes) and the presence of 11 serotype O:6,31 from various sources. Serotype O:9, the second most common human isolate in Europe,[16] was not found in our study and has not been reported in the U.S. The significance of some other serotypes, particularly the atypical *Y. enterocolitica* in the environment and the intestinal flora of man and animals is not clear. Some consider the presence of atypical *Y. enterocolitica* to be normal in the intestinal flora,[19] however, others are speculative.[20,21] In our studies both typical and atypical *Y. enterocolitica* based on Brenner et al.[10] DNA relatedness groups were isolated from humans, animals, and the environment. Further in-depth studies are required in order to better understand the pathogenicity and extent of yersiniosis in humans.

REFERENCES

1. **Gilbert, R.,** Interesting cases and unusual specimens, Annu. Rep., Division of Laboratories and Research, New York State Department of Health, Albany, 1933, 57.
2. **Schleifstein, J. I. and Coleman, M. B.,** An unidentified microorganism resembling *B. lignieri* and *Past. pseudotuberculosis,* and pathogenic for man, *N.Y. State J. Med.,* 39, 1749, 1939.
3. *Yersinia enterocolitica* Outbreak — New York, Center for Disease Control Morbidity and Mortality Weekly Report, 26, 53, 1977.
4. **Black, R. E., Jackson, R. J., Tsai, T., Medvesky, M., Shayegani, M., Feeley, J. C., MacLeod, K. I. E., and Wakelee, A. M.,** Epidemic *Yersinia enterocolitica* infection due to contaminated chocolate milk, *N. Engl. J. Med.,* 298, 76, 1978.
5. **Shayegani, M., Menegio, E. J., McGlynn, D. M., and Gaafar, H. A.,** *Yersinia enterocolitica* in Oneida County, New York, *Contrib. Microbiol. Immunol.,* 5, 196, 1979.
6. **Highsmith, A. K., Feeley, J. C., and Morris, G. K.,** *Yersinia enterocolitica:* a review of the bacterium and recommended laboratory methodology, *Health Lab. Sci.,* 14, 253, 1977.
7. **Niléhn, B.,** Studies on *Yersinia enterocolitica* with special reference to bacterial diagnosis and occurrence in human acute enteric disease, *Acta Pathol. Microbiol.,* 206, 1, 1969.
8. **Wauters, G.,** Contribution a l'etude de *Yersinia enterocolitica,* Ph.D. thesis, Vander, Louvain, Belgium, 1970.
9. **Knapp, W. and Thal, E.,** Differentiation of *Yersinia enterocolitica* by biochemical reactions, in *Contributions to Microbiology and Immunology*, Vol. 2, Winblad, S., Ed., S. Karger, Basel, 1973, 10.
10. **Brenner, D. J., Steigerwalt, A. G., Falcao, D. P., Weaver, R. E., and Fanning, G. R.,** Characterization of *Yersinia enterocolitica* and *Yersinia pseudotuberculosis* by deoxyribonucleic acid hybridization and by biochemical reactions, *Int. J. Syst. Bacteriol.,* 26, 180, 1976.
11. **Brenner, D. J.,** Biotyping of Enterobacteriacea in the clinical laboratory, in *Biotyping in the Clinical Microbiology Laboratory,* Balows, A. and Isenberg, H. D., Eds., Charles C Thomas, Springfield, Ill., 1978, 12.
12. **Keet, E. E.,** *Yersinia enterocolitica* septicemia, source of infection and incubation period identified, *N.Y. State J. Med.,* 74, 2226, 1974.
13. **Toma, S.,** Survey on the incidence of *Yersinia enterocolitica* in the province of Ontario, *Can. J. Public Health,* 64, 477, 1973.
14. **Toma, S. and Lafleur, L.,** Survey on the incidence of *Yersinia enterocolitica* in Canada, *Appl. Microbiol.,* 28, 469, 1974.
15. **Pai, C. H., Mors, V., and Toma, S.,** Prevalence of enterotoxigenicity in human and nonhuman isolates of *Yersinia enterocolitica, Infect. Immun.,* 22, 334, 1978.
16. **Bottone, E. J.,** *Yersinia enterocolitica:* a panoramic view of a charismatic microorganism, *CRC Crit. Rev. Microbiol.,* 5(2), 211, 1977.
17. **Bissett, M. L.,** *Yersinia enterocolitica* isolates from humans in California 1968–1975, *J. Clin. Microbiol.,* 4, 137, 1976.
18. **Weaver, R. E. and Jordan, J. G.,** Recent human isolates of *Yersinia enterocolitica* in the United States, in *Contributions to Microbiology and Immunology*, Vol. 2, Winblad, S., Ed., S. Karger, Basel, 1973, 120.
19. **Kapperud, G.,** *Yersinia enterocolitica* and Yersinia like microbes isolated from mammals and water in Norway and Denmark, *Acta Pathol. Microbiol. Scand. Sec. B,* 85, 129, 1977.
20. **Bottone, E. J., Chester, B., Malowany, M., and Allerhand, J.,** Unusual *Yersinia enterocolitica* isolates not associated with mesenteric lymphadenitis, *Appl. Microbiol.,* 27, 858, 1974.
21. **Bottone, E. J. and Robin, T.,** *Yersinia enterocolitica.* Recovery and characterization of two unusual isolates from a case of acute enteritis, *J. Clin. Microbiol.,* 5, 341, 1977.

Chapter 17

YERSINIA ENTEROCOLITICA INFECTIONS IN CANADA 1966 TO AUGUST 1978

Sandu Toma and Lucette Lafleur

TABLE OF CONTENTS

I.	Introduction	184
II.	Materials and Methods	184
III.	Results	184
	A. Human Isolates	184
	B. Nonhuman Isolates	185
IV.	Discussion	186
References		189

I. INTRODUCTION

Data on the first human infections with *Y. enterocolitica* in Canada were published by Lafleur et al. in Quebec, between 1966 and 1972.[1-5] Schieven reported the first *Y. enterocolitica* isolation in Ontario.[51] Fifty-nine *Yersinia* cultures, isolated in different laboratories throughout Ontario, were studied by Toma.[6-9] Other cases of human yersiniosis were reported from Quebec,[10-16] British Columbia,[17-19] Ontario,[20-22] and also summarized in an editorial.[23] Isolations of *Y. enterocolitica* from animals and environmental sources were recorded in British Columbia,[24] Ontario,[25-30] and Quebec[16] and two papers were presented on experimental pathogenicity of *Y. enterocolitica* strains.[31,32]

Due to an increasing number of *Y. enterocolitica* infections in humans, the Ontario Ministry of Health, in collaboration with the Laboratory Center for Disease Control, Department of National Health and Welfare, Ottawa, established in 1974 a Canadian National Reference Center for *Yersinia* located in the Toronto Public Health Laboratory. As a direct result of the activity of the National Reference Center, three surveys on the incidence of *Y. enterocolitica* infections in Canada were published between 1974 and 1976[33-35]

The present paper deals with laboratory and epidemiological aspects relating to 1980 *Y. enterocolitica* cultures isolated in Canada from 1966 to August 1978 (1485 human and 495 nonhuman strains).

II. MATERIALS AND METHODS

The majority of *Yersinia* cultures reported in this paper were referred to the Canadian National Reference Center for confirmation and bio-sero-phage typing. Some of the cultures isolated in the Province of Quebec were identified and bio-serotyped at the Quebec Provincial Public Health Laboratory and the results were forwarded to the National Reference Center.

The methods used for isolation of *Y. enterocolitica* from fecal or environmental sources were described previously.[5,7,14,16,26,28,29] Identification of cultures was carried out by standard methods.[36] Several of the tests were modified or supplemented as recommended by Niléhn[37] and Wauters.[38] Biotyping was done according to Wauters' scheme.[38] Serotyping was performed by slide agglutination using cultures incubated for 24 to 48 hr at 22° C and 34 absorbed and unabsorbed O antisera prepared in rabbits, according to Wauters and Winblad.[38-40] Phage typing was performed by the *Yersinia* Center at the Institut Pasteur, Paris (H.H. Mollaret) and, later, for *Y. enterocolitica* serotype O:3 only by the Canadian National Reference Center for *Yersinia*, using phages supplied by the Institut Pasteur and the methodology described by Nicolle et al.[41]

III. RESULTS

A. Human Isolates

The majority of *Y. enterocolitica* strains (1325 or 89%) were isolated from two provinces only, Quebec and Ontario (Table 1). With the exception of Prince Edward Island, which did not refer any cultures, 160 strains were isolated in the remaining seven Canadian provinces. Most of the strains (1233 or 83%) were recovered from fecal specimens in clinical cases of acute diarrhea. Twenty cultures were isolated from blood specimens, thirteen from appendices, mesenteric lymph nodes, or deep-seated digestive abscesses, twelve from the respiratory tract (sputum, throat swabs) and twelve from miscellaneous sites (urine, mucocutaneous superficial infections). In 191 cases the source

Table 1
DISTRIBUTIONS OF 1485 HUMAN ISOLATES OF
Y. ENTEROCOLITICA IN CANADA
1966 TO AUGUST 1978

Province	No. of cultures	Serotypes O							Nontypable
		3	5,27	6,30	8	5	4,32	Others[a]	
Quebec	780	663	20	21	2	6	1	34	33
Ontario	545	423	26	20	8	16	1	19	32
Br. Columbia	48	3	4	1	21	—	11	6	2
Alberta	39	1	10	4	12	2	—	6	4
New Brunswick	38	31	4	—	—	—	—	1	2
Nova Scotia	20	3	3	4	2	3	—	4	1
Newfoundland	7	4	—	1	1	—	—	1	—
Manitoba	5	—	2	—	3	—	—	—	—
Saskatchewan	3	—	—	—	2	—	—	1	—
Total	1485	1128	69	51	51	27	13	72	74

[a] Serotype O: 6,31 (eleven); 7,8 (ten); 7,13 (ten); 16 (seven); 21 (seven); 34 (seven); 9 (four); 16,29 (three); 4,33 (two); 8,19 (two); 1 (one); 2 (one); 1,2,3 (one); 11 (one); 11,24 (one); 12,25 (one); 15 (one); 17 (one); 20 (one).

of isolation was not stated. The majority of patients (958) were children or adolescents; 226 were adults and in 301 cases the age was not stated.

Y. enterocolitica, indole-negative, serotype O:3, biotype 4, was the predominant serotype (76%), followed by indole-positive serotypes O:5,27 (4.7%); 6,30 (3.4%); and 8 (3.4%). Other serotypes, seldom encountered, were 5; 4,32; 6,31; 7,8; 7,13; 16; 21; 34; 9; 16,29; 4,33; and 8,19. Seventy-four cultures (15%) were nontypable with the 34 O antisera. Four serotype O:9 strains, the second most predominant serotype in Europe, South Africa, and Japan, were isolated from two humans and two monkeys.

Four hundred and eighty-six *Y. enterocolitica* cultures, serotype O:3, isolated in Ontario, Quebec, New Brunswick, Newfoundland, Nova Scotia, Alberta, and British Columbia were phage type IXb (the "Canadian" phage type). Seven cultures, serotype O:3, were phage type VIII (the "European" phage type), all but one of which were isolated from patients who had just arrived from Europe. None of our serotype O:3 cultures belonged to phage type IXa (the "South African" phage type). One hundred and seventy indole-positive cultures, of different O serotypes, belonged to subgroups of phage group X.

B. Nonhuman Isolates

Four hundred and ninety-five strains were referred from Ontario (345), Quebec (119), New Brunswick (19), British Columbia (6), Alberta (4), and Saskatchewan (2) (Table 2). From water specimens, 180 cultures were isolated, 149 from milk and milk product specimens, and 104 from various food products (hamburger, sausage, mushroom salad, oysters, fish). Most of these environmental strains were indole-positive, biochemically atypical, and serologically nontypable (56%) or belonging to serotypes which were seldom encountered in human infections. Thirty-four cultures were isolated from swine (cecal content of slaughtered animals) during a survey[26] and 28 cultures were isolated from chinchillas (nine), beavers (five), poultry (three), pets (two), monkeys (two), and from a guinea pig, a racoon, a Pekin robin, a Canada goose, and pigeon droppings.

Table 2
SOURCES OF 495 NONHUMAN ISOLATES OF Y. ENTEROCOLITICA IN CANADA 1966 TO AUGUST 1978

Source	No. of cultures	Serotypes										Nontypable
		5	6	4	17	16	7,13	21	3	5,27	Others[a]	
Water	180	33	19	8	13	3	6	7	—	—	14	77
Milk and milk products	149	9	11	13	2	5	2	—	—	—	8	99
Food products	104	8	4	9	3	5	1	3	—	—	3	68
Swine	34	1	1	—	—	—	1	—	13	7	3	8
Other animals and birds	28	—	1	3	2	—	1	—	1	10	4	6
Total	495	51	36	33	20	13	11	10	14	17	32	258

[a] Serotypes O: 14 (six); 7,8 (five); 18 (five); 11,24 (four); 34 (four); 1,2,3 (two); 9 (two); 8,19 (one); 12,26 (one); 15 (one); 20 (one).

IV. DISCUSSION

Most human cultures (1233 or 83%) were isolated from cases of acute gastroenteritis in which the predominant symptoms were diarrhea, abdominal discomfort, varying degrees of malaise, and fever. In cases where the clinical history was available, the symptoms were similar to those experienced with acute gastroenteritis caused by *Salmonella* or *Shigella*. Thirteen were isolated from patients with acute iliac fossa syndrome. In all likelihood, only a small number of the 191 cultures with no site of isolation stated were recovered from extra-intestinal sources. Therefore, the true percentage of *Y. enterocolitica* strains isolated from cases of acute gastroenteritis was, in all probability, greater than 90%. All age groups of the population were susceptible to *Yersinia* infection although most of the patients were children or adolescents (64.5%). Twenty cultures were isolated from cases of septicemia in elderly debilitated patients and 41 cultures were isolated from various clinical conditions and anatomical sites.

Considering that 83% (and probably > 90%) of *Y. enterocolitica* strains were isolated from fecal specimens associated with acute enterocolitis, the practical conclusion for medical laboratories is that their capability in isolating *Yersinia* is related to their general proficiency in the enteric bacteriology area. Failure to isolate *Yersinia* from fecal specimens is related to lack of experience and/or use of improper media and techniques rather than to the absence of *Y. enterocolitica* in a particular geographic area. Some difficulties in isolating *Y. enterocolitica* from fecal specimens may be overcome by the following recommendations:

1. Avoid those selective agar media which contain sucrose, xylose, or salicin (e.g., Hektoen enteric agar [H-E] or xylose-lysine-deoxycholate [XLD] medium); these carbohydrates are metabolized by some strains of *Yersinia* which yield *Yersinia* colonies resembling lactose fermenters, and, therefore, a lack of fermenting colonies.
2. Use fresh, and where possible, at least two different selective agar media. The plates should preferably be less than 3 days old when used and definitely no older than 5 days. They should be stored in protective containers such as plastic bags and refrigerated.

3. Use selective Agar media with a final pH of 7.2 to 7.4 rather than 7.0 to 7.2. A pH lower than 7.2, in association with the atmospheric acidity from a regular incubator plus the acid background produced by lactose fermenting colonies, makes it difficult to see a scant presence of minute, nonlactose fermenting, colonies of *Yersinia*. This safeguard is also valid for the isolation of other enteric pathogens, especially *S. sonnei* (slow lactose fermenter) or *S. typhi* (small colonies).
4. Use quality-controlled media. Every batch of enteric selective plate media must be quality-controlled to ensure the ability of the medium to support growth of even a few cells of *Yersinia*. Plates should be pretested using a stock culture of *Y. enterocolitica* O:3, and *Y. pseudotuberculosis* if possible. A dilute inoculum, e.g., one small loop of an overnight incubated broth culture, is diluted in 4 mℓ of broth medium and streaked onto the medium to be tested and on a nonselective agar medium. Quantitative evaluation of a controlled inoculum is even more desirable.
5. Use of a plate microscope (stereomicroscope, dissecting microscope), with or without oblique illumination[7,38] for examination of the selective plates was found to be essential for isolation of *Y. enterocolitica* from fecal specimens. The plate microscope ensured good examination of colony size, shape, structure, shades of color, periphery, surface, and consistency.*

The use of fluid enrichment media, especially the cold enrichment techniques,[14,26,28,42] is not essential for isolation of *Y. enterocolitica* in the acute stage of enterocolitis when large numbers of *Yersinia* are present. Cold enrichment techniques are very useful in the recovery of *Yersinia* from specimens containing small numbers of bacilli, as in late convalescent stages, carriers, and different surveys of population or from environmental specimens.

Practically all *Y. enterocolitica* strains isolated from acute gastroenteritis, septicemia, or deep-seated infections were biochemically typical and belonged to one of the previously mentioned predominant O serotypes. *Y. enterocolitica* which were biochemically atypical, or serologically nontypable, were very seldom isolated from humans; when they were, they were mainly from mucocutaneous, urinary, or nosocomial infections or they appeared to be part of the transient flora in different cavities of the human body.

There are many schemes available for biochemical characterization and biotyping of *Y. enterocolitica* cultures[37,38,44] and, undoubtedly, each scheme has its merits. We found Wauters' scheme[38] appropriate for biotyping of clinical isolates of *Y. enterocolitica*, especially considering the indole, xylose, and salicin reactions. Biochemically typical clinical strains which are indole-negative (after 4 days at 30°C or 8 days at room temperature), xylose- and salicin-negative are, in most cases, the pathogenic *Y. enterocolitica*, biotype 4, serotype O:3. Cultures having the same features, but xylose-positive, are usually biotype 3 and, although agglutinating with O:3 antiserum, they will probably belong to serotype O:1 or O:2. *Y. enterocolitica*, serotype O:8, are biotype one (indole-positive in 1 day at 30°C or 36°C) and most of these have the particular feature of being salicin- and aesculin-negative. Strains belonging to serotypes O:5,27 or O:9 are biotype two (indole-late-positive). Biochemical characterization of *Y. enterocolitica* DNA relatedness groups,[44] very valuable in characterization of environmental strains or

---

* In the experience of one of the authors, routine use of a plate microscope in all areas of clinical bacteriology is one of the major technical factors responsible for the isolation in our laboratory of over 2000 *Bordetella pertussis* strains over the past 5 years, and high isolation rates of *Shigella, Campylobacter, N. gonorrhoeae, H. vaginalis,* and *Mycoplasma* species. Routine use of the plate microscope also produced a significant decrease in the percentage of contaminated cultures for biochemical and susceptibility testing.

for taxonomic purposes, was of little value in bio-grouping of our human clinical isolates as less than 1% of *Y. enterocolitica* isolated from humans in Canada were rhamnose-positive. A similar distribution was found in the Institut Pasteur, Paris, where, out of 4800 *Y. enterocolitica* studied, only approximately 1% were rhamnose-positive.[45]

Serotyping of *Y. enterocolitica* cultures in conjunction with complete biochemical identification proved to be a very valuable method for differentiation of clinical pathogenic strains of environmental strains and for epidemiological purposes. *Y. enterocolitica* serotype O:3 was the predominant serotype in Canada (76%), followed by serotype O:5,27 (4.7%), O:6,30 (3.4%), and O:8 (3.4%). Several other infrequently encountered serotypes (112 strains) and nontypable strains (74 strains) represented 12.5% of the total number of human isolates. These general figures for the serotype incidence of *Y. enterocolitica* in Canada did not correspond to the serotype incidence in each province. In four western provinces the indole-negative strains, serotype O:3, were seldom isolated (4%). The predominant strains were indole-positive, serotype O:8 (40%), 5,27 (17%), 4,32 (12%), and 26 strains were either other serotypes or nontypable. It is difficult to interpret at this time the differences in the serotype predominance either in different Canadian provinces or in different countries. Serotype O:8, the predominant type in the western provinces, was also the most predominant serotype in the U.S. although infrequently isolated in central and eastern Canadian provinces (0.9%) and practically never isolated on any other continent. The small number of *Yersinia* strains isolated in the western provinces (95 cultures) was not statistically significant; therefore no conclusion could be reached. Significant evidence of serotype predominance in a given geographical area should be based on the consistent isolation of *Y. enterocolitica* from its commonest clinical syndrome, i.e., acute enterocolitis.

Phage typing of indole-negative strains of *Y. enterocolitica*, serotype O:3, was a valuable procedure for typing "Continental" distribution. Four hundred and eighty-six strains were phage type IXb. Nine *Y. enterocolitica* serotype O:3 strains were phage type VIII; they were isolated either from patients who had arrived from Europe or referred from the U.S. To the best of our knowledge, the "Canadian" phage type IXb has never been identified in the U.S. or in any other country. Phage typing of indole-positive *Y. enterocolitica* is not yet highly developed, and does not appear to be of epidemiological significance.

Nonhuman strains of *Y. enterocolitica* may be conveniently divided into environmental (waterborne, milk, food products) and animal strains. Most environmental strains were indole-positive, biochemically and serologically atypical, and significantly different from the strains isolated from human infections. Rhamnose- and/or Simmons' citrate-positive reactions were frequent biochemical features of these strains. Many were serologically nontypable (65%) and 26% belonged to serotypes O:5, O:4, and O:6. The strains O:5 (5A or fast indole-positive, different from 5,27 which are slow indole-positive) were biochemically typical but the strains of type O:4 or O:6 were often rhamnose-positive and had crossings between factors 32 and 33 of the serotypes O:4,32 and O:4,33 or of the factors 30 and 31 of the serotypes O:6,30 and O:6,31. The predominant human serotypes O:3; O:5,27 and O:8 were never isolated in Canada from environmental sources and serotype O:6,31, biochemically typical, was isolated in a few instances only. Several strains of *Y. enterocolitica*, serotype O:3, phage type IXb have been found in the U.S.

Swine were the only nonhuman source which yielded serotypes O:3 (13 strains) and O:5,27 (7 strains), representing 59% of the total number of isolates from this source. The strains O:3 isolated from swine belonged to the same phage type IXb as similar cultures isolated from humans.[26] In Japan, Zen Yoji et al.[46] reported 2254 *Y. enterocolitica* strains isolated from swine (15.3% isolation rate) and about 70% of these strains were serotype

O:3 and O:5,27, matching the prevalence of human strains. Pederson in Denmark[47] collected throat swabs from bacon pigs at slaughter houses and isolated 84 strains of *Y. enterocolitica* serotype O:3 (30% isolation rate). Wauters[48] swabbed the surface of pigs' tongues bought during 1974, 1976, and 1977 in butcher shops from various areas in Belgium. He isolated 165 strains of serotype O:3 and three strains of serotype O:9 (55.6% isolation rate) and proved that *Y. enterocolitica* serotype O:3 was practically a normal inhabitant or contaminant of the oral cavity of pigs in Belgium. All these data constitute sound evidence that swine are a reservoir of human serotypes of *Yersinia*. The link between swine as a reservoir and transmission to humans remains still unclear.

The strains isolated from animals other than swine and birds were an intermediate group between humans and environmental strains. Eleven cultures from this group (46%) belonged to serotypes O:5,27, O:9, and O:3 and they were isolated mainly from animals which have contact with humans, i.e., cat, dog, camel, chinchilla, and monkey. Kaneko[49] isolated *Y. enterocolitica* from 25 dogs (5% isolation rate) and serotypes O:3, O:5,27, and O:9 represented 76% of the isolated *Yersinia*.

According to our data and the findings of Mollaret[50] there are obviously two different groups of *Y. enterocolitica*. The first group is represented by strains which are (a) stable and they are well determined bio-sero-phage types, (b) well adapted to a specific human or animal host, and (c) producing typical pathological disorders. The second group is made up of *Yersinia enterocolitica*-like strains which are (a) widespread in the environment, (b) extremely variable in their biochemical and antigenic characteristics, and (c) only occasionally present in humans or animals of limited pathogenic potential, and may infrequently produce minor and nonspecific infections.

REFERENCES

1. **Lafleur, L.**, *Yersinia enterocolitica* outbreak in children, paper presented at the 70th Ann. Meet. American Society for Microbiology, Boston, April 26 to May 1, 1970.
2. **Albert, G. and Lafleur, L.**, *Yersinia enterocolitica* in children, Can. J. Public Health 62, 70, 1971.
3. **Lafleur, L., Martineau, B., and Chicoine, L.**, *Yersinia enterocolitica*, Union Med. Can., 101, 2407, 1972.
4. **Lafleur, L.**, Yersiniosis: an introduction, Can. J. Public Health, 64(Abstr.), 82, 1973.
5. **Delorme, J., Laverdiere, J. M., Martineau, B., and Lafleur, L.**, Yersiniosis in children, Can. Med. Assoc. J., 110, 281, 1974.
6. **Toma, S., Lior, H., Quinn-Hill, M., Sher, N., and Walker, W. A.**, *Yersinia enterocolitica* infection: report of two cases, Can. J. Public Health, 63, 433, 1972.
7. **Toma, S.**, Survey on the incidence of *Yersinia enterocolitica* in the Province of Ontario, Can. J. Public Health, 64, 477, 1973.
8. **Toma, S.**, *Yersinia enterocolitica* infection: bacteriological and serological studies, Can. J. Public Health, 64(Abstr.), 82, 1973.
9. **Toma, S.**, *Yersinia enterocolitica* gastroenteritis, Abstr. 73rd Ann. Meet. American Society for Microbiology, Miami Beach, May 6 to 11, 1973, 113.
10. **Abramovitch, H. and Butas, C. A.**, Septicemia due to *Yersinia enterocolitica*, Can. Med. Assoc. J., 109, 1112, 1973.
11. **Rodgers, B. and Karn, G.**, *Yersinia enterocolitica*, J. Pediatr. Surg., 10, 497, 1975.
12. **Lafleur, L., Pai, C., Hammerberg, O., and Delage, G.**, *Yersinia enterocolitica* in a pediatric population, paper presented at the 3rd Int. Symp. on *Yersinia*, Mont Gabriel, Quebec, Saranac Lake, New York, September 25 to 28, 1977.
13. **Marks, M. I., Pai, C. H., Lafleur L., Hammerberg, O., Lackman, L., and Chicoine, L.**, *Yersinia enterocolitica* gastroenteritis in children, paper presented at the 2nd Ped. Research Meet., New York, April 28, 1978.
14. **Sorger, S., Pai, C. H., Lafleur, L., Lackman, L., Hammerberg, O., and Marks, M. I.**, Is cold enrichment necessary for isolation of *Yersinia enterocolitica?*, paper presented at Interscience Conf. on Antimicrobial Agents and Chemotherapy, Atlanta, October 2, 1978.
15. **Hammerberg, O., Lafleur, L., Marks, M. I., Lackman, L., Auclair, J., and Pai, C. H.**, Serological response in children with *Yersinia enterocolitica* (YE) gastroenteritis and their household contacts, Abstr. presented at 18th Interscience Conf. on Antimicrobial Agents and Chemotherapy, Atlanta, October 2, 1978.

16. **Caprioli, T., Drapeau, A. J., and Kasatiya, S.**, *Yersinia enterocolitica:* serotypes and biotypes isolated from humans and the environment in Quebec, Canada, *J. Clin. Microbiol.,* 8, 7, 1978.
17. **Lawrence, M. R., Ting, S. K., and Neilly, S.**, Furuncle caused by *Yersinia enterocolitica, Can. Med. Assoc. J.,* 112, 1289, 1975.
18. **Spiller, G. W.**, Yersiniosis due to *Yersinia enterocolitica, B. C. Med. J.,* 15, 338, 1973.
19. **Crichton, E. P.**, Suppurative conjunctivitis caused by *Yersinia enterocolitica, Can. Med. Assoc. J.,* 118, 22, 1978.
20. **Schieven, B. C. and Randall, C.**, Enteritis due to *Yersinia enterocolitica, J. Pediatr.,* 84, 402, 1974.
21. **Randall, C. and Bannatyne, R. M.**, Experience with *Yersinia enterocolitica* at The Hospital for Sick Children, 1972–74, *Can. Med. Assoc. J.,* 113, 542, 1975.
22. **Narasimhan, S. L., Schieven, B. C., and Campsall, E. W. R.**, Septicemia caused by *Yersinia enterocolitica, Can. Med. Assoc. J.,* 118, 682, 1978.
23. The spectacular rise of *Yersinia enterocolitica.* Editorial, *Can. Med. Assoc. J.,* 108, 1097, 1973.
24. **Langford, E. V.**, *Yersinia enterocolitica* isolated from animals in the Fraser Valley of British Columbia, *Can. Vet. J.,* 13, 109, 1972.
25. **Hacking, M. A. and Sileo, L.**, *Yersinia enterocolitica* and *Yersinia pseudotuberculosis* from wildlife in Ontario, *J. Wild. Dis.,* 10, 452, 1974.
26. **Toma, S. and Deidrick, V. R.**, Isolation of *Yersinia enterocolitica* from swine, *J. Clin. Microbiol.,* 2, 478, 1975.
27. **Toma, S. and Deidrick, V. R.**, Cold enrichment methods used to isolate *Yersinia enterocolitica* and *Y. pseudotuberculosis* from feces specimens — swine survey, Abstr. 75th Ann. Meet. American Society for Microbiology, New York, 1975, 34.
28. **Schiemann, D. A. and Toma, S.**, Isolation of *Yersinia enterocolitica* from raw milk, *Appl. Environ. Microbiol.,* 35, 54, 1978.
29. **Schiemann, D. A.**, Association of *Yersinia enterocolitica* with the manufacture of cheese and occurrence in pasteurized milk, *Appl. Environ. Microbiol.,* 36, 274, 1978.
30. **Schiemann, D. A.**, Isolation of *Yersinia enterocolitica* from surface and well waters in Ontario, *Can. J. Microbiol.,* 24, 1048, 1978.
31. **Pai, C. H. and Mors, V.**, Production of enterotoxin by *Yersinia enterocolitica, Infect. Immun.,* 19, 908, 1978.
32. **Pai, C. H., Mors, V., and Toma, S.**, Prevalence of enterotoxigenicity in human and nonhuman isolates
33. **Toma, S. and Lafleur, L.**, Survey on the incidence of *Yersinia enterocolitica* infection in Canada, *Appl. Microbiol.,* 28, 469, 1974.
34. **Toma, S. and Deidrick, V. R.**, Incidence of *Yersinia enterocolitica* and *Y. pseudotuberculosis* in Canada; 1975 semi-annual report, *Can. Med. Assoc. J.,* 114, 16, 1976.
35. **Toma, S., Lafleur, L., and Deidrick, V. R.**, Canadian experience with *Yersinia enterocolitica* 1966–1977, paper presented at the 3rd Int. Symp. on *Yersinia,* Mont Gabriel, Quebec, Saranac Lake, New York, September 25 to 28, 1977.
36. **Sonnenwirth, A. C.**, *Yersinia,* in *Manual of Clinical Microbiology,* 2nd ed., American Society Microbiology, Washington, D. C., 1974, 222.
37. **Nilehn, B.**, Studies on *Yersinia enterocolitica* with special reference to bacterial diagnosis and occurrence in human acute enteric disease, *Acta Pathol. Microbiol. Scand. Suppl.,* 206, 1, 1969.
38. **Wauters, G.**, Contribution à l'étude de *Yersinia enterocolitica,* Ph.d. thesis, Vander, Louvain, Belgium, 1970.
39. **Wauters, G., Le Minor, L., Chalon, A. M., and Lassen, J.**, Supplément au schéma antigenique de *Yersinia enterocolitica, Ann. Inst. Pasteur Paris,* 122, 951, 1972.
40. **Winblad, S.**, Studies on the O-serotypes of *Yersinia enterocolitica, Contrib. Microbiol. Immunol.,* 2, 27, 1973.
41. **Nicolle, P., Mollaret, H. H., and Brault, J.**, Nouveau résultats sur la lysotypie de *Yersinia enterocolitica* portant sur plus de 4,000 souches d'origines diverses, *Rev. Epidemiol. Sante Publique,* 24, 479, 1976.
42. **Paterson, J. S. and Cook, R.**, A method for the recovery of *Pasteurella pseudotuberculosis* from faeces, *J. Pathol. Bacteriol.,* 85, 241, 1963.
43. **Knapp, W. and Thal, E.**, Differentiation of *Yersinia enterocolitica* by biochemical reactions, in *Contributions to Microbiology and Immunology,* Vol. 2, Winblad, S., Ed., S. Karger, Basel, 1973, 10.
44. **Brenner, D. J., Farmer, J. J., Hickman, F. W., Asbury, M. A., and Steigerwalt, A. G.**, Taxonomic and Nomenclature Changes in Enterobacteriaceae, HEW Publ. No. CDC 78-8356, 1977.
45. **Alsonso, J. M., Bejot, J., Bercovier, H., and Mollaret, H. H.**, Sur en group de souches de *Yersinia enterocolitica* fermentant le rhamnose, *Med. Mal. Infect.,* 5, 490, 1978.
46. **Zen-Yoji, H., Sakai, S., Maruyama, T., and Yanagawa, Y.**, Isolation of *Yersinia enterocolitica* and *Yersinia pseudotuberculosis* from swine, cattle and rats at an abattoir, *Jpn. J. Microbiol.,* 18(1), 103, 1974.

47. **Pedersen, K. B.**, Studies on the prevalence of *Yersinia enterocolitica* infection in pigs in Denmark, paper presented at the 3rd Int. Symp. on *Yersinia*, Mont Gabriel, Quebec, Saranac Lake, New York, September 25 to 28, 1977.
48. **Wauters, G.**, Carriage of *Yersinia enterocolitica* serotype 3 by pigs as a source of human infection, paper presented at the 3rd Int. Symp. on *Yersinia*, Mont Gabriel, Quebec, Saranac Lake, New York, September 25 to 28, 1977.
49. **Kaneko, K., Hamada, S., and Kato, E.**, Occurrence of *Yersinia enterocolitica* in dogs, *Jpn. J. Vet. Sci.*, 39, 407, 1977.
50. **Mollaret, H. H.**, Contribution à l'étude épidemiologique des infections à *Yersinia enterocolitica*, *Méd. Mal. Infect.*, 6, 442, 1976.
51. **Schieven, B. C.**, personal communication, 1971.

Chapter 18

YERSINIA ENTEROCOLITICA IN SOUTH AFRICA

Roy M. Robins-Browne, A. R. Rabson, and H. J. Koornhof

TABLE OF CONTENTS

I.	Introduction	194
II.	Bacteriology	194
III.	Epidemiology	194
IV.	Bacterial Synergy	195
V.	Isolation Technique	196
VI.	Clinical Picture	197
VII.	*Y. enterocolitica* Gastroenteritis	197
VIII.	Generalized *Y. enterocolitica* Infection	198
IX.	Conclusion	201
Acknowledgments		201
References		202

I. INTRODUCTION

Ex Africa semper aliquid novi

Pliny

The first confirmed isolation of *Yersinia enterocolitica* in South Africa was made in 1966 from the peritoneal fluid of a man with mesenteric lymphadenitis and ascites.[1] This discovery aroused keen local interest in *Y. enterocolitica* infections which have subsequently been diagnosed in patients presenting with a variety of clinical conditions.[2-6] In South Africa investigations of *Y. enterocolitica* have, for the most part, centered around a number of specific problems pertaining to the clinical, laboratory, and epidemiological aspects of yersiniosis encountered in this country. In this article our experience in South Africa with this unusually versatile pathogen is reviewed with particular attention to aspects of yersiniosis peculiar to this region.

II. BACTERIOLOGY

Until very recently only one serotype of *Y. enterocolitica* had been detected in South Africa during the past 12 years. The strain unique to our region is serotype O:3, phage type IXa. Biochemically it corresponds to biotype 4 of Nilehn and Wauters, being negative for rhamnose, xylose, salicin, esculin, lecithinase, and indole.[7] It thus closely resembles *Y. enterocolitica* O:3 reported from Europe and North America. Recently two strains of *Y. enterocolitica* serotypes O:5 and O:6 were isolated for the first time from human material in South Africa. Both strains conformed to biotype 1.

III. EPIDEMIOLOGY

Despite our interest in and consequent awareness of *Y. enterocolitica* the number of isolations in routine bacteriology laboratories in and around Johannesburg remains low. An average of approximately 10 strains are recovered from the more than 10,000 fecal specimens submitted for bacteriological examination each year. During the past 12 months, however, the number of isolations increased to 40. This increment almost certainly reflects a heightened consciousness of the organism rather than a genuine increase. Successful isolations of *Y. enterocolitica* are made throughout the year but with a peak prevalence during late summer and autumn (Figure 1).

An unusually large proportion of human cases have originated in a predominantly agricultural district within a 50-mile radius west and northwest of Johannesburg. Patients infected in this region frequently volunteer a history of close contact with farm animals, most notably pigs.

Although *Y. enterocolitica* has been recovered from a variety of animal species in different parts of the world,[7] in South Africa only the pig has regularly been shown to carry the same serotype and phage type as that found in man. To date no definite association between infection in man and pigs has been documented, although we have encountered one patient, a laborer on a pig farm, who developed generalized yersiniosis following close contact with infected animals.[3] The organism recovered from the liver and spleen of this patient was serotype O:3, phage type IXa, the same as that found in 6 of 12 pigs with which he had experienced contact. This finding stimulated interest in the association and possible transmission of *Y. enterocolitica* between pigs and man. In two separate studies, *Y. enterocolitica* was isolated from 4 of 250 pigs in which feces, mesenteric lymph glands, or cecal contents were examined. Evidence of animal to man

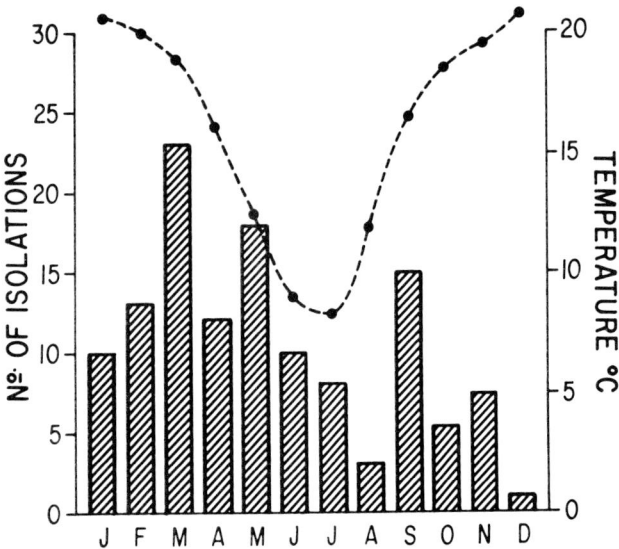

FIGURE 1. Distribution of 125 isolates of *Y. enterocolitica* recovered in Johannesburg each month between 1966 and 1977. The interrupted line indicates the mean daily temperature in Johannesburg.

transmission was lacking, however, as no fecal carriers and only one seroreactor was detected among 130 abattoir workers in close contact with infected material.

In an effort to obtain a clearer picture of the prevalence of *Y. enterocolitica* in South Africa we have been measuring *Y. enterocolitica* antibodies in serum specimens reaching our laboratory for typhoid, *Brucella*, and Weil-Felix investigations. Approximately 6% of some 3500 sera examined during the past 2 years displayed *Y. enterocolitica* agglutinins in a titer of 1 in 50 or higher. Of the 211 positive sera, 127 (60%) contained H-agglutinins alone, 28 (13%) O-agglutinins alone, and the remaining 56 (27%) both O- and H-agglutinins. Although some cases of acute yersiniosis were detected in this way, agglutinins were generally present in low titer (below 1 in 200 anti-O) in patients with illnesses unrelated to yersiniosis. These data suggest that a substantial proportion of South Africans have had previous exposure to this organism. Several isolates of *Y. enterocolitica* O:3, as well as the O:5 and O:6 strains alluded to previously, have been obtained from healthy carriers. Subclinical infections of this kind may be largely responsible for the high prevalence of seroreactors in the population.

IV. BACTERIAL SYNERGY

Y. enterocolitica is frequently isolated from fecal specimens which also contain other enteropathogens, most notably *Salmonella*. In addition, many patients with elevated antibody titers to *Y. enterocolitica* also have *Salmonella* agglutinins in their sera. Although it is likely that most of these cases indicate mixed infections originating from a common source, we considered the possibility that there may be a synergistic relationship between *Salmonella* and *Y. enterocolitica*. In order to test this hypothesis, we infected small groups of Balb/c mice intragastrically with from 5×10^2 to 5×10^5 colony-forming units (cfu) of *S. typhimurium* W118[8] in tenfold dilutions. Forty-eight hours later the same mice were similarly infected with from 10^{10} to 10^6 cfu of a locally isolated strain of *Y. enterocolitica* O:3 (*Y. enterocolitica* 6809 in the Institut Pasteur collection), and

Table 1
MORTALITY OF MICE INFECTED ORALLY WITH VARYING NUMBERS (cfu) OF Y. ENTEROCOLITICA 6809 AND SALMONELLA TYPHIMURIUM W118

S. typhimurium (cfu)	Yersinia Enterocolitica (cfu)					
	10^{10}	10^9	10^8	10^7	10^6	0
5×10^5	6/6	1/6	1/6	0/6	0/6	0/12[a]
5×10^4	2/6	0/6	2/6	0/6	2/6	0/6
5×10^3	2/6	1/6	0/6	0/6	1/6	0/6
5×10^2	1/6	0/6	1/6	0/6	0/6	0/6
0	0/12[a]	1/6	0/6	0/6	0/6	0/6

Note: All deaths occurred between the second and fourth weeks after infection.

[a] Half the mice in these groups received heat-killed bacteria.

thereafter observed for 1 month. The results summarized in Table 1 demonstrate that while *Y. enterocolitica* 6809 given alone is avirulent for mice, its pathogenicity is considerably enhanced by prior exposure of animals to *S. typhimurium*. Culture of liver, spleen, and heart blood undertaken in ten mice dying of infection yielded *Salmonella* and *Yersinia* alone in three animals each, and mixtures of both bacteria in the remaining four.

In a separate experiment we were unable to demonstrate any enhancement of pathogenicity when mice were inoculated intraperitoneally with suspensions containing varying proportions of *S. typhimurium* and *Y. enterocolitica*. These findings suggest that local gut immunity is important in conferring resistance to *Y. enterocolitica* on mice, and that penetration of the intestinal mucosa by *Salmonella* may facilitate subsequent invasion by *Y. enterocolitica*.

V. ISOLATION TECHNIQUE

In order to improve our yield of *Y. enterocolitica* from stool cultures, we have examined various enrichment and selective cultural procedures with a view to incorporating one of these into routine bacteriological investigation. In one experiment tenfold dilutions of a saline suspension containing from 2.5×10^6 to 25 cfu of a recently isolated strain of *Y. enterocolitica* were well mixed into six fresh stool samples obtained from healthy people. The cultural procedures investigated were (1) enrichment in selenite broth at 37°C for 18 hr, or at 4°C for 96 hr; (2) enrichment in modified Rappaport's broth[9] at 4°C for 96 hr; (3) direct plating on to MacConkey agar incubated for 18 hr at 37°C followed by 24 hr at room temperature, with or without a centrally placed filter paper disc containing 100 μg carbenicillin; (4) direct plating on to *Salmonella-Shigella* agar incubated for 96 hr at 4°C followed by 24 hr at room temperature; (5) direct plating on to SSD agar (SS agar supplemented with 20 g/l desoxycholate[9]) incubated at room temperature for 96 hr; and (6) direct plating on to Rappaport's agar (modified Rappaport's broth with 1.5% agar) incubated at room temperature for 96 hr. Methods necessitating prolonged cold enrichment were excluded because of their unsuitability for routine bacteriological investigations.

The results indicated that of the methods investigated direct plating on to SSD agar was the most satisfactory, yielding twice as many isolations as did the next best method, enrichment in selenite broth at 37°C. The only other procedure whereby positive results were obtained was direct plating on to MacConkey and Rappaport's agar. SSD agar was uniformly successful only when the stool contained 2.4×10^6 cfu per gram, although in one instance a stool containing as few as 1.9×10^3 cfu per gram was also positive.

In a field trial involving stool culture of 195 abattoir workers and 49 pigs, the superiority of selenite broth enrichment and SSD agar over the other isolation techniques was confirmed, inasmuch as four isolates were obtained by using at least one of these two media.[9] These results demonstrate that cold enrichment does not necessarily improve the yield of *Y. enterocolitica* from clinical specimens. The isolation techniques discussed here are not suited to all strains of *Y. enterocolitica*, the growth of especially the rhamnose-positive environmental strains often being inhibited by selenite.[7,9] We can not be emphatic, moreover, that the growth of all human strains will be encouraged to an equal extent in selective media. Consequently our search for an ideal isolation technique is continuing.

VI. CLINICAL PICTURE

The clinical manifestation of *Y. enterocolitica* infections in South Africa resemble those described elsewhere. Diarrhea, frequently accompanied by vomiting, is by far the most common presenting feature, especially in children under the age of 5 years. In older children and adults, mesenteric adenitis and acute terminal ileitis, with or without associated diarrhea, are commonly encountered. Although the infection does not regularly involve the colon, two unusual cases presenting with longstanding bloody diarrhea have been observed. Sigmoidoscopy revealed shallow rectal ulcers, from which the organism was cultivated. Normal bowel function was restored to both patients after tetracycline therapy.

Late sequelae of *Y. enterocolitica* infection have been noted on a number of occasions in South Africa and include both erythema nodosum and polyarthritis. Although initial reports did not include cases with either of these complications, at least 49 patients with histories highly suggestive of post-yersinia arthritis and/or erythema nodosum have now been documented. All cases have been diagnosed serologically by the finding of elevated titers primarily of H- but also of O-agglutinins to *Y. enterocolitica* antigens, prepared from a local isolate of serotype O:3 and phage type IXa. In most cases arthritis was preceded by fever and gastrointestinal symptoms. At the time joint involvement was manifest, however, the organism was undetectable in stool cultures, although no use was made of specific enrichment procedures which could have revealed low numbers of bacteria. The failure to isolate *Y. enterocolitica* from these cases raises the problem of whether the South African strain is truly arthritogenic.[7] The typical history, clinical picture, and response to therapy, however, is identical to the syndrome reported from other countries, and, when taken together with the absence of other detectable causes of arthritis, and the presence of raised (occasionally enormously raised) titers of antibodies, strongly suggests that these cases are indeed due to underlying yersiniosis.

VII. *Y. ENTEROCOLITICA* GASTROENTERITIS

For many years acute summer diarrhea of infants has posed a serious problem for South African pediatricians and microbiologists, largely because in many cases microbiological examination of the stool failed to yield enteropathogens.[10] In view of the well-known association of *Y. enterocolitica* with acute enteric infection,[7] we were interested to know whether *Y. enterocolitica* could be implicated in the causation of infantile diarrhea. Examination of stool specimens from several hundred patients suffering from enteritis, however, failed to reveal *Y. enterocolitica* in any, pointing to the relative unimportance of this organism in the etiology of this condition.

Recently a great deal of attention has been devoted to investigations of the pathogenic mechanism of gastroenteritis.[11] During the course of a study designed to examine the enteropathogenicity of *Y. enterocolitica*, we showed that a number of local strains produce an enterotoxin.[12] Subsequent experiments have revealed enterotoxin production by 39 of 71 local isolates. Fifty of these bacteria had been stored at room temperature for periods varying from 1 to 10 years, and only 19 were positive. In contrast, all but 1 of 21 fresh isolates were shown to be enterotoxigenic. The latter included three strains from pigs and two from healthy human carriers. Enterotoxin production is not restricted to *Y. enterocolitica* originating in South Africa, nor is it exclusive to serotype O:3.[13,14] One strain each of serotypes O:5 and O:6 recovered locally, as well as one each of O:4 and O:6 originating in France, and two strains of serotype O:8 recovered in the U.S. have all been shown to be enterotoxigenic. Of interest is that one of the enterotoxin-producing American isolates used in our studies was the WA strain of *Y. enterocolitica* which has been utilized in a number of studies of pathogenicity.[15-17]

The enterotoxin of *Y. enterocolitica* strongly resembles the heat-stable enterotoxin (ST) of *Escherichia coli* in being highly resistant to heat and acid. In addition, both toxins are active in the suckling mouse assay[18] but do not stimulate adenyl cyclase activity, a property shared by *Vibrio cholerae* enterotoxin and the heat-labile toxin of *E. coli*.[19] On the other hand, *Yersinia* enterotoxin differs substantially from ST with respect to the cultivation conditions required for its production. Thus synthesis of toxin is restricted to growth in air at temperatures between 18 and 28°C. As the enterotoxin of *Y. enterocolitica* is not produced at 37°C nor under anaerobic conditions, it is difficult to envisage an *in vivo* role for this entity. It is possible, however, should the ingestion of preformed *Yersinia* enterotoxin be shown to produce diarrhea, that the presence of toxin in food may conceivably give rise to food poisoning.

VIII. GENERALIZED *Y. ENTEROCOLITICA* INFECTION

Between the years 1966 and 1967, 14 patients with generalized *Y. enterocolitica* infection were encountered in South Africa.[3,5,20] Patients with this form of yersiniosis could be divided into two groups according to their clinical presentation. Eight patients, four blacks and four whites, ranging in age from 2 to 80 years, presented with an acute illness with rigors, in which the differential diagnosis was usually typhoid fever or malaria. The remaining six patients, all of whom were elderly black males, had a subacute illness with a history of malaise, anorexia, and weight loss during the preceding weeks. Abdominal pain, most marked in the right upper quadrant, was a common complaint and was associated with tender hepatomegaly. In these patients the clinical picture bore a close resemblance to amebic hepatitis.

The major clinical features of these two groups of patients are summarized in Figure 2. The most striking difference between the groups concerned the outcome of infection, which was very much worse in patients with the subacute form, and was associated with a tendency of the infection to localize as hepatic or splenic abscesses.

Autopsy examination of three patients with subacute yersiniosis revealed marked deposits of hemosiderin in the viscera, a finding pathognomonic of "Bantu siderosis".[21] This condition was until recently fairly common in South Africa, particularly in men who consumed excessive quantities of alcoholic beverages, containing large amounts of ionizable iron.[22] Altered drinking habits associated with freer access to commercially

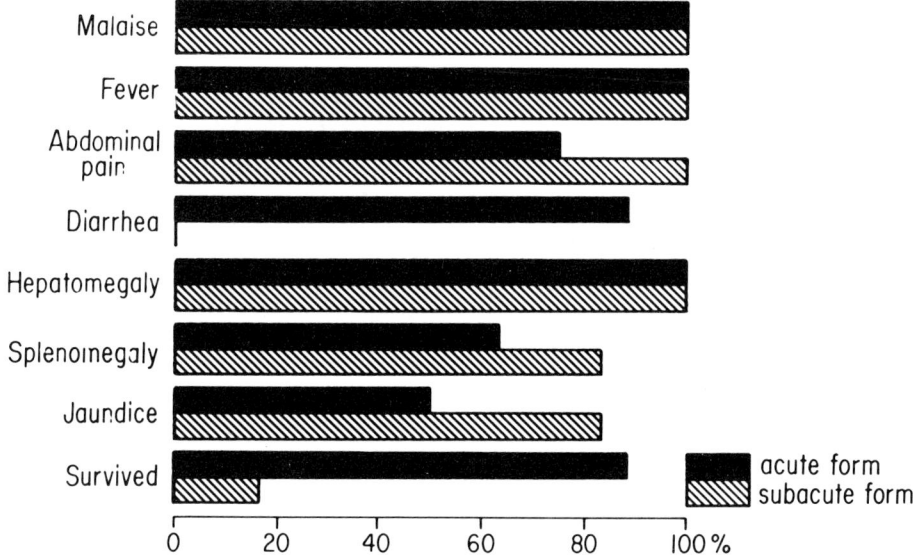

FIGURE 2. Major clinical features in eight patients with the acute septicemic and six patients with the subacute localizing forms of infection with *Y. enterocolitica*.

prepared liquor and a consequent decline in the use of iron utensils for domestic brewing, has led to a decreased prevalence of this disorder.[23]

The question of iron overload and its role in bacterial infections has been the subject of two recent reviews.[24,25] Evidence has been provided that the presence of excess iron in tissues may adversely effect the outcome of a variety of bacterial infections including a number caused by bacteria closely related to *Y. enterocolitica*. Thus iron overload may play a role in increasing the host's susceptibility to infections with *Y. pestis*,[26] *Y. pseudotuberculosis*,[27] *Pasteurella septica*, and *Escherichia coli*.[28] Experiments conducted in our laboratory indicate that iron overload may, in similar fashion, aggravate the clinical course of infections with *Y. enterocolitica*.[20]

Fresh pooled human serum is bactericidal for most strains of *Y. enterocolitica*, and the bactericidal activity is mediated by antibody and complement. Agammaglobulinemic, hypocomplementemic, and heat inactivated sera all permit the growth of *Y. enterocolitica*. In addition, the lethal effect of serum inactivated by heat is readily restored by the addition to such serum of fresh guinea pig serum, which per se displays no antiyersinia activity. The addition of ferric iron to serum also reduces or abolishes its bactericidal capacity, especially when the serum transferrin has been fully saturated by exogenous iron (Figure 3).

In separate experiments we have demonstrated that the administration of parenteral iron to mice results in a heightened susceptibility to fatal infection with *Y. enterocolitica*. Small groups of mice were given a single i.p. injection of ferric ammonium citrate containing 200, 100, or 50 μg of iron. Control animals received either ammonium citrate or saline. Twenty-four hours later from 10^9 to 10^5 cfu of *Y. enterocolitica* were administered as a second i.p. injection. Six of nine mice which received 200 μg of iron and 10^9 *Y. enterocolitica* died, compared with 3 of 32 mice given the same dose of iron but smaller doses of bacteria (p = 0.0025 using Fisher's exact test) as shown in Table 2. In contrast, only 6 of 35 mice died after receiving 10^9 bacteria and 100 μg or less of iron (p = 0.014). Histological examination of animals dying 11 or more days after infection,

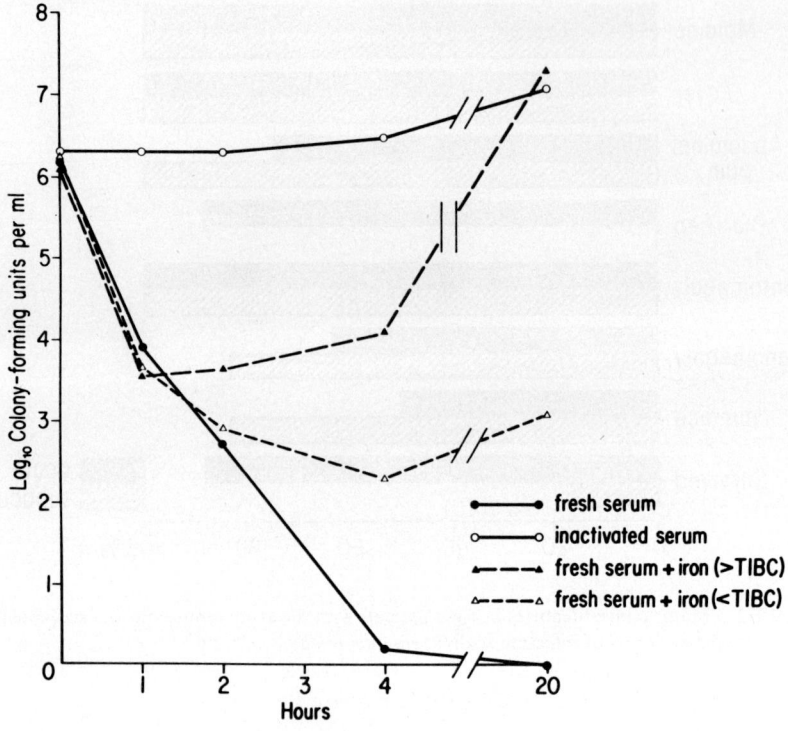

FIGURE 3. Fate of *Y. enterocolitica* serotype O:3 incubated with fresh pooled human serum, serum inactivated at 56°C for 30 min, and fresh serum supplemented with exogenous iron either below or exceeding the serum total iron binding capacity (TIBC).

Table 2
MORTALITY OF MICE WITHIN 28 DAYS OF BEING INJECTED INTRAPERITONEALLY WITH VARYING DOSES OF *Y. ENTEROCOLITICA* AND IRON

Y. enterocolitica (cfu)	Dose of iron (µg/mL)			
	0	50	100	200
10^9	3/17	3/9	0/9	6/9
10^8				0/8
10^7				1/8
10^6				1/8
10^5				1/8

revealed that two had hepatitis with patchy liver necrosis resembling that found in human patients (Figure 4), and that two others had changes consistent with septicemia.[9] *Y. enterocolitica* antigen was demonstrated in the livers of infected animals by means of immunofluorescent examination (Figure 5).

These experiments provide evidence that the presence of excess iron in the tissues can have a deleterious effect on the course of infection with *Y. enterocolitica*, and suggest that the high prevalence of siderosis in South Africa may have accounted for the frequency with which generalized yersiniosis was encountered in this country. Additional

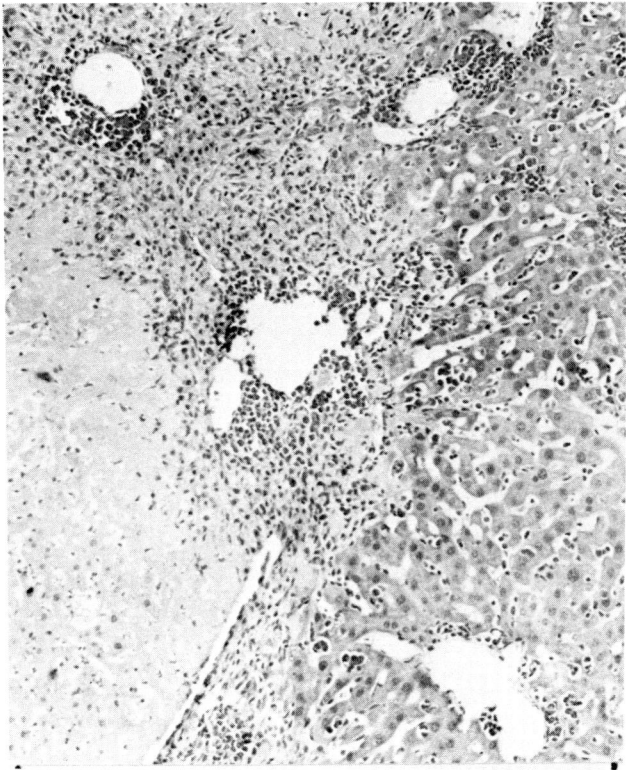

FIGURE 4. Inflammation and necrosis in liver of mouse which died 12 days after i.p. injection of 10^9 cfu of *Y. enterocolitica* and 200 µg of iron. (H&E × 135.)

circumstantial evidence for an association between siderosis and yersiniosis stems from the recent decline in prevalence of subacute yersiniosis which parallels the decline in Bantu siderosis in this country.

IX. CONCLUSION

Despite the relative infrequency with which *Y. enterocolitica* is encountered during the conduct of routine bacteriology, it remains a microorganism of considerable interest. In particular the paucity of knowledge regarding its role in infections ranging from subclinical to fulminating, its association with autoimmune states, its interrelationships with other bacteria and iron, and its ability to survive in nature and to infect new hosts is likely to preoccupy both contemporary and future microbiologists for some time.

ACKNOWLEDGMENTS

We are indebted to Professor H. H. Mollaret of the Institut Pastuer, Paris, for his invaluable advice and assistance during the past few years. Thanks are also due to Drs. P. B. Carter of the Trudeau Institute, Saranac Lake, New York, S. B. Formal of the Walter Reed Army Institute of Research, Washington, D. C., and A. C. Mauff of Johannesburg for providing some of the bacteria used in these studies; and to Carol Still and Marianne Miliotis for excellent technical assistance. These investigations were supported in part by a grant from the South African Medical Research Council.

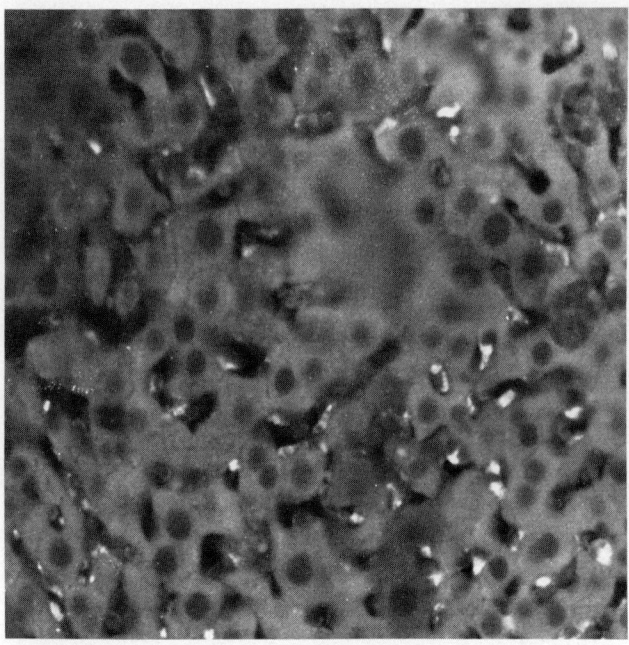

FIGURE 5. Immunofluorescent staining of *Y. enterocolitica* in liver of mouse which died 11 days after receiving 10^9 cfu of *Y. enterocolitica* and 200 µg of iron. (Magnification × 360.)

REFERENCES

1. **Koornhof, H. J.**, Pasteurella "X" *(Yersinia enterocolitica)* infection in man. A report of two cases, *Med. Proc.*, 13, 352, 1967.
2. **Rabson, A. R. and Koornhof, H. J.**, *Yersinia enterocolitica* infections in South Africa, *S. Afr. Med. J.*, 46, 798, 1972.
3. **Rabson, A. R., Koornhof, H. J., Notman, J., and Maxwell, W. G.**, Hepatosplenic abscesses due to *Yersinia enterocolitica, Br. Med. J.*, 4, 341, 1972.
4. **Finlayson, M. H., Coldrey, N. A., Street, B., and Brede, H. D.**, *Yersinia enterocolitica* in the western Cape, *S. Afr. Med. J.*, 47, 111, 1973.
5. **Rabson, A. R., Hallett, A. F., and Koornhof, H. J.**, Generalized *Yersinia enterocolitica* infection, *J. Infect. Dis.*, 131, 447, 1975.
6. **Thomas, A. F., Solomon, L., and Rabson, A.**, Polyarthritis associated with *Yersinia enterocolitica* infection, *S. Afr. Med. J.*, 49, 18, 1975.
7. **Bottone, E. J.**, *Yersinia enterocolitica*: a panoramic view of a charismatic microorganism, *CRC Crit. Rev. Microbiol.*, 5, 211, 1977.
8. **Gianella, R. A., Formal, S. B., Dammin, G. J., and Collins, H.**, Pathogenesis of salmonellosis. Studies of fluid secretion, mucosal invasion, and morphologic reaction in the rabbit ileum, *J. Clin. Invest.*, 52, 441, 1973.
9. **Wauters, G.**, Improved methods for the isolation and the recognition of *Yersinia enterocolitica*, in *Contributions to Microbiology and Immunology*, Vol. 2, Winblad, S., Ed., S. Karger, Basel, 1973, 68.
10. **Koornhof, H. J., Robins-Browne, R. M., Richardson, N. J., and Cassel, R.**, Etiology of infantile enteritis in South Africa, *Isr. J. Med. Sci.*, 14, 341, 1979.
11. **DuPont, H. L.**, Etiologic diagnosis of acute diarrhea, *Ann. Intern. Med.*, 88, 707, 1978.
12. **Robins-Browne, R. M., Jansen van Vuuren, C. J., Still, C. S., Miliotis, M. D., and Koornhof, H. J.**, The pathogenesis of *Yersinia enterocolitica* gastroenteritis, in *Contributions to Microbiology and Immunology*, Vol. 5, Carter, P., Lafleur, L., and Toma, S., Eds., S. Karger, Basel, 1979, 324.
13. **Feeley, J. C., Wells, J. G., Tsai, T., and Puhr, N.**, Enterotoxin detection in *Yersinia enterocolitica*, Proc. 3rd Int. Symp. Yersinia, Montreal, Canada and Saranac Lake, N.Y., September 1977.

14. **Pai, C. H. and Mors, V.**, Production of enterotoxin by *Yersinia enterocolitica, Infect. Immun.,* 19, 908, 1978.
15. **Carter, P. B. and Collins, F. M.**, Experimental *Yersinia enterocolitica* infection in mice: kinetics of growth, *Infect. Immun.,* 9, 851, 1974.
16. **Carter, P. B.**, Pathogenicity of *Yersinia enterocolitica* for mice, *Infect. Immun.,* 11, 164, 1975.
17. **Carter, P. B.**, Oral *Yersinia enterocolitica* infection of mice, *Am. J. Pathol.,* 81, 703, 1975.
18. **Dean, A. G., Ching, Y.-C., Williams, R. G., and Harden, L. B.**, Tests for *Escherichia coli* enterotoxin using infant mice: application in a study of diarrhea in children in Honolulu, *J. Infect. Dis.,* 125, 407, 1972.
19. **Sack, R. B.**, Human diarrheal disease caused by enterotoxigenic *Escherichia coli, Annu. Rev. Microbiol.,* 29, 333, 1975.
20. **Robins-Browne, R. M., Rabson, A. R., and Koornhof, H. J.**, Generalized infection with *Yersinia enterocolitica* and the role of iron, in *Contributions to Microbiology and Immunology*, Vol. 5, Carter, P. B., Lafleur, L., and Toma, S., Eds., S. Karger, Basel, 1979, 277.
21. **Wainwright, J.**, Siderosis in the African, *S. Afr. J. Lab. Clin. Med.,* 3, 1, 1957.
22. **Bothwell, T. H., Seftel, H., Jacobs, P., Torrance, J. D., and Baumslag, N.**, Iron overload in Bantu subjects. Studies on the availability of iron in Bantu beer, *Am. J. Clin. Nutr.,* 14, 47, 1964.
23. **Isaacson, C.**, The changing pattern of liver disease in South African Blacks, *S. Afr. Med. J.,* 53, 365, 1978.
24. **Weinberg, E. D.**, Iron and infection, *Microbiol. Rev.,* 42, 45, 1978.
25. **Bullen, J. J., Rogers, H. J., and Griffiths, E.**, Role of iron in bacterial infection, *Curr. Top. Microbiol. Immunol.,* 80, 1, 1978.
26. **Jackson, S. and Burrows, T. W.**, The virulence-enhancing effect of iron on non-pigmented mutants of virulent strains of *Pasteurella pestis, Br. J. Exp. Pathol.,* 37, 577, 1956.
27. **Bradley, J. M. and Skinner, J. I.**, Isolation of *Yersinia pseudotuberculosis* serotype V from the blood of a patient with sickle-cell anaemia, *J. Med. Microbiol.,* 7, 383, 1974.
28. **Bullen, J. J. and Rogers, H. J.**, Bacterial iron metabolism and immunity to *Pasteurella septica* and *Escherichia coli, Nature, (London),* 224, 380, 1969.

Chapter 19

EPIDEMIOLOGIC ASPECTS OF YERSINIOSIS IN JAPAN

Hiroshi Zen-Yoji

TABLE OF CONTENTS

I.	Introduction	206
II.	The Outbreaks of Yersiniosis	206
	A. Outbreak A	206
	B. Outbreak B	207
	C. Outbreak C	208
	D. Outbreak D	208
	E. Outbreak E	210
	F. Outbreaks F and G	210
III.	The Incidence of Sporadic Human Yersiniosis	211
	A. *Y. enterocolitica* Infection	211
	B. *Y. pseudotuberculosis* Infection	211
IV.	*Yersinia* Isolations from Animals	212
V.	Conclusion	214
	References	214

I. INTRODUCTION

Although a large number of cases of yersiniosis have been reported in European countries and both the clinical features and epidemiologic aspects of the disease were mostly clarified during the last two decades, *Yersinia* infections had not been noted at all in Japan until 1972 when Zen-Yoji and Maruyama[1] reported the first isolation of *Yersinia enterocolitica* from sporadic cases with enteritis and from the intestinal contents of two autopsy cases with death due to infection. Since then, however, in Japan yersiniosis has been gradually recognized as an important enteropathogenic disease and a considerable number of sporadic cases have been found up to now.

Comparison of the occurrence of yersiniosis in Japan with its occurrence in other countries, especially in Europe and North America, showed remarkable differences in the epidemiology. The first difference to be noted was the occurrence in Japan of large scale outbreaks of enteritis due to *Y. enterocolitica* infection. In countries other than Japan, reported cases were sporadic, with a few family outbreaks involving several patients. Contrary to this, in Japan in the 8-year period from 1972 to the present, seven outbreaks of enteritis due to infection by *Y. enterocolitica* occurred at seven separate schools, six primary and one junior high, involving several hundred patients in each. The second difference would be that yersiniosis cases with sequelae, such as arthritis, erythema nodosum, and eye inflammation were quite rare as compared with the clinical features presented by Ahvonen.[2] In addition, only one case with septicemia has been known in Japan.

In this country most human *Yersinia* infections were due to *Y. enterocolitica*, and those with *Y. pseudotuberculosis* infection were extremely small in number. Thus this review deals primarily with the epidemiologic aspects of seven outbreaks of abdominal illness at schools caused by *Y. enterocolitica* and secondarily with the incidence of sporadic yersiniosis and the results of *Yersinia* isolations from animals surveyed in the search for the infection source to humans.

II. THE OUTBREAKS OF YERSINIOSIS

A. Outbreak A

The first outbreak of enteritis due to *Y. enterocolitica* occurred at a primary school and the kindergarten located on the site of this school, which is situated in a rural area of Shizuoka Prefecture, during the period from January 31 through February 26, 1972.[3] In the outbreak, 182 children and one teacher among 390 persons in the primary school and 6 of 51 children in the kindergarten suffered from an acute intestinal illness resembling bacterial food poisoning. Out of 189 cases, 172 occurred during the first 11 days of the outbreak, and the remaining cases continued to occur sporadically until February 26 (Figure 1). With the possible exception of a junior high school boy whose brother was involved in the outbreak, no secondary cases occurred in the families of the affected children.

The major symptoms observed in the patients included abdominal pain (85.7%), fever (60.1%), nausea (23.5%), and vomiting (4.0%). Some patients were appendectomized upon suspicion of appendicitis. There were no deaths and most of the patients recovered within 2 days.

At the onset of the outbreak, only the known enteropathogenic bacteria were examined because *Y. enterocolitica* was not considered, and no causative agent was isolated from stools of the patients. Thus a total of 113 fecal specimens consisting of stools from 44 earlier cases, which were kept at −20°C, and from 69 patients with acute abdominal symptoms were examined for *Yersinia*; *Y. enterocolitica* was found in 42.4%

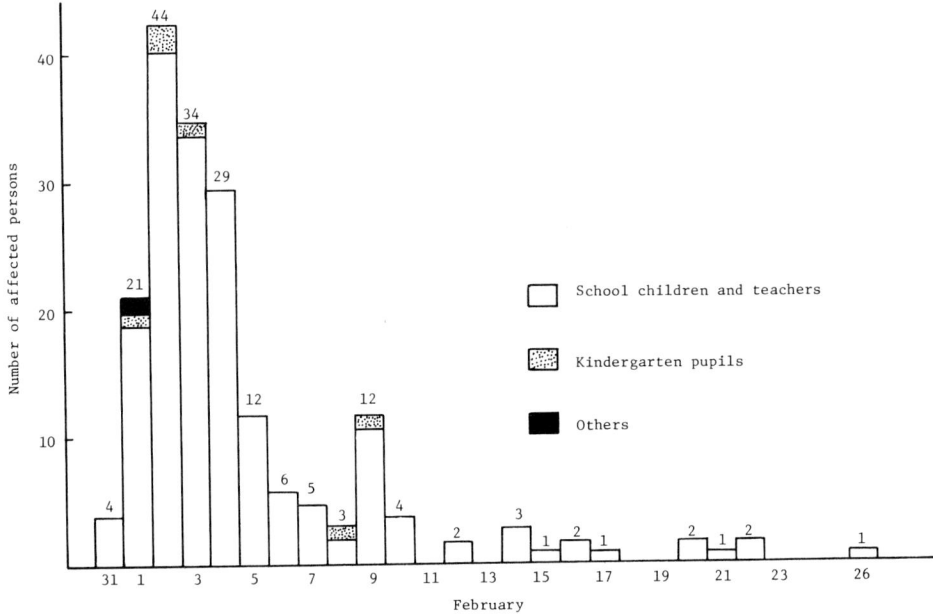

FIGURE 1. Daily occurrence of cases of *Y. enterocolitica* infection during an outbreak at a primary school and kindergarten.

of these specimens. Later, 20.5% of 175 convalescent cases examined on March 11, 40 days after the outbreak started, were also confirmed to have excreted the bacterium. In addition, the examination on 234 specimens from symptomless pupils and teachers revealed that nine persons (3.8%) were positive for the bacterium. All the isolates in the outbreak were bacteriologically identified as *Y. enterocolitica*, serogroup O:3. From the fact that *Y. enterocolitica* was isolated from the feces of most of the cases and no other enteropathogenic bacteria were detected, this bacterium was presumed with great certainty to have been the causative agent in the outbreak.

In the search for the source of infection and the vehicle of transmission, school-water, fecal specimens from pet animals kept in the school, and rats caught near the school were examined for *Y. enterocolitica* with negative results. Although epidemiologically it was suspected that the outbreak was due to food- or water-borne infection, the source and mode of infection remained unknown.

B. Outbreak B

Between July 11 and July 22, 1972, the second explosive outbreak due to *Y. enterocolitica*, serogroup O:3, occurred among children in a primary school in another district in Shizuoka Prefecture, 15 km distance from the school involved in the first outbreak.[3] Of 993 children and 49 adults at risk, 544 pupils (52.6%) had intestinal illness. Of these, 522 cases (95.9%) occurred within 4 days, but sporadic cases continued to occur until July 22.

Similar to the first outbreak, the main symptoms were abdominal pain (63.9%), fever (49.6%), diarrhea (32.4%), and vomiting (10.6%). Although 30% of 270 ill persons with fever had a temperature of 39°C or higher and watery stools were observed in the majority of cases, most patients had relatively mild symptoms and recovered within 1 day; neither appendectomized cases nor deaths were present. Despite a great effort to determine the source and mode of infection in this outbreak, neither could be clarified.

C. Outbreak C

Coincidental to the occurrence of the second outbreak, the third epidemic of *Y. enterocolitica* infection took place between July 14 and July 21, 1972 at a junior high school in a rural area in Tochigi Prefecture[4] which is located 200 km from Shizuoka.

In this outbreak, a total of 198 out of 1086 pupils suffered from intestinal illness for a duration of 8 days, and the peak of occurrence after onset of illness was between July 17 and 19. The major symptoms in these patients consisted of abdominal pain (76.3%), fever (60.6%), diarrhea (35.8%), and vomiting (12.2%). In the case of abdominal pain, it was noticed that half the patients were seized with severe pain in the umbilical region or the right lower quadrant of the abdomen. Among the patients with severe abdominal pain, three had appendectomies. The body temperature ranged from 38 to 39°C. Diarrhea was generally mild with a frequency of about three times per day. In all cases, neither erythema nodosum, arthritis, nor other complications were observed. Most patients recovered within several days. Of 198 patients, 195 fecal specimens obtained within 7 days of onset of their illness were available for bacteriological examination, and *Y. enterocolitica*, serogroup O:3, O:5, and O:9 were isolated, totaling 111, 2, and 5 specimens, respectively, as shown in Table 1.

One month after onset of the outbreak, to determine the excretion period of the bacterium, an examination was performed of the fecal specimens from 195 patients, 117 patients positive and 78 patients negative for the bacterium on the first isolation attempts, and from 868 pupils and 48 teachers who were symptomless at the time of the outbreak. The results demonstrated that the feces of 20 out of 117 patients (17.9%), 5 out of 78 patients (6.41%), 16 out of 868 pupils (1.84%), and one out of 48 teachers were positive for *Y. enterocolitica* (Table 1). After that, feces from 42 persons who were positive on the second isolation attempt were reinvestigated 2 months after onset of outbreak and 14.28% of them were positive for the bacterium. Since most of the patients in the outbreak were not treated with any antibiotics, the results described suggested that some patients with *Y. enterocolitica* infection might continue to excrete bacterium for over 2 months, if not given antibiotic therapy.

In this outbreak also, the infection source and route could not be elucidated, although based on the fact that most of patients were taken ill within 3 days and that there was no evidence of secondary spread in households, it was presumed to be an infection spread from a source such as food and school water contaminated with the bacterium.

D. Outbreak D

Between April 22 and April 26, 1974, the fourth outbreak due to *Y. enterocolitica* occurred at a primary school situated in a rural area in Kyoto Prefecture.[5] In the outbreak a total of 296 patients, 291 out of 779 pupils and 5 out of 35 teachers, suffered from intestinal illness complaining of fever. The illness was again characterized by fever (90.0%), diarrhea (72.7%), abdominal pain (70.7%), vomiting (38.2%), and malaise (30.9%). Similar to the cases in the outbreaks previously described, the body temperature ranged from 38 to 39°C in most of the cases. Out of all the cases, seven patients with severe symptoms were admitted to hospital, and the remaining 59 (19.9%) had severe symptoms, 160 (54.1%) relatively severe, and 70 (23.6%) mild. None of the patients had appendectomies. Since it was known from treatment of sporadic cases as well as some patients in the past outbreaks that antibiotic therapy was effective for *Yersinia* infections, most of the patients were administered chloramphenicol and recovered within 2 or 3 days.

Bacterial examination of fecal specimens from all the pupils and teachers was carried out for a causal agent. Of the pupils, 291 patients and 488 symptomless children, 187 (64.3%) and 87 (17.8%) were positive for *Y. enterocolitica*, serogroup O:3, respectively.

Table 1
SUMMARIZED RESULTS ON THE RECOVERY OF *Y. ENTEROCOLITICA* DURING AN OUTBREAK AT A JUNIOR HIGH SCHOOL[4]

Date of collection of specimens	Subject tested	No. of subjects	No. of positives	No. of negatives	No. of strains isolated	Serogroup and biovar of the isolates		
						O:3 (biovar 4)	O:5 (biovar 2)	O:9 (biovar 2)
July 21, 1972 (1st attempts)	Patients	195	117[a]	78	118	111	2	5
August 22, 1972 (2nd attempts)	Patients positive on 1st attempt	117	20	97	20	20		
	Patients negative on 1st attempt	78	5[b]	73	5	5		
	Pupils passed symptomless	868	16	852	16	15		1
	Teachers passed symptomless	48	1	47	1	1		
Total of isolates						132	2	6

Note: Specimens utilized for the isolation attempts were feces. Parenthesis indicates the biovars of Niléhn.[41]

[a] One specimen contained two serogroups of O:3 and O:9.
[b] One specimen contained serogroup O:5 bacterium at the time of the first isolation attempt.

In teachers, out of 5 symptomatic and 30 symptomless persons, 5 (40.0%) and 2 (6.7%), respectively, were also positive for serogroup O:3. Again the outbreak infection source and route remained unknown.

E. Outbreak E

Following four previous outbreaks, another caused by *Y. enterocolitica*, took place in June 1975 at a primary school in a rural area in Miyagi Prefecture.[6] The outbreak, in which a total of 198 (29.8%) out of 486 pupils suffered from acute abdominal illness, commenced on June 14 and lasted 8 days with a peak of cases between June 15 and 17. Most patients had the same abdominal symptoms and fever as those in the outbreaks described before. In the search for the causal agent, fecal specimens from 19 patients randomly selected from among many cases with typical symptoms were examined bacteriologically, and *Y. enterocolitica*, serogroup O:3, was isolated from all of them. Although the number of specimens examined was small, the bacterium was suspected to be a causal agent in the outbreak because no other pathogens were detected from the specimens and almost all children affected had symptoms quite similar to those of the typical cases. In this outbreak also, source and mode of spread of the infection were not established.

F. Outbreaks F and G

The other two outbreaks due to *Y. enterocolitica*, serogroup O:3, occurred in January and November of 1979 at primary schools in Miyagi and Hiroshima Prefectures. In the outbreak at the school in Miyagi Prefecture which occurred on January 27, about 200 patients out of 700 suffered from intestinal illness complaining of fever. The main symptoms were the same as those of Outbreak E.

The outbreak at the school in Hiroshima Prefecture took place on November 17. A total of 184 patients out of 954 complained of abdominal illness; abdominal pain (61.5%), diarrhea (36.5%), and fever (88.5%).

Since the details of epidemiological aspects of both outbreaks have not been made public yet the infection sources, incubation periods, and courses of illness remain obscure. However, all patients in the outbreaks seem to recover within 10 days by antibiotic therapy.

In countries other than Japan, community outbreak of *Y. enterocolitica* infections had not been described until 1976 when an outbreak of intestinal illness caused by the consumption of chocolate milk contaminated with *Y. enterocolitica*, serogroup O:8, occurred among children in five Oneida County schools in New York State. In Japan, on the other hand, during the period from 1972 to 1979, seven explosive outbreaks of *Y. enterocolitica* infection had occurred among children in seven separate primary and junior high schools. These outbreaks were characterized by occurrence in schools all situated in rural areas. For this reason, the spread of the infection was at first suspected to be due to drinking water contaminated with the bacterium from wildlife reservoirs, but in the outbreaks neither *Y. enterocolitica* nor coliform organisms were detected in the water. In addition, various foods provided as school lunch were bacteriologically examined early in the outbreak, and were found to be uncontaminated by *Y. enterocolitica* as well as the known enteropathogens. Ultimately, the reason why these outbreaks occurred only in the schools in rural areas remained obscure.

Another characterization was that *Y. enterocolitica*, serogroup O:3, the most prevalent serogroup in Japan, was causative in all the outbreaks. As reported by Une,[7-9] Une et al.,[10] and Lee et al.,[11] *Y. enterocolitica* of serogroups O:3, O:5, and O:9 have an ability to invade the epithelial cells. Subsequently, enterotoxin production has been recognized among serotype O:3 and O:5 strains.[40] Since among these serogroups,

O:3 bacterium seems to have the most intensive ability to invade, it seems that some patients with this serogroup infection could continue to excrete the bacterium for 2 months or more if not treated with antibiotics.

III. THE INCIDENCE OF SPORADIC HUMAN YERSINIOSIS

A. *Y. enterocolitica* Infection

In Japan the first report of *Y. enterocolitica* infection in humans was presented by Zen-Yoji and Maruyama in 1972.[1] During a 1-year period, March 1971 through February 1972, these investigators attempted the isolation of *Yersinia* from a total of 442 specimens consisting of 43 pieces of ileum resected from autopsy cases which exhibited pathological changes in their intestines, 292 fecal samples from patients hospitalized with symptoms suggestive of dysentery, 22 from sporadic cases of diarrhea in children, and 85 appendix vermiformis excised with suspicion of appendicitis. From this material, Zen-Yoji and Maruyama isolated 12 cultures of *Y. enterocolitica*: 2 from the 43 autopsy cases (4.7%), 5 from excised appendix vermiformis (5.9%), 2 from hospitalized patients (0.7%), and 3 from children with diarrhea (13.6%). In this survey, two autopsy cases, a 57-year-old male and a 15-year-old male, showed conspicuous hemorraghic necrosis and edema confined to about 1m of the ileocecum, but no other abnormalities in the intestine. Such evidence of limited pathological changes in the intestine might suggest a possibility that there was a correlation between the limiting of the change in length and mildness of diarrhea generally observed in older children and adults with *Y. enterocolitica* infection. In this survey, bacterium-positive patients had abdominal pain, fever, and mild diarrhea, but no sequelae were observed. As for the serogroup, of 12 isolates 11 were O:3 and 1 was O:5.

Following this report, Ohkawa et al.[12] reported the case of a 1-month-old baby with *Y. enterocolitica*, serogroup O:3 infection, whose main symptom was frequent bloody diarrhea which lasted for about 1 month. The rise in body temperature was not significant. For 2 years, from November 1972 through October 1974, Shinozuka et al.[13] also isolated 18 cultures of *Y. enterocolitica* and 2 of *Y. pseudotuberculosis* from fecal specimens of 1156 patients with diarrhea at the pediatric out-patient clinic of a Tokyo Metropolitan Hospital. Of 18 isolates 15, 2, and 1 were serogroup O:3, O:5, and O:9, respectively. In addition to the abdominal symptoms, sore throat was found in about one half of the patients; skin rash, rhinorrhea, conjunctival hyperemia, and bloody stool were seen only in some patients younger than 5 years old. After that, Kato et al.[14] reported the findings of radiological examination of 12 patients with acute terminal ileitis due to *Y. enterocolitica* infection, and pointed out the importance of *Yersinia* isolations for differentiating the disease from the acute stage of Crohn's disease.

Although there were no patients with sequelae such as erythema nodosum and arthritis in these reports, recently Kanazawa et al.[15] reported for the first time two adult female cases of *Y. enterocolitica*, serogroup O:3 infection hospitalized with major symptoms of cervical lymphadenitis, erythema nodosum, fever, and arthritis. Except for these two cases, *Y. enterocolitica* infection with such complications has not been found in Japan.

As described above, almost all the patients reported so far were infected with *Y. enterocolitica*, serogroup O:3, the predominant serogroup in Japan as well as in Europe, Canada, and South Africa, but in Japan their symptoms seemed to be relatively mild as compared to those of patients in other countries.[2,16-18] In addition, the bacterium of serogroup O:8 has, so far, not been isolated in Japan, although this serotype is well known as an occasional cause of septicemia associated with a high fatality rate.[19,20]

B. *Y. pseudotuberculosis* Infection

In Japan, the incidence of *Y. pseudotuberculosis* infection has been significantly rare not only in animals, but also in humans, as compared to Europe where human and animal infections with the bacterium occurred frequently.[21-24] To date, only a few cases have been reported. In 1913 Saisawa[25] had already succeeded in isolating *Y. pseudotuberculosis* from a man who died of septicemia. However, no isolation of the bacterium had been made from humans in the ensuing 50 years.

The first successful isolation in recent years was reported by Tsubokura et al.,[26] who isolated *Y. pseudotuberculosis*, serovar IB, from the inflamed appendix vermiformis excised from a boy. Following this report, Aikawa et al.[27] isolated *Y. pseudotuberculosis*, serovar I, on three occasions from the blood of a girl with severe septicemia hospitalized for 44 days. Kanazawa et al.[28] reported the isolation of *Y. pseudotuberculosis*, serovar III, from the stool of a woman with acute terminal ileitis. Subsequently, two cultures, serovar III and V, were isolated by Shinozuka et al.[13] from feces of two patients among 1156 cases with diarrhea. Until now only six patients with *Y. pseudotuberculosis* infection, including Saisawa's reported case, have been found in Japan.

IV. YERSINIA ISOLATIONS FROM ANIMALS

Although there were a few reports concerning the infection of *Y. pseudotuberculosis* in animals in Japan since the 1930s,[29-31] a long-term epidemic in animals with yersiniosis, which was confirmed to be virtually due to this bacterium, was recently reported by Hirai et al.[32] Between 1964 and 1972, this epidemic occurred spontaneously among patas monkeys kept in an outdoor compound at the Japan Monkey Center in Aichi Prefecture. During the outbreak, a total of 41 monkeys died, 3 to 7 animals each year, from severe watery or bloody diarrhea. Necropsy showed pathological findings characteristic of diphtheriod and necrotic enteritis in all the animals that died. *Y. pseudotuberculosis*, serovar IB, was isolated from the viscera of three monkeys necropsied in November 1972. The clinical symptoms and pathological findings of the three animals were quite similar to those in 38 other monkeys that died over the past 9 years. For detection of carrier animals, an additional 154 feces collected randomly from the ground of the compound were examined in February 1973, and *Y. pseudotuberculosis*, serovar IB, was isolated from eight of them. It was assumed from the facts that the epidemic was due to *Y. pseudotuberculosis* and continued to occur among the group of monkeys over 9 years by means of fecal-oral transmission.

In Japan the isolation of *Yersinia* from animals was attempted in order to clarify the source and mode of transmission of the infection (rather than for ecological studies), because of the occurrence of five explosive outbreaks of human *Yersinia* infections since 1972. The first attempt at *Yersinia* isolation from livestock for this purpose was carried out by Tsubokura et al.[33] They successfully isolated from 299 specimens of the cecal contents of swine at a slaughterhouse 13 *Y. enterocolitica* consisting of strains of serogroup O:3, 5 of O:5, 1 of O:10 and O:12, and 2 of nontypable strains.

Coincidentally, Zen-Yoji et al.[34] also attempted the isolation of *Yersinia* from swine and cattle collected from rats inhabiting an abattoir in Tokyo. In addition to assessing the carrier rates of *Yersinia* in these animals, this survey was carried out in summer and winter in order to examine the seasonal fluctuation in the frequency of isolation. As indicated in Table 2, it was noticed that both the carrier rate in swine and the frequency of detection of *Y. enterocolitica* and *Y. pseudotuberculosis* was higher during winter months than in summer months. In cattle, on the other hand, neither *Y. enterocolitica* nor *Y. pseudotuberculosis* was recovered from any of the specimens examined. Thus the survey strongly suggests that swine are of great importance as a source of *Yersinia* infections in humans, whereas cattle are probably not an important reservoir of the bacterium in Japan. On the other hand, rats inhabiting the abattoir did carry *Y.*

Table 2
ISOLATION OF *Y. ENTEROCOLITICA* AND *Y. PSEUDOTUBERCULOSIS* FROM SWINE AND RATS: SEROGROUPS AND BIOVARS OF THE ISOLATES[34]

Material	Time of investigation	No. positives	No. of positives (%)	*Y. enterocolitica* O-serogroup Biovar													*Y. pseudotuberculosis* No. of positives (%)	Serovar		
				3	5A	5B	6	7	7	9	12	14	16	25	UT		I	III	UT[b]
				4	1	2	1	1	3	1	2	1	1	1	1				
Swine																			
Cecal contents	Aug.—Sept. 1972	798	67[a] (8.4)	54				10	1						3	14 (1.8)		14	
Cecal contents	Feb. 1973	998	153[a] (15.3)	79	4	12	4	3			51				5	27 (2.7)	2	24	1
Mesenteric lymph nodes	Aug.—Sept. 1972	917	5 (0.5)	4		1										0			
Subtotal		2713	225	137	4	23	5	3	3		51				8	41	2	38	1
Rat																			
Cecal contents	Mar.—June 1973	165	58[a] (35.2)		10		25	16		1		3	2	5	13	0			
Total		2878	283	137	14	23	30	19	3	1	51	3	2	5	21	41	2	38	1

[a] Several specimens were positive for strains of different O serogroups.
[b] UT = untypable.

enterocolitica in a high percentage — 35.2, but not serogroups O:3 and O:5B which are prevalent in human infections as major causal serogroups. Although rats are assumed to play a role in the secondary contamination of carcasses and meats, it seems unlikely, on the basis of the results of this survey, that the animals were important as a reservoir of infection to humans. Quite recently, Yanagawa et al.,[35] the author's co-workers, have attempted additional isolation of *Yersinia* from dogs and cats and found that of 704 dogs and 373 cats, 42 dogs (6.0%) and 11 cats (2.9%) carried *Y. enterocolitica*, and 13 dogs (1.8%) and 12 cats (3.2%) carried *Y. pseudotuberculosis*. About 73.0% of the isolates of *Y. enterocolitica* in this survey were identified as serogroup O:3. In view of the results of the survey and a report by Gutman et al.[20] (in which a dog assumed to be carrying *Y. enterocolitica* was presumed to be the common infection source in the occurrence of *Yersinia* infections among the members of two families) and also a case report of *Y. enterocolitica* infection in an infant associated with infection in household dogs which was presented by Wilson et al.,[36] it could be said that pet animals such as dogs and cats should not be ignored as an infection source.

Tsubokura et al.[37,38] also isolated 87 (4.3%) cultures of *Y. enterocolitica* and 28 (1.4%) of *Y. pseudotuberculosis* from the cecal contents of 2041 swine at separate abattoirs in three cities and demonstrated that *Yersinia* was detected more frequently in winter than in summer, as reported by Zen-Yoji et al.[34] After that, Tsubokura et al.[39] carried out an ecological survey for the isolation of *Yersinia* from dogs, rats, frogs, mud-snails, fresh-water fish, fresh-water mussel, and snakes, and found 2 isolates of *Y. enterocolitica* from 115 dogs and 5 isolates from 65 rats, but no *Yersinia* from the other animals. As yet, little is known of the ecology of *Yersinia* in Japan.

V. CONCLUSION

An interesting feature of the occurrence of yersiniosis in Japan as compared to that in other countries is that seven explosive outbreaks of abdominal illness due to *Y. enterocolitica*, O:3, have occurred among children in seven schools during the 8 years from 1972 to 1979. For this reason, in Japan *Yersinia* infections have been studied exclusively in the fields of bacteriology and public health, and it has still been scarcely noticed by general physicians and gastroenterologists. Thus far, therefore, a relatively small number of sporadic cases of infection due to *Y. enterocolitica*, as well as to *Y. pseudotuberculosis* are reported. However, as the number of physicians interested in yersiniosis seems to be gradually increasing, many such cases will be found in Japan in the future. Only then will the question as to whether or not yersiniosis accompanied by complications such as erythema nodosum and arthritis is virtually rare be clarified.

REFERENCES

1. **Zen-Yoji, H. and Maruyama, T.,** The first successful isolations and identification of *Yersinia enterocolitica* from human cases in Japan, *Jpn. J. Microbiol.*, 16, 493, 1972.
2. **Ahvonen, P.,** Human yersiniosis in Finland. II. Clinical features, *Ann. Clin. Res.*, 4, 39, 1972.
3. **Asakawa, Y., Akahane, S., Kagata, N., Noguchi, M., Sakazaki, R., and Tamura, K.,** Two community outbreaks of human infection with *Yersinia enterocolitica, J. Hyg.*, 71, 715, 1973.
4. **Zen-Yoji, H., Maruyama, T., Sakai, S., Kimura, S., Mizuno, T., and Momose, T.,** An outbreak of enteritis due to *Yersinia enterocolitica* occurring at a junior high school, *Jpn. J. Microbiol.*, 17, 220, 1973.
5. **Uchida, K., Furui, S., and Ibuki, H.,** An outbreak of human infection with *Yersinia enterocolitica, Annu. Rep. Kyoto Pref. Inst. Publ. Health*, 19, 25, 1974 (in Japanese).
6. **Yuda, K.,** An outbreak of human infection with *Yersinia enterocolitica* occurred in Miyagi Prefecture, *Med. Biol.*, 92, 493, 1976 (in Japanese).
7. **Une, T.,** Studies on the pathogenicity of *Yersinia enterocolitica*. I. Experimental infection in rabbits, *Microbiol. Immunol.*, 21, 349, 1977.

8. **Une, T.,** Studies on the pathogenicity of *Yersinia enterocolitica.* II. Interaction with cultured cells in vitro, *Microbiol. Immunol.,* 21, 365, 1977.
9. **Une, T.,** Studies on the pathogenicity of *Yersinia enterocolitica.* III. Comparative studies between *Y. enterocolitica* and *Y. pseudotuberculosis, Microbiol. Immunol.,* 21, 505, 1977.
10. **Une, T., Zen-Yoji, H., Maruyama, T., and Yanagawa, Y.,** Correlation between epithelial cell infectivity in vitro and O-antigen groups of *Yersinia enterocolitica, Microbiol. Immunol.,* 21, 727, 1977.
11. **Lee, W. H., McGrath, P. P., Carter, P. H., and Eide, E. L.,** The ability of some *Yersinia enterocolitica* strains to invade HeLa cells, *Can. J. Microbiol.,* 23, 1714, 1977.
12. **Ohkawa, S., Matsumura, Y., Kawakami, H., and Ishi, S.,** An infant case with the main symptom of bloody diarrhea due to *Yersinia enterocolitica* infection, *Pediatr. Clin.,* 27, 1052, 1974 (in Japanese).
13. **Shinozuka, T., Kouno, Y., Imai, T., Zen-Yoji, H., and Maruyama, T.,** Yersiniosis in children, with special reference to clinical symptoms, *Acta Pediatr. Jpn. Overseas Ed.,* 80, 342, 1976.
14. **Kato, Y., Hattori, T., Oh-Ya, H., Yoshino, S., Kato, H., and Nishikawa, H.,** Acute terminal ileitis and *Yersinia enterocolitica* infection, *Gastroenterol. Jpn.,* 12, 36, 1977 (in Japanese).
15. **Kanazawa, Y., Kubo, M., and Fujimaki, S.,** Two human cases with erythema nodosum due to *Yersinia enterocolitica* infection, *Clin. Bacteriol.,* 4, 49, 1977 (in Japanese).
16. **Rabson, A. R. and Koornhof, H. J.,** *Yersinia enterocolitica* infections in South Africa, *S. Afr. Med. J.,* 46, 798, 1972.
17. **Rabson, A. R., Hallett, A. F., and Koornhof, H. J.,** Generalized *Yersinia enterocolitica* infection, *J. Infect. Dis.,* 131, 447, 1975.
18. **Larsen, J. H.,** *Yersinia enterocolitica* infections and arthritis, in *Infection and Immunology in the Rheumatic Diseases,* Dumonde, D. C., Ed., Blackwell Scientific, London, 1976, 133.
19. **Mollaret, H. H.,** *Yersinia enterocolitica* infection: a new problem in pathology, *Ann. Biol. Clin.,* No. 1, 7, 1972.
20. **Gutman, L. T., Ottesen, E. A., Quan, T. J., Noce, P. S., and Katz, S. L.,** An inter-familial outbreak of *Yersinia enterocolitica* enteritis, *N. Engl. J. Med.,* 288, 1372, 1973.
21. **Mair, N. S., Mair, H. J., Stirk, E. M., and Corson, J. G.,** Three cases of acute mesenteric lymphadenitis due to *Pasteurella pseudotuberculosis, J. Clin. Pathol.,* 13, 432, 1960.
22. **Hnatko, S. I. and Rodin, A. E.,** *Pasteurella pseudotuberculosis* infection in man, *Can. Med. Assoc. J.,* 88, 1108, 1963.
23. **Mair, N. S.,** Sources and serological classification of 177 strains of *Pasteurella pseudotuberculosis* isolated in Great Britain, *J. Pathol. Bacteriol.,* 90, 275, 1965.
24. **Mair, N. S.,** Yersiniosis in wildlife and its public health implications, *J. Wildl. Dis.,* 9, 64, 1973.
25. **Saisawa, K.,** Über die Pseudotuberkulose beim Menschen, *Zschr. Hyg. Inf. Krkh.,* 73, 353, 1913.
26. **Tsubokura, M., Otsuki, K., Itagaki, K., Kiyotani, K., and Matsumoto, M.,** Isolation of *Yersinia pseudotuberculosis* from an appendix in man, *Jpn. J. Microbiol.,* 17, 427, 1973.
27. **Aikawa, T., Kubo, S., and Masuko, S.,** A case of septicemia due to *Yersinia pseudotuberculosis* in man, *Rep. Hokkaido Inst. Publ. Health,* 24, 84, 1974 (in Japanese).
28. **Kanazawa, Y., Ikemura, K., Sasagawa, I., and Shigeno, N.,** A case of terminal ileitis due to *Yersinia pseudotuberculosis, Jpn. J. Infect. Dis.,* 48, 220, 1974 (in Japanese).
29. **Ikegaki, R.,** The isolation of a new strain of *Pasteurella pseudotuberculosis* from an animal, *Jpn. J. Bacteriol.,* 483, 341, 1936 (in Japanese).
30. **Katayama, N., Kishi, H., Okazaki, Y., and Ueda, T.,** An outbreak of yersiniosis due to *Pasteurella pseudotuberculosis* occurring at a zoological gardens in Tokuyama City, *J. Zool. Aquar.,* 9, 1, 1969 (in Japanese).
31. **Tsubokura, M., Itagaki, K., and Kawamura, K.,** Studies on *Yersinia pseudotuberculosis.* I. Source and serological classification of the organism isolated in Japan, *Jpn. J. Vet. Sci.,* 32, 227, 1970.
32. **Hirai, K., Suzuki, Y., Kato, N., Yagami, K., Miyoshi, A., Mabuchi, Y., Nigi, H., Inagaki, H., Otsuki, K., and Tsubokura, M.,** *Yersinia pseudotuberculosis* infection occurring spontaneously in a group of patas monkeys *Erythrocebus patas, Jpn. J. Vet. Sci.,* 36, 351, 1974.
33. **Tsubokura, M., Otsuki, K., and Itagaki, K.,** Studies on *Yersinia enterocolitica.* I. Isolation of *Y. enterocolitica* from swine, *Jpn. J. Vet. Sci.,* 35, 419, 1973.
34. **Zen-Yoji, H., Sakai, S., Maruyama, T., and Yanagawa, Y.,** Isolation of *Yersinia enterocolitica* and *Y. pseudotuberculosis* from swine, cattle, and rats at an abattoir, *Jpn. J. Microbiol.,* 18, 103, 1974.
35. **Yanagawa, Y., Maruyama, T., and Sakai, S.,** Isolation of *Yersinia enterocolitica* and *Y. pseudotuberculosis* from apparently healthy dogs and cats, *Microbiol. Immunol.,* 22, 1978, in press.
36. **Wilson, H. D., McCormik, J. B., and Feely, B. C.,** *Yersinia enterocolitica* infection in a 4-month-old infant associated with infection in household dogs, *J. Pediatr.,* 89, 767, 1976.
37. **Itagaki, K., Yamaoka, K., and Wakatsuki, M.,** Studies on *Yersinia enterocolitica.* II. Relationship between detection from swine and seasonal incidence, and regional distribution of the organism, *Jpn. J. Vet. Sci.,* 38, 1, 1976.

38. **Tsubokura, M., Ohtsuki, K., Fukuda, T., Kubota, M., Imamura, M., Itagaki, K., Yamaoka, K., and Wakatsuki, M.,** Studies on *Yersinia pseudotuberculosis*. IV. Isolation of *Y. pseudotuberculosis* from healthy swine, *Jpn. J. Vet. Sci.,* 38, 549, 1976.
39. **Tsubokura, M., Fukuda, T., Otsuki, K., Kubota, M., and Itagaki, K.,** Isolation of *Yersinia enterocolitica* from some animals and meats, *Jpn. J. Vet. Sci.,* 37, 213, 1975.
40. **Pai, C. H. and Mors, V.,** Production of enterotoxin by *Yersinia enterocolitica, Infect. Immunol.,* 19, 908, 1978.
41. **Niléhn, B.,** Studies on *Yersinia enterocolitica, Acta Pathol., Microbiol. Scand. Suppl.,* 206, 1, 1969.

INDEX

A

Abdomen
  mass, right lower quadrant, 108
  pain, 99
Abscess
  in childhood yersiniosis, 99
  perianal, 108
  splenic, 198
Acetoin, see Acetylmethylcarbinol
Acetylmethylcarbinol, production, 22
Acidification, of fermentable substrates, 23—25
  Y. pseudotuberculosis and, 35
Aeromonas hydrophila, biochemical reaction, 13
Agar media, inoculation of, 12
Age, in erythema nodosum, 128
Agglutination reaction
  in erythema nodosum, 126
  in thyroid disease, 141
Agglutinins, 195
6-Amino-penicillanic acid, see 6-APA
Ampicillin
  inhibition of β-lactamases, 59
  minimum inhibitory concentration of, 57
Animal(s)
  isolation of *Y. enterocolitica* from, 146—149
  laboratory, yersiniosis in, 74—77
Animal isolates
  Canada, 185—186, 188—189
  Japan, 212—214
  New York State, 179
  Oneida County, N.Y., during outbreak, 176—177
Animal model, 88
Animal pathogenicity tests, 37—38
Ankylosing spondylitis, 108
  in Reiter's disease, 116
Antibiotic(s)
  β-lactam
    chromosomal resistance to, 56—70
    inhibition of β-lactamase by, 59
    mean inhibitory concentration of, 57
    resistance to, 89
  in childhood gastroenteritis, 101
  in *Yersinia* arthritis, 121
  resistance to, β-lactamases in, 59, 61
Antibiotic therapy, efficacy of, 89
Antibody(ies)
  anti-*Yersinia*, in thyroid disease, 137—141
    geographic distribution, 137
  cross-reacting *Yersinia*, in thyroid disease, demonstration of, 141
  humoral, in syphilis, 117
  titers, in erythema nodosum, 128
Antigen(s), 42—52
  cross-reacting, 46—48
    Yersinia, in thyroid disease, 138
  flagella (OH), 126
  H, 49—51
  histocompatibility, disease association, proposed mechanisms, 116
  HLA-B27, 87, 108, 114
  K, 48—49
  O, 42—48
  somatic (O), 126
  thyroid, 136
  "unheated," 50
  V, 78
  W, 78
Antigenic scheme
  O-antigens, 43—46
  of *Y. pseudotuberculosis*, 36
Antistreptolysin titer, in erythema nodosum, 128—129
6-APA
  and induction of β-lactamase production in strain H66, 60
  minimum inhibitory concentration of, 57
Aphthous ulcers, 109, 110
Appendicitis, 86
Arthritis, 99
  in HLA-B27-negative individual, 114—115
  in HLA-B27-positive individual, 114—115
  peripheral, 108
  reactive
    carditis, in, 119—120
    clinical features, 119—121
    diagnosis, 118
    differential diagnosis, 118
    genetics, 114—115
    joint symptoms, 119
    laboratory findings, 118
    pathogenesis, 115—118
    sequelae, 120—121
    treatment, 121
  in South Africa, 197
Auramine-agglutinability, O-antigens, 42—44

B

Bacteremia, 87
  in childhood yersiniosis, 99
Bactericidal activity, serum, 199
Bacteriology
  childhood gastroenteritis, 96—97
  diagnostic, 88
  in South Africa, 194
Bantu siderosis, 198—201
Beef, *Yersinia* in, 163—164
Benzylpenicillin, minimum inhibitory concentration, 57
Biochemical characterization, 2—4, 7, 187
  typical *Y. enterocolitica*, 20—25
  Y. psuedotuberculosis, 34—35
Biochemical reactions, 2—4, 7, 13
Biotypes, 24
  of serotypes from New York State, 179—180

Biotyping, 187
  correlation with β-lactamase of strains, 65, 67—70
  of isolates from Oneida County, NY, 175
  schemes for, 5
Blood culture, 10
  isolation of Y. pseudotuberculosis from, 33
Broth media, inoculation of, 12
Brucella species, antigenic relationship with Y. enterocolitica, 46—48
Bushbaby, yersiniosis in, 150

C

Canada
  serotypes in, 181
  yersiniosis in, 184—189
Canadian National Reference Center for Yersinia, 184
Canadian phage type, 185
Carbenicillin
  inhibition of β-lactamases, 59
  minimum inhibitory concentration, 57
Carbohydrate utilization, 23
  of Yersinia species, 37—38
  Y. pseudotuberculosis, 35
Cardiac sequelae, in Yersinia arthritis, 121
Carditis, 119—120
Carrier state, 85
  in animals, 212
Cary-Blair transport medium, 10
Cattle, isolation of Yersinia from, in Japan, 212
Cefamandole
  inhibition of β-lactamases, 59
  minimum inhibitory concentration, 57
Cefoxiitn
  inhibition of β-lactamases, 59
  minimum inhibitory concentration, 57
Cell-mediated immunity
  in rheumatic disease, 117
  in thyroid disease, 136
Cephalexin, minimum inhibitory concentration, 57
Cephaloridine, minimum inhibitory concentration, 57
Cephalosporin, minimum inhibitory concentration, 57
Cephalothin, minimum inhibitory concentration, 57
Cerebrospinal fluid culture, 10
Cervical adenopathy, 108
Chemical preservation, and survival of Y. enterocolitica in food, 168
Childhood infections
  clinical manifestations, 97—99
  food-borne yersiniosis, 174—177
  gastroenteritis, 96—102
Chinchillas, yersiniosis in, 149
Chocolate milk, yersiniosis and, 152, 163, 174—177
Chromatography, in separation of β-lactamases from strain W222, 56, 58
Chromosome, mediation of β-lactamase production, 56
Chromobacter violaceum, biochemical reaction, 13
Cirrhosis, Y. enterocolitica sepsis in, 87
Citrobacter diversus, biochemical reaction, 13
Classification, 6—7
  biochemical characterization, 2—4
  DNA relatedness, 5—6
  guanine-cytosine content, 5
Clinical specimens, see Isolation techniques
Cloxacillin
  inhibition of β-lactamases, 50
  minimum inhibitory concentration, 57
Cold enrichment, 10, 187
  for isolation of Y. pseudotuberculosis from fecal culture, 34
Colitis, 109
Colonic mucosa, sigmoidoscopic findings, 86
Colony characteristics, 21
  of Y. pseudotuberculosis, 32—33
Conjunctivitis, 87, 120
Convulsions, in childhood gastroenteritis, 99
Crohn's disease, 87, 106
  clinical features, 108
  differentiation from Yersinia enteritis, 108—110
  endoscopic findings, 109
  pathology, 109—110
  pathophysiology, 110
  radiologic findings, 108—109
Cross-agglutination, 36
Cross-reactions
  and auramine-agglutinability in O-antigens, 42—44
  with other Gram-Negative bacilli, 46—48
Culture, see Isolation techniques

D

Dairy products, Y. enterocolitica in, 162—163
Diagnostic bacteriology, 88
Diarrhea, 85, 98
  in yersiniosis in South Africa, 197
DNA hybridization groups, 18
DNA-relatedness, 5—6
  of isolates from New York State, 179—180
Dog, yersiniosis in, 152

E

Endoscopy, in differentiation of Crohn's disease and Yersinia enteritis, 109
Enrichment cultivation, 126
Enteritis, 85
  clinical features, 108
  differentiation from Crohn's disease, 108—110

endoscopic findings, 108
pathology, 109—110
pathophysiology, 110
radiologic findings, 108—109
regional, see Crohn's disease
Enterobacter, biochemical reactions, 13
Enterobacteriaceae, differentiation of *Yersinia* species, 36—39
Enteropathogenicity, 198
Enterotoxin, 78, 198
Environmental biotypes, 163, 166—167
Environmental isolates
from Canada, 185—186, 188
from New York State, 179
Oneida County, 177
virulance testing, 77
Environmental specimens, see Isolation techniques
Epidemiology, 84
in New York State, 174—181
in South Africa, 194—195
Erythema nodosum, 85, 87
age distribution, 128
antibody titers in, 128
antistreptolysin titer in, 128—129
chest X-ray in, 130
comparison with control group, 127—130
diagnostic methods, 126—127
erythrocyte sedimentation rate in, 128—129
historical aspects, 126
joint pain in, 131—132
related infections, clinical symptoms, 130—131
sex distribution of cases, 128
South Africa, 197
tuberculin reaction in, 129—130
Yersinia arthritis, 120
Erythrocyte sedimentation rate (ESR)
in erythema nodosum, 128—129
in *Yersinia* arthritis, 118
European phage type, 185
Exudative pharyngitis, 86

F

Fecal culture, 10, 88
in erythema nodosum, 126—127
isolation techniques, recommendations, 186—187
in South Africa, 196—197
Y. pseudotuberculosis in, 34
Fistulas, perianal, 108
Fluorescent antibody test, for *Y. pestis*, 37—38
Food
isolates from, from Canada, 185—186
specimens, collection, transport, and preparation of, 11—12
Y. enterocolitica in, significance of, 166—167
survival and control of, 167—168
Freezing, and survival of *Y. enterocolitica* in food, 167—168

Freezing, and survival of *Y. enterocolitica* in food, 167—168

G

β-Galactosidase activity, 23
Gastroenteritis, 85—86
Canada, 186
childhood, 96—102
age in, 97
bacterial, 96—97
clinical manifestations, 97—99
communicability, 100—102
incidence, 96—97
seasonal distribution, 97, 98
serologic response, 99—100
treatment, 101
communicability, 100—102
familial outbreaks, 100—102
pathogenesis, 101
South Africa, 197—198
Gel filtration, in estimation of molecular weight of β-lactamases, 56, 60
Genetics, of reactive arthritis, 114—115
Geographic distribution, of anti-*Yersinia* antibodies in thyroid disease, 137
Geographic variation, in isolation rates, 88
Glomerulonephritis
acute proliferative, 87
in *Yersinia* arthritis, 120
Goats, yersiniosis in, 151
Gram-Negative bacilli, antigenic relationship with *Y. enterocolitica*, 46—48
Granulomas, 110
Graves' disease, 136
Great Britain, anti-*Yersinia* antibodies in, in thyroid disease, 140—141
Guanine-cytosine content, 5
Guinea pigs, yersiniosis in, 74

H

H-antigens, 49—50, 51
Y. pseudotuberculosis, 36
H66 strain, β-lactamases from, 56
Hashimoto's thryoiditis, 136
Heating, and survival of *Y. enterocolitica* in food, 168,
HeLa cells, invasiveness of, 77
Hemachromatosis, *Y. enterocolitica* spesis in, 87
Hemorrhage, intestinal, 86
Histocompatibility antigens, see also HLA-B27 antigen
disease association, proposed mechanisms, 116
HLA-B27 antigen, 87
disease association with, 114
frequency, in reactive arthritis, 114

peripheral arthritis, 108
Humoral antibodies, in syphilis, 117
Humoral response, 79

I

Iliac fossa syndrome, 86, 186
Immune complexes, 117
Immune response, 79
  in syphilis, 117
  in thyroid disease, 136
Incidence
  of childhood gastroenteritis, 96
  seasonal, 85
Incubation
  of selective plating media, 12
  of *Y. pseudotuberculosis*, 33
Incubation period, 85
Indirect immunofluorescence, of thyrotoxic thyroid epithelial cells, 138
Indole production, 20, 22
  rhamnose-fermenting strains, 26
Infections, erythema nodosum and, 130—131
Invasiveness
  in laboratory animal models, 77—78
  of HeLa cells, 77
IP97 strain
  β-lactamase inducibility in, 63—64
  mutants of, 63, 65, 66
Iritis, 120
Iron
  and susceptibility to yersiniosis, 78
  overload, and yersiniosis, 199
Irradiation, and survival of *Y. enterocolitica* in food, 168
Isoelectric focusing, 65, 68—70
Isolates
  animal, 146—149
  from Canada, 184—186
  from New York State, 177—179
    Oneida County, 175—177
  from South Africa, 194
Isolation techniques, 10—14, 88—89
  beef, 164
  cultural characteristics, 18—20
    of *Y. pseudotuberculosis*, 32—33
  culture results, in erythema nodosum, 128
  dairy products, 163
  erythema nodosum, 126—127
  fecal cultures, recommendations, 186—187
    in South Africa, 196—197
  for food specimens, 165—166
  inoculation, incubation, and examination of selective plating media, 12—13
  mixed flora specimens, 34
  nutrient supplementation, 10
  pork, 164
  sample preparation, 11—12
  selection of suspect colonies, 13—14
  selective plating media, inoculation, incubation, and examination of, 12—13
  specimen collection and transport, 10—11
  specimens from Oneida County, N.Y., 174—175
  Y. pseudotuberculosis, 33—34
Israel, anti-*Yersinia* antibodies in, in thyroid disease, 139—140

J

Japan, yersiniosis in, 206—214
Joint symptoms
  in *Yersinia* arthritis, 119
  pain, in erythema nodosum, 131

K

K-antigens, 48—49
Klebsiella, biochemical reactions, 13

L

Laboratory identification, see Isolation techniques
β-Lactam antibiotics, see Antibiotics
β-Lactamases
  antibiotic resistance, 59, 61
  cellular location, 61—62
  inhibition of, by β-lactam antibiotics, 59
  isoelectric points of, in grouping strains, 65, 68—70
  molecular weight, estimation of, 56, 60
  other strains of O:3 and O:9, 59
  production, 56
    6-APA induction of, in strain H66, 60
  selective release of, by osmotic shock and spheroplast formation, 62
  serotype O:5b strains, 63
  strain H66, 56
  strain W222, 56
    substrate profiles, 58
  type B, inducibility of, 63—64
Lactose fermentation, 23
Lamb, *Y. enterocolitica* in, 163
Lecithinase activity, 22
Leukocyte-migration inhibition assay, in thyroid disease, 138
Lymph node, isolation of *Y. pseudotuberculosis* from, 34
Lymphoid reaction, 110.

M

Magnesium chloride-malachite green-carbenicillin broth, 12
Mean inhibitory concentration (MIC), of β-lactam antibiotics, 57

Meat, *Y. enterocolitica* in, 163—165
Mesenteric adenitis, 86, 109
Mesenteric lymphadenitis, isolation of *Y. pseudotuberculosis* in, 34
Methicillin
   inhibition of β-lactamases, 59
   minimum inhibitory concentration, 57
Microbiology, of *Y. pseudotuberculosis*, 32—39
Microscopic morphology, 18
Molecular mimicry hypothesis, 117
Molecular weight, of β-lactamases, 56
Monkeys, yersiniosis in, 76, 150, 212
Morganella (Proteus) morganii
   antigenic relationship, 48
   biochemical reactions, 13
Morphology, of *Y. pseudotuberculosis*, 32
Mortality rate, in yersinial gastroenteritis, 86
Motility
   temperature and, 18
   Yersinia species, 27, 38
   Y. pestis, 32
   Y. pseudotuberculosis, 32
Mouse
   immune response in, 79
   yersiniosis in, 74—76
Mutant immune response gene, 117
Mutants, of strain IP97, 63, 65, 66
MZ phenotype, 116

N

Natural reservoir, 84, 153
   for serogroups, 24
New York State
   Oneida County, food-borne yersiniosis in, 174—177
   Y. enterocolitica isolates in, 177—179
O-Nitrophenol-β-D-galactoside (ONPG), 23
Nutrient supplementation, of cultures, 10

O

O-antigens, 42—48
   auramine-agglutinablity and cross-reactions, 42—44
   biochemical nature, 48
   factors of, 43—46
   scheme for, 43—46
   Y. pseudotuberculosis, 36
Ocular disorders, see also specific disorder
   in *Yersinia* arthritis, 120
Oligoarthritis, 87
Oneida County, N.Y., see New York State
Osmotic shock, and selective release of β-lactamases, 62
Oxacillin, minimum inhibitory concentration of, 57

P

Pathogenic mechanisms, 77—79
Pathogenicity
   documentation of, 85
   for laboratory animals, 74—77
Pathologic findings, antemortem, in terminal ileitis, 87
pH, and growth, 168
Phage types, of isolates from New York State, 180—181
Phage typing
   Canadian serotypes, 188
   isolates from Oneida County, NY, 175
Pharyngitis, 108
Phenotype, MZ, 116
Physiotherapy, 121
Plasmid RY3, 70
Plasmids, resistance, see Reistance (R) plasmids
Plate microscope, 187
Pleural fluid culture, 10
Pork, *Y. enterocolitica* in, 164—165
Poultry, *Y. enterocolitica* in, 165
Proteus (Morganella) morganii
   antigenic relationships, 48
   biochemical reactions, 13
Proteus rettgeri, biochemical reactions, 13
Providencia, biochemical reactions, 13
Pyoderma gangrenosum, 108

R

R-antigen, 42—43
   Y. pseudotuberculosis, 36
Rabbits, yersiniosis in, 76
Radiology
   chest X-ray, in erythema nodosum, 130
   in differentiation of Crohn's disease from *Yersinia* enteritis, 108—109
   in follow-up of *Yersinia* arthritis, 121
Rash, in childhood yersiniosis, 99
Rats
   isolation of *Yersinia* from, in Japan, 212—214
   yersiniosis in, 77
Reactive arthritis, see Arthritis
Refrigeration, and survival of *Y. enterocolitica* in food, 167
Regional enteritis, see Crohn's disease
Reiter's disease
   ankylosing spondylitis in, 116
   family study, 115
Resistance (R) plasmids, 70
   and antibiotic therapy, 89
Rhamnose fermentation, 23
Rhamnose-fermenting strains, 25
   biochemical profiles and characteristics, 26—27
Rhamnose-positive strains, 2
Rheumatic fever, differentiation from *Yersinia*

arthritis, 118
Right iliac fossa syndrome, 86

S

Salmonella
  arthritis, see Arthritis, reactive
  bergen, antigenic relationship, 48
  typhimurium, growth parameters, temperature and, 152—153
  urbana, antigenic relationship, 48
  yersiniosis, 195—196
Scandinavia
  anti-*Yersinia* antibodies in, in thyroid disease, 137—138
  erythema nodosum in, 126
Seafood, *Y. enterocolitica* in, 165
Seasonal incidence, 85
  childhood gastroenteritis, 97, 98
Selective plating media, inoculation, incubation, and examination, 12
Sereny reaction, 78
Sereny test, 163
Serodiagnosis, 89
  childhood gastroenteritis, 99—100
  erythema nodosum, 126
Serogroup O:5,27, natural reservoir, 24
Serogroup O:8, natural reservoir, 24
Seronegative spondyloarthropathies, 114
Serotype(s), 19
  animals, 147—149
  biochemical characterization, 25
  isolates from New York State, 179—180
  natural reservoirs, 24
  O:1, 84
  O:3
    antibiotic resistance in, 59, 61
    antibodies to, in thyroid disease, 137—141
    childhood gastroenteritis, 96—97
    erythema nodosum and, 127
    geographic distribution, 181
    in Japan, see Japan
    laboratory animal infection, 76
    β-lactamases from strains of, 59
    strain H66, 56
    strain W222, 56, 62
  O:5, 181
  O:5b, β-lactamases from, 63
    strain IP97, 63—64
  O:6, 30, 181
  O:8, antibodies to, in thyroid disease, 139—140
    and childhood gastroenteritis, 96—97
    geographic distribution, 84
    immune response to, 79
    laboratory animal infection, 75
    Oneida County, N.Y., 175—176
  O:9, antibiotic resistance in, 59, 61
    antibodies to, in thyroid disease, 139—140
    childhood gastroenteritis, 96—97
    erythema nodosum and, 127

geographic distribution, 84
    laboratory animal infection, 76
    β-lactamases from strains of, 59
Serotyping
  correlation with β-lactamases of strains, 65, 67—70
  in diagnostic bacteriology, 50—52
  isolates from Canada, 188
  isolates from Oneida County, N.Y., 174—175
  Y. pseudotuberculosis, 35—36
Serratia, biochemical reactions, 13
Serum bactericidal activity, 199
Serum complement (C3), in *Yersinia* arthritis, 118
Sex predilection, of erythema nodosum, 128
Shigella
  and acute terminal ileitis, 107
  arthritis, see Arthritis, reactive
Sigmoidoscopy, 86
Soil samples, collection, transport, and preparation, 11
Sonication, in demonstration of cross-reacting *Yersinia* antibody in thyroid disease, 142
South Africa
  bacteriology of *Y. enterocolitica*, 194
  Yersinia gastroenteritis, 197—198
  yersiniosis in, clinical manifestations, 197
  epidemiology, 194—195
South Africa phage type, 185
Specimens, see Isolation techniques
Spheroplast, formation of, and selective release of β-lactamases, 62
Steroids, in *Yersinia* arthritis, 121
Substrate, fermentable, acidification of, 23—25
  Y. pseudotuberculosis and, 35
Sucrose fermentation, 23
Sucrose-negative strains, 2, 27
  differential characteristics, 28
Swine
  and human yersiniosis, 194
  isolation of *Yersinia* from, in Japan, 212—214
  serotypes in, 188—189
  yersiniosis, in 150—151
Synergistic infection, *Salmonella* and, 195—196
Synovial fluid, in *Yersinia* arthritis, 118
Syphilis, immune responses in, 117
Systemic yersiniosis, in South Africa, 198—201

T

T-cell abnormalities, in rheumatic disease, 117
Taxonomy, 2, 146
Temperature
  and acetoin production, 22
  and growth, 88, 152—153
    of *Y. pseudotuberculosis*, 33
    surface antigen phenotypic expression and, 141
  and lecithinase activity, 22
  and motility, 18

of *Y. pseudotuberculosis,* 32
and survival of *Y. enterocolitica* in food, 167—168
and urease activity, 23
and virulence of strains, in animals, 77
of incubation, see also Incubation, 2
Terminal ileitis, 87
  historical perspectives, 106—108
Tetrathionate broth, 12
Throat culture, 10
Thyroid disease
  anti-*Yersinia* antibodies in, 136—143
    geographic distribution, 137
  cross-reacting *Yersinia* antibodies in, demonstration of, 141
  immune responses in, 136
  pathogenesis, 137
Thyroiditis, subacute, *Yersinia* in, 138
Tissue culture, 10
Transmission
  aerosol, 87
  animal-to-human, 84, 152—155
  human-to-human, 85
  swine-to-human, 194
Trehalose-negative strains, 2
Tuberculin reaction, in erythema nodosum, 129—130

U

Ulcers, aphthous, 109
"Unheated" antigen, 50
United States, anti-*Yersinia* antibodies in, in thyroid disease, 138—139
Urease
  activity, 23
  production, 37
Urine culture, 10
Urologic disorders, in *Yersinia* arthritis, 120
Uveitis, acute anterior, 116

V

V-antigens, 78
Vacuum packaging, and survival of *Y. enterocolitica* in food, 168
Vegetables, *Y. enterocolitica* in, 165
Vibrio cholerae
  antigenic relationship, 48
  biochemical reactions, 13
Virulence, of strains, temperature and, 77
Voges-Proskauer test, see Acetylmethylcarbinol (acetoin) production

W

W-antigens, 78
W222 strain
  β-lactamases from, 56

W238 strain, selective release of β-lactamases from, 62
Water samples, collection, transport, and preparation of, 11
Wound exudate, culture of, 10

Y

Yersinia
  arthritis, see Arthritis, reactive *enteritidis,* 4
  enterocolitica, biochemical characteristics, 7
    biochemical reactions, 2—4
    classification, 1—7
    cross-reacting antigens, 46—48
    differentiation from *Y. pseudotuberculosis,* 37—38
    survival and control of, 167—168
    human infection, laboratory models for, see also Yersiniosis, 74—79
    in food, significance of, 166—167
    isolation techniques, 10—14
    laboratory identification, DNA hybridization groups, 18
    trehalose-negative strains, 2
    typical, biochemical characteristics, 20—25
      differentiation from rhamnose-fermenting strains, 27
      differentiation from sucrose-negative strains, 28
  enterocolitica-like organisms, biochemical characterization, 2—4, 7, 13
  frederiksenii, biochemical characteristics, 7, 19
    biochemical reactions, 13
    DNA homology group, 19
    representative serotypes, 19
  intermedia, biochemical characteristics, 7, 19
    biochemical reactions, 13
    DNA homology group, 19
    representative serotypes, 19
  kristensenii, biochemical characteristics, 7, 19
    biochemical reactions, 13
    DNA homology group, 19
    serotype, 19
  pestis, biochemical reactions, 13
    differentiation from *Y. pseudotuberculosis,* 37—38
    motility of, 32
  pseudotuberculosis, in animals, 212—214
    antigenic relationship, 48
    antigenic scheme, 36
    biochemical characteristics, 34—35
    biochemical reactions, 4, 13
    cultural characteristics, 32—33
    differentiation from other organisms, 36—39
    differentiation from other *Yersinia* species, 37—38
    differentiation from rhamnose-fermenting strains, 27
    differentiation from sucrose-negative strains, 28
    erythema nodosum and, 127

incidence in Japan, 211—212
isolation from clinical specimens, 33—34
microbiology, 32—39
serotyping, 35—36
Yersiniosis, see also specific infection
animal model, 88
in animals, see Zoonotic yersiniosis
bacterial synergy in, 195—196
Canada, 184—189
childhood, 97—99
clinical observations, 84—89
clinical presentation, 85—88
complicated, in Sweden, 131—132
diagnosis, 88—89
epidemiology, 84
erythema nodosum and, 126—132
food-borne, 162—168
  in Oneida County, N.Y., 163, 174—177
frequency, 84
immune responses, 79
incubation period, 85
Japan, 206—214
laboratory models for, 74—79
noncomplicated, in Sweden, 131—132

pathogenic mechanisms, 77—79
pathophysiology, 88
seasonal incidence, 85
serotypes causing, 25
Siderosis, 198—201
South Africa, clinical manifestations, 197
  epidemiology, 194—195
sporadic, in Japan, 211—212
susceptibility to, iron and, 78
synergistic infection, 195—196
systemic, in South Africa, 198—201
transmission see also Transmission, 84—85
treatment, 89
water-borne, 152—153
zoonotic, 146—155

Z

Zoonotic yersiniosis, 146—155
  clinical manifestations, 149—152
  isolation of organism, 146—149
  transmission, 152—155